DAVID WALTNER-TOEWS | JAMES J. KAY | NINA-MARIE E. LISTER

EDITORS

THE ECOSYSTEM APPRO

Complexity, Uncertainty, and Managing for Sustainability

T0176203

The Ecosystem Approach

Complexity in Ecological Systems Series

The Ecosystem Approach

Complexity, Uncertainty, and Managing for Sustainability

Edited by David Waltner-Toews, James J. Kay, and Nina-Marie E. Lister

COLUMBIA UNIVERSITY PRESS NEW YORK

Columbia University Press
Publishers Since 1893
New York   Chichester, West Sussex

Library of Congress Cataloging-in-Publication Data

    The ecosystem approach : complexity, uncertainty, and managing for sustainability /
David Waltner-Toews, James J. Kay, and Nina-Marie Lister, editors
    p.   cm. — (Complexity in ecological systems series)
Includes bibliographical references and index.
  ISBN 978-0-231-13250-3 (cloth : alk. paper) / ISBN 978-0-231-13251-0 (pbk. : alk. paper)
  ISBN 978-0-231-50720-2 (ebook)
1. Ecological integrity   2. Ecosystem management   I. Waltner-Toews, David,
1948–   II. Kay, James J.   III. Lister, Nina-Marie E.   IV. Title   V. Series.

  QH541.15.E245E38 2008
  333.72—dc22

                                                                          2008014964
⊚

Columbia University Press books are printed on permanent and durable acid-free paper.
This book is printed on paper with recycled content.
Printed in the United States of America

c 10 9 8 7 6 5 4 3 2 1
p 10 9 8 7 6 5 4 3 2 1

Contents

A Preface

David Waltner-Toews, Nina-Marie E. Lister, and Stephen Bocking

The universe we live in is confusing, complex, and sometimes opaque to our queries of it. As scientists, we simplify it to try to understand it. Systems thinking is one approach to simplification that has proven useful for answering certain sets of questions. The ecosystem approach as described in this book is an attempt to bring together an ecological understanding of the world with a desire to make the world a convivial place for our species.

The distinction between social systems and ecological systems, and the linking of them into social-ecological systems, is a useful simplification for teasing apart difficult problems. Insofar as our species interacts with other species and the landscapes we live in, we are ecological beings; insofar as we consume and excrete nutrients and use energy, we are members of ecosystems. An urban landscape is certainly a social system. It is also as much an ecosystem as any rural landscape or wilderness. Just as the restructuring of landscapes by cattle, elephants, or coral do not change scientific abilities to describe those landscapes in ecosystemic terms, just so urban restructuring by people does not change the essential ecological nature of a city.

Pandemics of avian influenza and salmonellosis reflect the sad fact that the ecological nature of urban settlements is often forgotten or neglected. This neglect results in outbreaks, epidemics, and pandemics caused by infectious agents that take advantage of various feedback loops in the ecosystems we inhabit. For instance, there is evidence that the pandemic of human disease with *Salmonella enteritidis* in the 1980s and 1990s was attributable to eradication of the poultry pathogen *Salmonella pullorum* (the cause of fowl plague) from commercial chickens (Rabsch et al. 2000).

This pandemic (like that of avian influenza) can be understood as a function of social systems, ecological systems, linked social-ecological systems, and fully integrated socioecological or ecosocial systems. Each of these systemic constructions is useful for answering a different set of questions.

For the poultry industry, the *Salmonella* pandemic can be usefully explored as a function of a social system constructed to feed urban populations. Biologists can learn a great deal about ecological roles and the global circulation and adaptation of microbial populations by looking at the pandemic as a function of an ecological system. Linking the social and ecological systems yields useful insights into policy alternatives that might satisfy several conflicting criteria related to keeping down consumer prices and preventing foodborne diseases. Finally, by considering the interactions between different species of mammals, birds, and bacteria, using different criteria (organism, animal, flock; individual, population) at different scales (microscopic, bioregional, global), one can begin to ask questions related to the sustainability of different efforts to promote global human health and nutrition.

The elements that an investigator chooses to put into a particular systemic model also reflect the questions being asked. These might include the molecular structure of a microbe, its genetic history, eating habits of various urban consumers (which reflect in part cultural histories), shedding rates of different bacteria by chickens under different housing conditions, and patterns of global trade.

It is the premise of the work in this book that the reality humanity inhabits can only be known through our perceptual organs (primarily, the eyes and ears) and their technological extensions. We are inside the world and have evolved within it. We have no external observer to tell us when we have got it right. Scientific disciplines have generally been defined, and have made progress, by narrowing the field of permissible questions and agreeing on sets of rules that enable the group to answer those questions. For scientists that work in this philosophical tradition, which some have termed reductionist, other questions, other modes of inquiry, and other criteria for quality are often marginalized and sometimes denigrated.

The epidemiological literature, for instance, is rife with aspersions on ecological studies, which are seen as being primarily case studies and which are dismissed as the weakest form of study possible. The ecological fallacy—drawing conclusions about individuals based on studies of groups—is spoken of with a certain level of derision in the epidemiological literature. This pejorative use of ecological has arisen because epidemiologists have wanted to define risk factors for diseases in individuals that are universally true. What would be the point of a study that could only conclude that smoking caused lung cancer in one subset of individuals living in an area of Nepal, with a particular ecological and social history? Hence epidemiological studies have focused on large populations and relatively stable traits. These studies have been most successful at identifying genetic markers, toxic exposures, and individual behaviors that result in disease. They have been much less successful in determining relationships between human disease and changes in social and ecological systems. Most post-World War II epidemiological investigations have been deliberately designed to exclude these kinds of variables.

Acceptance of a wider range of modes of inquiry and criteria for evaluation of results has only recently changed when some epidemiologists decided to ask different kinds of questions, particularly with respect to relationships between human well being and global environmental and climate change. Similar changes have occurred in a subset of those scholars asking related questions related to environmental management. By collectively pooling our perceptions and openly challenging them according to multiple, agreed upon criteria, we, as a species, are likely to be more successful in solving the difficult problems of global sustainability than if we fall back into mutually exclusive paradigms. As the Millennium Ecosystem Assessment made clear, answering questions about how to manage for sustainability incorporates both science as we have come to know it and pushes its boundaries into realms of policy and philosophy.

According to Funtowicz and Ravetz (chapter 17), these new problems can be characterized by situations where the facts are uncertain, values are in dispute, the decision-making stakes are high, and there is a sense of urgency that decisions be made. In these situations, they have suggested that normal paradigm-driven science (in the Kuhnian sense) is insufficient and that there is a need to accommodate a much wider range of modes of constructing knowledge about the world. In this setting, paradigms do not replace each other. By expanding the peer group, they play off against each other to give us a richer understanding of the world. Funtowicz and Ravetz have called this post-normal science. Post-normal scientific inquiry, in their view, is an act of collaborative learning and knowledge integration. Expertise is collective, and the role of this expertise shifts from giving correct advice to sharing information about options and trade-offs. Reductionist forms of inquiry are not marginalized but drawn upon and embraced to inform approaches that are not so much holistic as they are transdisciplinary and comprehensive. Contradictory evidence and uncertainty from complex reality is not set aside, excused, or used as an excuse to privilege some information over others but rather is incorporated into a richer, albeit less predictive, understanding.

The theoretical constructs and the case studies from around the world that are presented in this book represent attempts to elaborate on such a science and to explore the ways in which a variety of systems constructions can be used to inform post-normal scientific inquiry.

Since the 1980s, the sustainable management of environments, societies, and economies has become the broadly accepted backdrop for policy and management decisions in most parts of the world. For many, this backdrop provides reassuring scenery against which we play out our daily lives and think about the future. The hard fact is, however, that in all regions of the planet (individually, collectively, nationally, regionally, and globally), people are having to make urgent and controversial decisions on public health, environmental, economic development, and agricultural issues with very little scholarly or practical guidance on how to integrate multiple perspectives across spatial and temporal scales. These decisions are made even more difficult by the fact that sustainability itself, being the capacity to create, test, and maintain adaptive capacity over time (Gunderson and Holling 2002), requires us to reconcile ecological, economic, and social imperatives (Dale 2001). Given the scholarly and policy interest in questions that link social and ecological activities, it should not be surprising that some of the key insights into how to structure inquiries to answer these questions should have come from ecologists and environmental managers.

The issues of this work of reconciliation are complex and interacting, where the decision-making stakes, the scientific uncertainty, and the conflicts over what constitutes legitimate knowledge are high. Even as we speak of managing for global sustainability, we are faced with ethical imperative to "deal with" ethnic pluralism, multiple epistemologies, and place-based biodiversity. Scientists are often at loggerheads over how to advise politicians and the public, quarrelling about data and their interpretation in front of television cameras. Having grown accustomed to a way of doing science that breaks out the pieces and gives them to experts, we are faced with making decisions that demand integrated and integrative answers, where the collective expertise is an emergent property of many different modes of inquiry. Furthermore, as Berkes and Davidson-Hunt (chapter 7) point out, we are increasingly aware that "if an ecosystem is not an object to be understood or managed by external agents, but a place of dwelling for a group of people who use it, then what are the implications of this for research and management?"

Indeed, after all the rhetoric of global conferences, we might ask, is managing for sustainability possible? If so, how? Without good theory, good management can only happen by accident. Without practice, all theories are suspect. Are there ways to combine science and management in a way that builds on the strengths of both? Can we learn our way into a sustainable (human) future on this planet? These issues cut across all areas of practice—including health, urban planning, natural resource conservation, agriculture, international development—and a wide variety of areas of scholarly inquiry.

The ecosystems we call home are diverse, complex, and dynamic. As a species, we have at once accelerated the speed and increased the scale of changes that were already nonlinear and full of surprises. As such, there is considerable scientific uncertainty surrounding ecosystem change and the management of human activities. The high degree of uncertainty, coupled with an urgent need for more for sustainable development necessitates a fundamentally different and more creative approach to decision-making. Traditional (disciplinary) science, while necessary, is not by itself sufficient for understanding and dealing with ecosystems, especially if these are understood to have embedded in systems and organizations created by that peculiar species, *Homo sapiens*. As ecologists and environmental managers were among the first to recognize, a new, broadened, and interdisciplinary approach bridging science and management is essential.

The history of environmental management in the twentieth-century has been described in some detail by Bocking (1997, 2004). What we describe in this book is what has emerged in the 1980s and took new forms in the 1990s and beyond. The new ecosystem approach we describe represents a synthesis between conventional ways of framing both ecological problems and environmental management and more recent theories of complex systems. This approach is not intended to displace the focused science that has been the backbone of biological inquiry for the past century but rather to embrace that work, to build on it, and to find ways to connect it to large temporal and spatial contexts.

Beginning in the early 1970s, confidence in ecosystem ecology's central role in ecological research and environmental policy declined. In part, this reflected the discovery that construction of realistic ecosystem models able to predict impacts of human activities is more difficult than first expected. Some also criticized the failure of ecologists and other scientists to explain their results in ways that society can comprehend. For example, early studies of climate change have been criticized for neglecting socioeconomic implications and focusing on atmospheric change to the exclusion of the economic and political variables that drive the current social-ecological system and can (perhaps) be manipulated to find solutions (Cohen et al. 1998).

Most importantly, the political context changed. In the United States, and to some extent in Canada, the belief in comprehensive management prevalent in the late 1960s was replaced by renewed reliance on processes more typical of a pluralistic political system, such as negotiation, compromise, and brokerage of competing interests, which are often conducted in an adversarial environment. Acceptance of a positive role for government in fostering society's interests was replaced by greater reliance on competition, private initiative, and individual interests. This implied a shift in the perceived role of science. No longer an alternative to "political" processes (as ecosystem ecology was once envisaged), science has instead become a participant, contributing knowledge that is useful in decision-making and dispute resolution and that is considered factual and value neutral. In this way, science was seen to impart an air of objectivity to the resulting decision (Bocking 2004; Jasanoff 1989; Nelkin 1992). Such a role placed a premium on quantifiable, precise predictions, which ecosystem ecologists were not immediately able to provide.

Although ecosystem ecology in its old, purely biophysical sense suffered a period of eclipse in this new political landscape, a variety of vigorous alternative views of ecosystems developed to take into account the rapidly changing nature of scientific inquiry and its role in society. Those who claimed ecosystems studies had disappeared were perhaps looking in the wrong places. New insights from fields as disparate as ecosystem studies, general systems theory, cybernetics, soft systems methodology, complex systems theories, hierarchy theory, thermodynamics, and chaos theory have brought together in new ways of understanding both ecosystems and our roles in them. Many of these new insights emerged from the work of scholars, such as Henry Regier, George Francis, Thomas Hoekstra, and Tim Allen, who were active in the scientific activities of International Joint Commission of the Great Lakes from the 1970s onward, as well as the Resilience Alliance of Holling and Gunderson. From these insights into the complex interactions within what now can only be viewed as hybrids of social and ecological systems (there being no pristine, nonhuman-influenced systems extant) have emerged new concepts of management and, indeed, new ways of thinking about science. These have, in turn, led to the notions of post-normal science and emergent complexity.

The ecosystem approach as described in this book reflects those new ways of thinking. Some of the pioneers in ecosystem studies might have difficulty recognizing the intellectual world inhabited by the ecosystem approach as described in this book, as they would no doubt have difficulty recognizing the biophysical and social world we now inhabit. Yet these new ways of thinking about our interactions with the natural world are deeply indebted to these pioneers and their ideas.

Scientific concepts rarely reflect simply an objective understanding of empirical reality. As the history of the ecosystem approach suggests, their evolution reflects not only our changing understanding of nature but our evolving sense of the role of science, and ultimately, of our place in the world. In describing nature, we describe ourselves. Since the 1930s, our understanding of ecosystems has been shaped not only by empirical observations but also by ecologists' changing visions of their discipline, of their place within society, and of the shared intellectual and cultural landscape we inhabit. The broader acceptance of ecosystem approach has been conditioned not only by empirical evidence or theoretical rigor, nor even by the reinvention of ecosystems studies themselves as described in this book, but also by how society views itself and the roles of individuals and institutions within it. By understanding this interdependence of ideas of nature, science, and society, we can better understand how the ecosystem approach can address the challenge of fostering respect for, and nurturing the sustainability of, the ecosystems that are our home.

This book presents an emerging integrative and innovative approach to managing for sustainability, not only in the midst of uncertainty and complexity but also in the midst of political, economic, and ecological turmoil. Integrating complex systems theories and participatory (and in some cases, collaborative) management, the ecosystem approach as we describe it in this book has grown out of, and feeds back into, case studies from around the world, ranging geographically from Canada and New Zealand to India, Latin America and Africa, and in focus from urban and community planning, and public health, to agriculture and management of natural areas.

All the authors represented in this book are struggling to find integrative and innovative solutions to the practical and theoretical challenges of understanding and nurturing sustainable, convivial, and just ways of living on this planet. Although the work of the late James Kay, which comprises a large part of the first section, informed much of the work, the relationships between the case studies and the theoretical ideas put forward by Kay were a kind of conversation and a debate. Rather than being illustrations of Kay's frameworks, they are perhaps best seen as arguments with it, from which all of us learned, and moved on to new place-based experiments. James Kay's untimely departure in 2004 left a large silent space in the argument around this scholarly table. We welcome our readers to engage in this discussion and to carry it to the streets and alleys, the mountain valleys and the swamps—wherever people live—and to join us in the daunting and exciting task of learning our collective way into a sustainable and convivial future.

References

Bocking, S. 1997. *Ecologists and Environmental Politics: A History of Contemporary Ecology*. New Haven, Conn.: Yale University Press.

Bocking, S. 2004. *Nature's Experts: Science, Politics, and the Environment*. New Brunswick, N. J.: Rutgers University Press.

Cohen, S., D. Demeritt, J. Robinson and D. Rothman. 1998. Climate change and sustainable development: Towards dialogue. *Global Environmental Change* 8(4):341–371.

Dale, A. 2001. *At the Edge: Sustainable Development in the 21st Century*. Vancouver, Canada: University of British Columbia Press.

Gunderson, L. H. and C. S. Holling (eds.). 2002. *Panarchy: Understanding Transformations in Human and Natural Systems*. Washington, D. C.: Island Press.

Jasanoff, S. 1989. The problem of rationality in American health and safety regulation. In *Expert Evidence: Interpreting Science in the Law*, eds. R. Smith and B. Wynne, 151–183. London, UK: Routledge.

Nelkin, D. (ed.). 1992. *Controversy: The Politics of Technical Decisions* (3rd ed.). Newbury Park, Calif.: Sage.

Rabsch, W., B. Hargis, R. Tsolis, R. Kingsely, K-H. Hinz, H. Tschäpe and A. Bäumler. 2000. Competitive exclusion of Salmonella enteritidis by Salmonella gallinarum in poultry. *Emerging Infectious Diseases* 6(5):443–448.

The Ecosystem Approach

PART I

Some Theoretical Bases for a New Ecosystem Approach

In this section (chapters 1–9), we cover the main theoretical and practical challenges of an ecosystem approach to managing for sustainability and some important possible responses, particularly as reflected in the ideas of the late James Kay and a few close colleagues. The intent of this section is not to provide an in-depth review of complex systems thinking but rather to identify those features that are deemed most important for the implementation of a reasonable scholarly and management response to the complexity of the world. Chapter 1 provides a basis for the more applied chapters that follow and that serve as a kind of argument, or conversation, with the theories as posited in Part I.

1

An Introduction to
Systems Thinking

James J. Kay

The Nature of the Beast

Environmental issues and sustainability have thwarted our society's scientific approach to dealing with the world. One need only contemplate global climate change to experience the frustration and confusion. In this book, we are using the term complexity as a concept that covers problematic situations that have eluded traditional scientific solutions. Complex situations involve uncertainty and surprise. They give the impression that there is no right way of looking at them and no right answer to the problems they raise. The problem is really the singularity of our concept of the "right answer." Complexity defies linear logic as it brings with it self-organization and feedback loops, wherein the effect is its own cause. Circular relationships between cause and effect require nonlinear logic, explanations in terms of morphogenetic causal loops where form is determined by and determines its own plans. In essence, complexity is characterized by situations where several different coherent future scenarios are possible, each of which may be desirable, all of which have an inherent irreducible uncertainty as to the likelihood for their actually coming about.

The differences between the above scenarios require a number of different perspectives at different scales of investigation. Understanding complex situations thus invokes alternative perspectives, which can be perplexing. Yet there is no avoiding our environmental concerns, and so we must take up the challenge of complexity. While not a panacea, systems thinking seems to offer some insights and approaches for dealing with complexity. As such, it holds the promise of helping us chart the course to sustainability.

Systems thinking is about patterns of relationships and how these translate into emergent behaviors. This section explores the notion of systems and its application to ecosystem thinking. Systems thinking provides us with a window on the world that informs our understanding of nature and our relationship to it. It provides us with a way of framing our investigations and a language for discussing our understanding. Translating systems thinking into action is what systems approaches are about. In this section the focus will be on systems thinking as it applies to biophysical systems.

Making Sense of Nonlinearity: Self-Organization

One of the puzzling observations about issues of sustainability is that everything seems to happen at once. Teasing apart causal links using conventional scientific techniques doesn't appear to help us answer the important questions. Systems thinking can help us by providing a language and conceptual tools for talking about the richness that comes with complexity.

Underlying systems thinking is the premise that systems behave as a whole and that such behavior cannot be explained solely in terms that simply aggregate the individual elements. This premise is, of course, the antithesis of prevalent reductionist thinking. Take, for example, evapotranspiration in a wetland. If one measures the evapotranspiration for the plants that make up the wetland when they are isolated in pots and add this to the evaporation for open pans of water (the classical experiment), one gets a higher value than the evapotranspiration of the plants and the open water when they are together in a wetland. One perspective on this is that when the plants transpire, they increase the humidity of the local atmosphere, thus decreasing the evaporation from the open water. Then again, does increased open water decrease plant evapotranspiration? The nonlinear causality in the loops typical of such systems makes distinguishing causal order impossible. Furthermore, this emergent property of wetlands cannot be deduced from more intensely studying their individual elements in isolation. Yet the dominant reductionist approaches are so entrenched that I have personally dealt with senior scholars who cannot accept that the evapotranspiration of a wetland is not simply the sum of the evapotranspiration of its component parts. There is a certain myopia in the dominant reductionist approaches, and it hinders our ability to deal with situations where emergence (i.e., the whole is more than the sum of the parts) is an important feature. Systems thinking is well suited to understanding such situations that require considerations of the whole as an emergent with its own properties.

An important emergent property of the whole is self-organization. We shall discuss this in more detail in a later chapter. However, it is important for us to introduce the notion here because self-organization is the phenomenon that gives us a sense that a system has an "identity" of its own. A simple example is a school of fish or a herd of wildebeest. The school as a whole seems to move of its own accord. Understanding or modeling this movement comes from understanding the relationship that is maintained between individual fish and wildebeest rather than from independent behavior of the individual itself. Self-organization is about how coherent patterns of relationships are internally structured and develop over time. How these relationships develop over time leads to a number of surprising and counterintuitive phenomena.

One of the manifestations of self-organization, which gives us a sense of a "whole" is the way in which systems deal with disturbance and, indeed, often incorporate disturbance as an important element of their dynamics. DeAngelis (1986) gives an example of this from southeastern Australia. The dominant trees are sclerophyllous eucalyptus, but the undergrowth consists of lush mesophytic vegetation. Normally, these circumstances would give rise to a temperate rain forest. However, these systems are subject to frequent fire, which would not occur if the mesophytic vegetation dominated. Fire increases soil leaching and sclerophylls are better adapted to poorer soils than mesophylls. Thus the dominance by sclerophyllous forest depends on fire and the occurrence of fire depends on the dominance by sclerophyllous forest. So fire has been incorporated as an integral element to the existence of the sclerophyllous dominant forest.

In a sense, the self-organization is a happenstance outcome of fire and the vegetation meeting. The components of the vegetation are significantly already evolved before the components of the new stable configuration ever came together. While the organisms in the forest are coded by DNA, the self-organization supersedes all that. There may be some microevolution that causes the components to line up in detail as they stabilize the emergent vegetation type. However, as with all self-organization, it comes down to flux and process; there is no plan or script for how the situation plays out.

This example also illustrates the importance of feedbacks and morphogenetic causal loops in understanding self-organization. In this case the feedback loop is that fires increase soil leaching, which increases the sclerophylls at the expense of the mesophylls, thus increasing the amount of forest fire. This would quickly get out of control, except that the mesophytic undergrowth limits the amount of fire, and so the whole system is in balance. It is such a balanced network of nonlinear causality that is referred to as a morphogenetic causal loop. The morphogenetic causal loop of sclerophyllous dominance, fire, and soil infertility obstructs the development of temperate rain forests and preserves the status quo.

The nonlinear causality of such systems gets us into trouble as environmental managers. An example is forest fire in temperate forests of North America. Forests are adapted to fire and are organized in such a way that normal fires cause only small areas of damage. The fire releases nutrients and makes openings for seedlings, promoting reproduction. Normal forest fires rejuvenate forests, keep the fuel level down, and prevent larger, more damaging fires and pest outbreaks. Suppressing forest fires prevents the rejuvenation process, allows fuel to accumulate and sets the stage for conflagrations, like the one that occurred in Yellowstone in 1988. Even in hindsight, researchers remain ambivalent about the Yellowstone fire, as there was in place a management regime that encouraged fire suppression. Later research indicated that there are huge fires every 400 years or so, of which the 1988 fire may be argued as an example. Suppressing forest fires usually makes forests less healthy! Indeed, anyone who depends on linear causal models as the basis for their management decisions will find the world a perplexing place.

Self-organizing systems have in their repertoire of behaviors a way of dealing with disturbance through their buffering capacity. In essence, one can substantially change the environmental context for such a system up to a point (a threshold or tipping point) with little apparent effect on the system. However, a slight change beyond the threshold and the system will suddenly change, that is, it reorganizes itself in a very dramatic and often unpredictable way. The effect of acid rain on lakes is an example of this phenomenon. The acidity of the precipitation running into lakes did not suddenly change; rather, it changed incrementally over decades. The pH of the lake water, however, did not change substantially, relatively speaking, over the same period (Stigliani 1988). The lakes maintained their organizational state (low pH) through a series of feedback loops that largely buffered the lake (in a chemical sense) from the environmental change. Eventually, the runoff from precipitation into the lake reached a level of acidity that exceeded the compensatory capacity of these loops. Once this happened, the effectiveness of the system decreased, which, in turn, decreased the capacity of the loops to compensate, which decreased the effectiveness of the system, and then quickly the organization unraveled and the system flipped to a different organizational state, in this case a "dead" acidified lake. The pH of the lakes dropped in a very short time period, less than one summer season. In some instances the change occurred in weeks.

Again, our linear thinking can get us in trouble when we make decisions regarding such systems. When Steve Carpenter began to work with human management of lacustrine systems, he was at first surprised by the flips of behavior he saw because there are not that many examples of them in basic science applied to lake systems. However, after a series of examples in managed systems, Carpenter is now surprised if he does not find such discrete jumps in the state of the system. Carpenter et al. (1999) report this as a common phenomenon. For quite a while our interaction with the system appears not to have any (deleterious) effect. As we increase what we are doing to the system, nothing appears to happen. Then suddenly, with little warning, a small change in our behavior causes the system to change dramatically, and too late we realize that we were impacting the system. The ability of systems to buffer themselves from external influences and to incorporate external disturbance as an integral part of their patterns of organization is part of what gives us our sense of them as a whole, a whole that is adapted to the situation that it is in.

The acid rain–lake interaction is also an example of important self-organizing phenomenon. Complex systems self-organize through feedback loops, and their openness predisposes them to dramatic reorganizations at critical points of instability (Nicolis and Prigogine 1977) (e.g., the dramatic "death" of an acidified lake, which is a flip to a plankton-dominated ecology). These instabilities and the resulting jumps or abrupt changes in the system are caused by self-amplified internal fluctuations mediated especially through positive feedback loops. These give rise to the spontaneous emergence of new structures and forms of behavior. Amplification is thus a source of new organization and complexity in the system. At the points at which these new structures emerge, the system may branch off into one of a number of quite different organizational states, often referred to as attractors. The existence of multiple stable states, multiple possibilities necessarily implies indeterminacy, as which path is taken depends on the system's history and various external conditions that can never be completely predicted (Nicolis and Prigogine 1989), thus the unpredictable nature of complex systems.

It is one thing to recognize such complexity in multiple case studies, but how do we use this information to help us make decisions about sustainability? In the 1930s, von Bertalanffy (1968) noted that open self-organizing systems exhibited common attributes regardless of the disciplinary domain of study. He called this property of systems "isomorphism " (Blauberg et al. 1977: chap. 2). The existence of isomorphisms allows us to make generalizations about open self-organizing systems, that is, to build a general theory about their behavior and characteristics. This is one of the premises and the impetus behind the development of von Bertalanffy's general systems theory as well as more recent advances in systems thinking. By furnishing us with a typology and description of the patterns of relationships that can occur, both within the system and between the system and its environment, and the types of behaviors that can emerge, systems thinking provides us with a language, questions, and techniques for thinking through the self-organizing aspects of systems.

A Brief History of Systems Theory

The origin of the modern systems movement is generally attributed to von Bertalanffy's work in evolutionary biology. He began his work in the 1920s, and his first major presentation was a series of lectures at the University of Chicago (1937–1938). However, his work became more widely known in the late 1940s after his arrival in Canada. His commonly known publications date back to the 1960s. Von Bertalanffy's general systems theory was one of the first schools of thought that provided alternative models and modes of inquiry to the reductionist methods of disciplinary science. General systems thinking emphasizes connectedness, context, and feedback. Research questions identify and explain interactions, relationships, and patterns. The essential properties of the parts of a system can only be understood from the organization of the whole, as they arise from the configuration of ordered relationships that are specific to that particular system (von Bertalanffy 1968). Understanding comes from looking at how the parts operate together rather than from teasing them apart.

The next major contribution is generally attributed to Wiener (1948), who developed the field of cybernetics. While "systems thinking" originated in fields associated with natural systems, those researching mechanical and human systems quickly adopted them. Early adopters of the systems ideas included Margaret Mead (see von Foerster 1952), Gerard (physiology), Rapoport (mathematical biology; Rapoport and Horvath 1959), and Boulding (1956). Churchman (1968) and Beer (1959) linked systems concepts into operations research and organizational cybernetics. In recent years, primarily through the work of Senge (1990) and Checkland (1981), systems concepts have been integrated into the management sciences.

Complex systems thinking is the grandchild of von Bertalanffy's general systems theory. It emerged in the wake of the new science of the 1970s: catastrophe theory, chaos theory, nonequilibrium thermodynamics and self-organization theory, Jaynesian information theory, complexity theory, etc. A number of authors have focused specifically on self-organizing systems (di Castri 1987; Jantsch 1980; Kay 1984; Nicolis and Prigogine 1977, 1989; Peacocke 1983; Wicken 1987).

Systems theory was first developed by von Bertalanffy in response to his sense that reductionist science was not sufficient to deal with biological systems. During the Second World War the development of systems theory was spurred on by the logistics (command and control) problems of assembly, delivery, and support of large numbers of men and machines to specific locations at specific times (e.g., amphibious assaults, bomber raids). The problems of tracking multiple moving targets also motivated much thinking about cybernetics and systems organization. During this time, aircraft also began to fly so fast that human response times became a major safety issue, thus motivating the development of ergonomics. The cold war had a similar impetus on systems thinking and approaches, the problem being the organization behind the operation and control of strategic bombers and intercontinental ballistic missiles (ICBMs). Similarly, the race to the moon of the 1960s motivated the development of systems. Much of the work on systems, over the years, was carried out in the Soviet Union, and this association led to ideological issues during the Reagan years, which saw the systems movement fall into disfavor in America.

However, with the publicity in America surrounding the Santa Fe Institute and its association with successful business management, systems thinking is once again being pursued. Currently, systems thinking is playing a major role in dealing with environmental issues, organization of global corporations, and computer networks. It is interesting to note that systems approaches have played a central role in some of the most technically and organizationally challenging activities of humans over the past half century. It is unfortunate that the impetus for the development of systems thinking and approaches has often been the problems posed by the most detestable of human activities—the waging of war. For more complete discussion of the development of systems thinking, I suggest Blauberg et al. (1977) and Flood and Jackson (1991).

Questioning Reality from a Systems Viewpoint

One well-known statement associated with systems thinking is that "everything is connected to everything else." This is an overstatement as the connections between things can be quite weak. In the systems we study, most potential connections are set to zero. So what does the adage mean to say? It is not that everything is connected, though technically true, but rather that we should expect unsuspected and surprising connections to be important with some regularity. Because there are so many potential ways that things may surprise us, there are several dialects of systems approaches that can be useful. Systems theories, particularly those associated with nonlinearity and complexity, are not suitable for all classes of problems.

Many problems or investigations can be completed successfully using the Newtonian worldview. Many of the discussions within the systems community focus on defining the contexts and problems best suited to the various approaches and worldviews. Weinberg (1975), following on the work of Weaver (1948), proposed the partitioning of problem situations based on their complexity and level of randomness.

Organized (not random), simple situations, with small numbers of interactions, are designated "small-number" problems. Two masses orbiting each other, the trajectory of propelled objects, and simple pendulums are examples. The behavior of these entities can be explained by Newtonian science and mechanistic explanation. This is because the interactions and relationships between objects in these situations are tightly constrained and can be written as simple solvable equations.

Highly unorganized complex situations, dominated by large numbers of random interactions and aggregate behavior, are designated "large-number" problems. Gas molecules in a room and large groups of people are examples. They can be adequately described by statistics and statistical mechanics. Averages mean something because there are large numbers of unconstrained (i.e., random) interactions between objects.

The remaining middle ground, with intermediate numbers of interactions and organized complexity with only a degree of unpredictability, are designated "middle-number" problems. Interactions between objects are loosely constrained and in sufficient number such that averages are not helpful and equations that can be written to describe the interactions are not uniquely solvable.

Human organizations, three masses orbiting each other, double pendulums, and ecosystems are all examples of situations where behavior cannot be explained by Newtonian mechanics or linear cause-and-effect explanations nor described in a useful way by statistical means. In middle number situations, prediction is not possible, but one can still get an answer to a slightly different question. Systems thinking is most applicable in situations where simple prediction fails. One must change the context to get an answer to the original question to one that is still useful. This idea of partitioning problem situations based on complexity, organization, and degree of constraint is key to understanding the domain of applicability of systems thinking. In middle number systems the constraints are ambiguous. As systems thinking changes the context by bounding the system in some new way, some reliable constraints come to the fore. There is a certain freedom in finding those helpful constraints, and it is the nature of the interactions between objects in a situation that determines which tools are appropriate.

The problem is that, because normal scientific concepts work for some classes of problems, we seem to think that they must be valid for all classes of problem.

We tend not to think through the connections that might lead us to a more systemic view. Environmental management projects, ranging from dams to agricultural development, have often resulted in unanticipated, usually negative, impacts. For example, large dams were seen as a panacea in development, but few thought through the ecological change they would cause. The Aswan dam on the Nile disrupted the annual flooding of the Nile, which provided the nutrients for the downstream agriculture. In order to maintain this agriculture, fertilizer has to be produced to replace the nutrients once provided by the flooding. The energy required to produce the fertilizer exceeds the energy produced by the dam, thus nullifying the energy-producing benefits of the dam. According to The World Commission on Dams (2000), many ecosystem impacts of large dams were unanticipated, even as late as the 1990s.

In deciding whether or not we are dealing with a simple, complicated, or complex problem, we need to ask what needs to be considered and what can be ignored. How far upstream and downstream from the dam must one think about the problem? And at what level of detail? These questions combine to ask what scale of analysis is appropriate? Need we be worried about individual farms or agriculture in general? This raises one of the fundamental conundrums of systems thinking and dealing with complexity. It is a given property of systems that everything is connected (at least weakly) to everything else. However, no scientist can look at everything at once. So any analyst must make decisions about what to include and what to leave out of the system to be studied. Scale, extent, and type of study must be selected, as discussed by Allen (chapter 3). Decisions on scale and type, while done in a systematic and consistent way, are necessarily subjective, reflecting the viewpoint of the analyst about which connections are important to the study at hand and which can be ignored. So, because of their very nature, the notion of an objective scientific observer is not applicable to the study of self-organizing systems, because the new level of emergence forces changes in the decisions as to how to bound the system.

Furthermore, this conundrum begs the question: Who gets to decide what is important and what is not? Whose values are used and why? These questions must be answered at the beginning of a system's investigation and are clearly political in nature. The inability, in principle, to have a unique objective systems description of a situation immediately moves systems thinking and systems approaches out of the domain of traditional scientific approaches and into the realm of post-normal science. Post-normal science applies when the stakes are high, the time is short, there is much intrinsic uncertainty, and values are in conflict. Post-normal science puts front and center the issue of who gets to decide, and it develops a stagecraft for finding better, not correct, answers to that question.

Systems thinking provides us with a heuristic tool and common language for framing situations and exploring self-organizing phenomena. It provides us with guidance about how to decide what is important to look at, and not look at, and how to describe a situation. It helps us to understand the self-organizing possibilities in a situation and thus to map out potential future scenarios. It provides a basis for synthesizing our understanding of a situation into narratives about how the future might unfold and the trade-offs that exist between choosing different paths. It also helps us understand what it is we don't understand.

Conclusions

In summary, then, the complexity of problems offered by asking questions about sustainability offers a major challenge to systems thinkers. Addressing sustainability means finding a way to deal with this complexity. It has become clear that systems explanations of social-ecological complexity require different types of perspectives and at different scales of examination. There is no single correct perspective. Rather, a diversity of perspectives is required for understanding. Such systems are self-organizing; their dynamics are largely a function of positive and negative feedback loops. Linear causal mechanical explanations of dynamics are insufficient to understand them. Emergence and surprise are normal phenomena in systems dominated by feedback loops. Inherent uncertainty and limited predictability are inescapable consequences of these system phenomena. Such systems organize about attractors. Even when the environmental situation changes, the system's feedback loops tend to maintain their current state. However, when change does occur, it can be very rapid and even catastrophic. Precisely when the change will occur and to what state the system will change are often not predictable. Frequently, in a given situation, there are several possible states (attractors) that are equivalent. Which state the system currently occupies is a function of its history. There is not a "correct" preferred state for the system.

This enhanced understanding of systems, as complex systems, forms the backdrop for navigating a path to sustainability. Moving toward sustainability involves long enough timelines for the actors and the context of the ecosystem to change. These insights into buffering capacity of self-organizing phenomena address that long timeline. By contrast, the conventional science approaches of modeling and forecasting are often so inflexible as to be inappropriate, as are prevailing explanations in terms of linear causality and homeostatic properties that underpin ecosystem management of the traditional sort. The new understanding of complexity leads to an approach that is different from traditional ecosystem approaches. The conventional approaches may be interdisciplinary and participatory in nature, but they focus on analysis, forecasting, and a single type of entity such as a watershed or forest community. Complex systems approaches go beyond interdisciplinary to transdisciplinary, which invokes emergence between the disciplines over merely working between them. The new approaches remain participatory but go on to be adaptive and multiscale in their focus. In short, the new approach is in the mode of post-normal science (Funtowicz and Ravetz 1992, 1993). At its heart is the portrayal of ecological systems as self-organizing. They are also seen as requiring multiple levels of organization and indeed link the twinned hierarchies of social and biophysical systems in a given place. Accordingly, the new approach has coined a neologism, generally accepted practitioners in this new endeavor. They are called self-organizing hierarchical open (SOHO) systems and will be discussed in some depth in chapter 4.

Having explored the rationale underlying systems thinking and its application to the problem of sustainability, it is time to turn our attention to some of the basic concepts, tools, techniques, principles, and considerations of systems thinking. Looking at a problem situation through the lens of systems thinking involves three phases. The first involves framing the situation by generating a systems description or map of what is involved and the important relationships that are used to define the system. In the second phase a description of the dynamics of the situation is developed. The third phase involves synthesizing the understanding gained from the first two phases into narratives about how the situation might or could unfold in the future. The discussion that follows in the next few chapters is not by any means a complete review of systems thinking. Rather, it is meant to establish some of the important aspects and point the reader to more detailed sources. Enough will be said to give the reader a basis for starting to apply systems thinking to the problem of sustainability.

References

Beer, S. 1959. *Cybernetics and Management*. London: English Universities Press.

Blauberg, I. V., V. N. Sadovsky and E. G. Yudin. 1977. *Systems Theory: Philosophical and Methodological Problems*. Moscow: Progress.

Boulding, K. E. 1956. General systems theory—The skeleton of science. *Management Science* 2:197–208.

Carpenter, S. R., D. Ludwig and W. Brock. 1999. Management of eutrophication for lakes subject to potentially irreversible change. *Ecological Applications* 9:751–771.

Checkland, P. 1981. *Systems Thinking, Systems Practice*. New York: Wiley.

Churchman, C. W. 1968. *The Systems Approach*. New York: Dell.

DeAngelis, D. L., W. M. Post and C. C. Travis. 1986. *Positive Feedback in Natural Systems*. New York: Springer-Verlag.

di Castri, F. 1987. The evolution of terrestrial ecosystems. In *Ecological Assessment of Environmental Degradation, Pollution, and Recovery*, ed. O. Ravera. New York: Elsevier Science, pp. 544–556.

Flood, R. L. and M. C. Jackson 1991. *Creative Problem Solving: Total Systems Intervention*. New York: Wiley.

Funtowicz, S. O. and J. R. Ravetz 1992. Three types of risk assessment and the emergence of post-normal science. In Krimsky, S., and D. Golding, eds, *Social Theories of Risk*. Westport, CT: Praeger.

Funtowicz, S. O. and J. R. Ravetz 1993. "Science for the post-normal age," *Futures*, 25(7): 735–755.

Jantsch, E. 1980. *The Self-Organizing Universe: Scientific and Human Implications of the Emerging Paradigm of Evolution*. New York: Elsevier.

Kay, J. 1984. Self-organization in living systems. Ph.D. thesis. Waterloo, Ontario, Canada: Systems Design Engineering, University of Waterloo.

Nicolis, G. and I. Prigogine. 1977. *Self-Organization in Non-Equilibrium Systems*. New York: Wiley.

Nicolis, G. and I. Prigogine. 1989. *Exploring Complexity*. New York: W. H. Freeman.

Peacocke, A. R. 1983. *The Physical Chemistry of Biological Processes*. Oxford, UK: Oxford University Press.

Rapoport, A. and W. J. Horvath. 1959. Thoughts on organizational theory. General Systems 4.

Senge, P. 1990. *The Fifth Discipline: The Art and Practice of the Learning Organization*. New York: Doubleday Currency.

Stigliani, W. M. 1988. Changes in valued "capacities" of soils and sediments as indicators of non-linear and time-delayed environmental effects. *International Journal of Environmental Monitoring and Assessment* 10:95–103

The World Commission on Dams. 2000. "Dams and Development: A New Framework for Decision-Making" (Sept. 7, 2004), at www.dams.org.

von Bertalanffy, L. 1968. *General Systems Theory*. New York: George Brazillier.

von Foerster, H.(ed.). 1952. *Cybernetics: Circular Causal and Feedback Mechanisms in Biological and Social Systems: Transactions of the Eighth Conference*. New York. Josiah Macy, Jr. Foundation.

Weaver, W. 1948. Science and complexity. *American Scientist* 36:537–544.

Weinberg, Gerald M. 1975. *An Introduction to General Systems Thinking*. New York: Wiley.

Wicken, J. S. 1987. *Evolution, Thermodynamics, and Information: Extending the Darwinian Program*. Oxford, UK: Oxford University Press.

Wiener, N. 1948. *Cybernetics: or Control and Communication in the Animal and the Machine*. New York: Wiley.

2

Framing the Situation
Developing a System Description

James J. Kay

Framing the situation, that is, identifying the key elements of the situation and the relationships between them, seems like the obvious starting point for any investigation. However, it is my observation that this is rarely done in a systematic way. In our experience, this phase alone is often sufficient to dramatically affect the problem situation. People's understanding can be significantly altered through the exercise of explicitly framing the situation; the resulting fresh perspective can make the path to a resolution quite clear. Three examples of the difficulties one may run into if one proceeds to intervene in a situation without thinking through how to frame it are provided.

A famous Canadian environmental impact assessment focused on the effects of a proposed pipeline on reproduction of caribou herds. The assessment was framed this way because many scientists have been trained to think of the standard biological "framing" of species investigation in terms of reproduction. Later it was noted that the major effect of the development was on the food source of the caribou, which was not studied as part of the impact assessment. The study focus was on the reproductive habitat of a single species; it ignored the broader food chain. In other words, steps were taken to preserve the integrity of the reproductive process and habitat but not the integrity of the food chain, thus making the steps that were taken moot relative to the ecosystem overall. The assessment had not been framed properly at the start.

A second example involves an energy conservation strategy of a large power utility in the main office of the utility. The staff decided on a strategy to decrease the electricity consumption by turning off unused office equipment and lights, particularly on the weekend. However, the staff did not study the energy system in the building. They were unaware that the building was designed to capture the waste heat from these office devices (and also people), store it in a large "swimming pool," and re-release it later when the heat was needed to maintain the building temperature, particularly on cold winter nights. In fact, the building had no source of heat other than the captured waste heat. Turning off the office equipment could result in a very cold building. Fortunately, this author (who was aware of the total energy system in the building) reviewed the conservation proposal and pointed out the obvious problem. Staff undertook the project using simple linear thinking: turn off a machine and energy is saved in the building. Consideration was not given to how energy flows through the whole building, that is, how turning off a machine affects the other elements of the energy system. The possibility of such a relationship was not even considered. Had the staff started off thinking about how to frame the situation to generate an energy system description, they would never have made the proposal in the first place.

The case study described by Murray et al. (chapter 13), in the Ucayali region of Peru, began with a literature review and discussions with researchers who focused on this region. Dozens of research agencies, for years, described the region in terms of livestock management, slash-and burn-agriculture, and deforestation. Research about the area had been framed in these terms, with particular emphasis on livestock management. The next step in the project was to generate a systems description of the situation. Three main findings emerged that described different forces shaping the development and dynamics of the region. First, timber extraction employs the largest percentage of people, generates the greatest amount of revenue, and represents Ucayali's main export product. Second, annual crops provide the majority of the staple foods consumed and household income, and fish is the key source of protein and other micronutrients. On the average farm, cattle do not play an important role in either food production or income generation. Third, coca, the invisible crop, is cultivated over an area larger than that of the two main legal crops—plantains and bananas—combined. The system description developed by the project team framed the situation in a very different way than previous work had suggested; in particular, viewing the development problems of the region in terms of livestock management, fish production, drug enforcement, and timber extraction results in very different sets of research and policy issues being examined.

These examples demonstrate that the first step in dealing with a complex situation is to explicitly frame the situation. We now discuss how a systems description is a means of doing this.

Generating a Systems Description

A systems description is always from the perspective of an observer, and the question or issues in which they are interested. Weinberg (1975) discusses this quite well. He notes that every system description is a partial description of a reality based on one observer's perspective. So when we talk about a system, we are not talking about a physical object but rather our limited mental representation of it. The system is not "out there" but "inside us." Some of the epistemological and ethical implications of this are discussed by Rosen (1991) and Maturana and Varela (1987). This suggests that any situation should be described by a number of systems descriptions, and in this way a richer polyocular (though still incomplete) description of the situation can be generated. Of course, this begs the question of which perspectives get used, a question that we return to several times in this book and that remains a fundamental challenge for working on issues of sustainable development.

The essence of a system is that we perceive a collection of things to be a whole. They are perceived to have an identity and to do something, either actively or passively in that they fulfill some function in the world around them. Systems are made up of components that are interconnected in a particular way. These interconnections are about the relationships between components. In physical systems these relationships and interconnections are about mass, energy, and information flow between the components. The particular way in which the components are interconnected constitutes the system structure. The components, configured together in the particular structure, constitute the system's organization, and it is this organization that allows the system to do something.[1]

Of particular interest in this regard are open systems: systems that have an environment that provides their context (the conditions in which they are embedded) and provides the source and sink for flows in and out of the system. When we look at an open system, we see its structure,[2] but not the processes. We see the consequences of the processes, usually in the form of inputs and outputs of the system or in the ability of the system to maintain itself in a state different from that of its environment.

The Starting Point

A systemic description identifies the important components, their interrelationships, our sense of the system as a whole and the system's relationship with its environment. Many interesting systemic descriptions have emerged from an undergraduate course at University of Waterloo entitled "Greening the Campus" (www.adm.uwaterloo.ca/infowast/watgreen/).

In one Greening the Campus project we sought to reduce the waste generated in a student residence cafeteria. The process began with background research about the type of system being studied, in this case, literature on student residence cafeterias and discussions with managers of such cafeterias. Some preliminary observations were then made about how the system was functioning, in this case, observation of the food that remained on plates after students had finished their meals. This revealed a large amount of untouched food (e.g., bananas and oranges), and unopened packages (e.g., milk cartons, orange juice) that ended up in the garbage. Surveys of students revealed that this was because meals were served on a fixed price, "all you can eat" basis. People were taking food on speculation so that they would not have to go through the line again. This then became the root issue that defined the system investigation, that is, orientated our investigation. Given this issue, we undertook to study the flow of food to the point that it became waste. Essentially, food for all the different cafeterias came into the university campus and was stored in a common facility. Food was then shipped to the appropriate kitchen, prepared, served in the servery, and consumed in the cafeteria. Cafeteria waste was removed in the dish cleaning room, and kitchen waste was removed in the pot cleaning room.

While this description looks quite straightforward, putting it together was not. It required identifying people who were involved in the handling of food and the production of meals, interviewing them, and actually following some food from its arrival on campus until it became "trash."

No one individual had a complete picture of the food production process. They only knew about their limited area, such as the kitchen or the scrape room. Indeed, in general, we have rarely found individuals who had a complete systemic understanding. Individuals know about the piece of the system in which they are most closely involved. It is our job as researchers and systems analysts to put the individual pieces together into an overall systems description.

Systems descriptions are deceptively simple and yet, in our experience, can be both challenging and profoundly change our understanding of the nature of a problematic situation. Consider one study that involved evaluating the indoor use of pesticides on a university campus (Cobean et al. 1995). The students started by attempting to describe the system but found much contradicting information during their conversations with various staff at the university. They realized, after months of effort and bewilderment, that, in fact, there were two pesticide application systems operating independently on the same campus. Two systems descriptions were generated, one for preventive activities and the other for complaint-driven actions. Pesticides were, in some instances, being doubly applied, resulting in a less than desirable situation. No one in the administration had stepped back far enough from the situation to see the big picture, i.e., the whole system.

Another example involves a Métis community. This community, which does not have indigenous people status, was perceived to be living on government handouts, whereas the community saw itself as living off the land. A complete system study of the material and economic flows in the community (Tobias and Kay 1994) was undertaken. This required a researcher to live in the community for two years and track, through various research instruments, all that was consumed by the community. All the fish caught were weighed and accounted for; all hunting, trapping, and harvesting were monitored; and all dollars were traced. The result was startling to all. The total harvest was 84.5 tonnes of edible meat or 0.342 kg per day for each of the 676 residents! Three tonnes of berries and 682 cords of fuel wood were also harvested. In fact, more than a third of the community's income came from bush harvest, i.e., the land. The remainder of the community income was divided between wages and various forms of government support payments. The major component of the community's income was from its land base. Government "handouts" represented a minor but significant fraction of the income. The study also identified and documented a number of species that were used by the community but that government officials insisted were not used. These examples demonstrate the way in which a systems description can alter our understanding of a situation and the difficulty that undertaking a system study can entail.

It has been our experience that there may be considerable information available about some elements and aspects of a system and virtually none about others. This is because a systems investigation asks very different (holistic) questions about a situation, questions that, previously, no one thought to ask. After undertaking systems investigations, I have been struck by how profound our ignorance of the biophysical world is. There are entire subsystems about which we simply have virtually no knowledge. This has left me humbled and with a profound sense of how little we actually know.

Suboptimization

So far we have only considered the most elementary of system considerations, tracing the flows through a system. Just doing this can provide significant new insight.[3] However, there are other considerations to explore as well. Returning to the student residence cafeteria example, after research, consultation, and discussion, a redesign of the cafeteria was undertaken so that students paid for each item they put on their plates. This reduced the waste scraped from plates after meals by 72 percent. The redesign seemed to be a big success. However, we monitored all the waste generated, from the time the food entered the university, through all the processing steps, until it was disposed of or consumed. The systems diagram was used as the basis for developing a monitoring program of the overall food system (see chapter 16 for a discussion of systems descriptions and monitoring). The changes in the overall food system required by the redesign actually increased the waste generated in some subsystems. However, when all the waste generated is taken into account, a 45 percent decrease was obtained.

This illustrates an important system principle, the principle of suboptimization. When one part of a system is optimized in isolation, another part will be moved farther from its optimum in order to accommodate the change. Generally, when a system is optimal, its components are themselves run in a suboptimal way. One cannot assume that imposing efficiency criteria on every component in a system will lead to the most efficient system overall. Frequently, it will not. If we had not monitored the whole system, we would have had a false sense of how well the redesign worked. This underlines the importance of how you frame a situation.

Boundaries

Implicit in the discussion, so far, have been the boundaries we choose for the student residence cafeteria system. It was explicitly decided to only consider food processing and consumption activities that occur on the campus. We did this because we had control over what happened on campus. The outside world was treated as external, a source of inputs (food) and outputs (food waste, including recyclables and compostables). We also did not look at energy costs associated with the change, even on campus. We left things out, for example, more "single-serve" prepackaged food was purchased in the new system, which meant suppliers would produce more waste. The new system also produced more recyclables, which has an effect on the waste stream handled off campus by the government. We could not look at everything. We had to bound the situation, decide what would be in the foreground versus the background.[4] This decision is arbitrary but not capricious.

Drawing boundaries may be based on science, but it is an art form and a political and ethical decision. Every person studying systems must explicitly specify the criteria for bounding the situation. How one draws boundaries and frames the situation can radically alter the conclusions one draws about the situation. We have noted this earlier in the examples of the Aswan dam and energy conservation in a building. Had we drawn the boundaries in our system description of the cafeteria to include only what happens in the meal consumption area, then we would have had a different sense of how effective the new program was at reducing waste.

In the year 2000, more than 2000 people became ill, and 7 died, in the small city of Walkerton, Ontario, when one of the city's wells became contaminated with *Campylobacter* and pathogenic *E. Coli* bacteria (O'Connor 2002). The farmer whose land was apparently the source of runoff into the city well had a well-designed and implemented environmental farm plan. The plan, of course, stopped at his farm boundary. The city of Walkerton, when digging that particular well, considered what was inside its own boundaries but did not consider the adjacent farm. Neither the farmer nor the city had prior concerns about climate change or the unprecedented rainfall of May 2000. The provincial government, in downloading responsibilities for monitoring water to the city (to save money), did not consider whether the city officials had the training or capacity to do so, nor did they consider communications pathways for reporting water monitoring results from the newly privatized laboratories to the public health officials. By drawing different boundaries, each of the players in this system saw, in a sense, a different system, judged by different criteria and, until the tragedy occurred, judged to be a success.

Often controversies about environmental issues reflect different framings of a situation, particularly with respect to boundaries. Thinking through the different unarticulated mental maps of a situation that proponents have through the exercise of developing a system description can help to clarify what the underlying issues are. This is particularly true when discussing issues of safety. Consider, for example, the safety of nuclear power plants. If we draw boundaries around the power plant and only consider what occurs inside the plant boundaries, we might conclude that nuclear power is safe, that is, that the plant can operate safely.[5] Others might choose their boundaries to include everything from mining to long-term disposal. This would probably lead to a different conclusion about the safety of nuclear power. The physical situation is the same, but different boundaries lead to different conclusions.

Drawing system boundaries is about values, about what people feel is important in a situation. As such, this step in the process of developing a system description can be politically volatile as participants debate what ought to be in the foreground and what can be in the background.

Holons and Nesting

In his *Ghost in the Machine*, Koestler (1967) coined a new word: "holon." The idea has developed since the original statement of forty years ago, but the central issue remains the same. Hierarchies used by biologists to classify organisms in a taxonomic scheme have levels such as species, genus, and family. Each of these levels, and all the other levels used to classify organisms, are collectively known as taxa, so that the species is a taxon. A holon is like a taxon, but it applies to any sort of hierarchical arrangement from the generals and soldiers in a military command to the cells, tissues, and organ systems in a human body. There are points of tension between what is meant by species as a general concept, a particular species, and the collection of all organisms in that species. These same points of tension surrounding the taxon also apply to holons.

Koestler's concept applies well to the general thrust of these opening chapters. Although Koestler was not explicit on the point, preferring to make holons more concrete, it is now generally understood that, since Allen and Starr (1982), what is inside a holon comes from the person who decides on the hierarchy and its intended uses. In other words, bounding the holon is a matter of taking responsibility for decisions and points of view. Koestler was particularly explicit that the holon is the boundary across which information and material are integrated for the components inside. That same boundary serves to integrate the information and material passage from inside the holon out to the rest of the universe.

Drawing boundaries is about stating what it is that was considered a "whole." In the example of the student residence cafeteria, the "whole" was the meal production system involving all the food processing steps on campus necessary to deliver a meal to a student. In the case of the Huron Natural Area (HNA), a relatively small, "natural" area in Kitchener, Ontario, the boundaries around the HNA are the legal property lines that define the area that is owned (see Lister and Kay (2000) and www.fes.uwaterloo.ca/u/jjkay/HNA/ for a more detailed description of the HNA and its management). It is ownership that gives the sense of a whole and defines the holon. However, the HNA must be considered in the context of the larger system of which it is part. From an aquatic perspective, surface water flow is very important. From this perspective the HNA is part of the Strasburg Creek subwatershed. All the water in the subwatershed drains into the HNA, and these drainage patterns set the stage for the streams and ponds in the HNA. Change the drainage patterns outside the HNA and you change the streams and ponds in HNA. The Strasburg Creek is referred to as the environment for the HNA. It provides the context within which the HNA exists. The HNA is nested within the Strasburg Creek subwatershed.

This identity of the HNA at two levels of analysis contributes to the complexity of the system as a management endeavor. This example introduces the systems concepts of holon, environment, nesting, and context. The upper level of the watershed context would not matter except that the HNA is an open system. The entity being focused on as an open system, a whole, is a holon. All holons are made up of holons and contained in holons. In Koestler's (1967, 1978) coining and usage of the term holon, he refers to this characteristic of systems as being like the Roman guardian of doorways, the god Janus, who is presented as having two faces, one looking forward and one looking backward. Similarly, each holon exists within a system that it is dependent on, sets the context for, and constrains the behavior of. Each holon is made up of other holons on which it is dependent and whose repertoire of behavior limits the holon's behavior. Holons are nested within holons and have holons nested within them. There are other sorts of hierarchical arrangement where the nesting of lower level holons inside upper level holons is not a requirement (e.g. pecking orders, food chains), but because we will be relying on certain thermodynamic properties of hierarchical relationships, nesting is more or less required to make even an informal bookkeeping tractable. Koestler (1978) referred to this nesting as a holarchy; Regier (see Appendix A) has proposed the term holonocracy as an alternative to make more explicit the mutual power relationships across scales and to deemphasize the residual notions of hierarchy that holarchy carries. A familiar example is the biological nesting of cells, within organs, within individuals, within populations, within species, within communities, etc.

This nesting of holons must be thought through when developing a systems description that will be using thermodynamic principles as guiding ideas. The surface water nesting for the Huron Natural Area is in fig. 2.1. The Huron Natural Area is a landscape made up of communities. Note that the nesting of plants in communities is not on the criteria of water flow inside watersheds. Even so, the whole HNA is part of a wider system, the Strasburg Creek subwatershed. Aggregation criteria are free to change because of nesting keeping things straight. The Strasburg Creek subwatershed is part of the Grand River watershed, which is the Huron Natural Area's environment. The Grand River, in turn, is part of the Great Lakes Basin, which is the Huron Natural Area's wider environment.

Scale

Nesting brings up another system's issue, that is, scale. Allen will discuss this in some detail in the following chapter. However, it is important to introduce some basic concepts here. When we think about the Huron Natural Area and its relationship with the Grand River, we are thinking about a very different set of relationships and concerns, at a very different scale of consideration than if we were examining the communities within the Huron Natural Area. Even our sense of the Huron Natural Area as a whole will change substantially, depending on the scale of consideration. Like boundaries, the scales chosen for a systems description have a profound influence on our understanding of a situation.

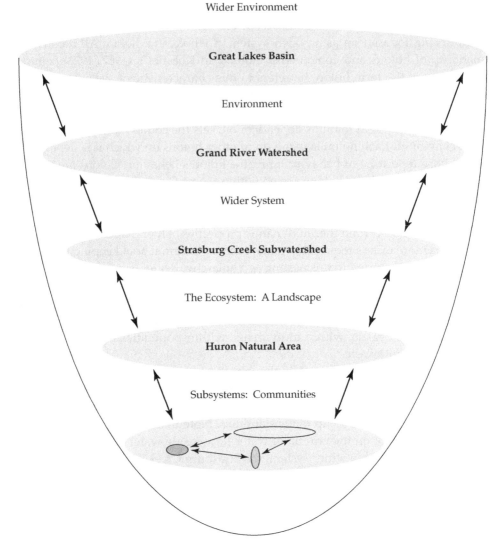

Figure 2.1 Huron Natural Area holarchy. Nesting of surface waters

To illustrate, consider a systems description of a bicycle. Most people would choose a scale where the relevant components would be perceived to be the seat, handlebars, wheels, pedals, etc. I assign this example in my classes. One year a member of the Canadian Olympic bicycle team was a student in the class. Her description focused on the scale of individual bearings, spokes, wires, and bolts. This level of scale is quite appropriate when a bicycle is studied as a means for maximizing speed but quite unnecessary when a bicycle is studied for its recreational use. The appropriate scales for consideration of a situation depend on the issues being explored and the questions being asked; that is, why we are looking at the situation.

Another example of the importance of scale comes from the HNA project (Lister and Kay, 2000). Currently, the aquatic ecosystem consists of a fast-moving, highly oxygenated cold-water stream in which brook trout thrive. However, another, incompatible regime is emerging: a slow-moving, low-oxygen, warm-water stream interrupted by ponds and small wetlands in which beaver and muskrat are dominant species that shape and maintain the habitat and its constituent communities. The current ecosystem trajectory is tending toward this regime because of several factors, including an invading population of beaver. The beaver dam the stream, thus providing more habitats for beaver. The dams slow the water flow, increase water temperature, and decrease dissolved oxygen. The emergence of this beaver-dominated regime was a complete and unpleasant surprise to the local managers of the natural area. However, managers operating at the much larger scale of the Grand River watershed basin, of which the natural area is part, were not surprised because, for years, they have followed the beaver migration toward the natural area. The surprise in the local management occurred because, much as between the campus emergency and preemptive pesticide delivers teams, there was no clear mechanism or opportunity to communicate the crucial information to local managers. This example illustrates how scale of consideration can significantly affect one's perspective on a situation.

When viewed from different scales, different relationships and considerations emerge. One needs to think about the situation from a variety of scales, to identify those scales that reveal the relationships and other considerations that aid in understanding the situation. It is very important to identify the scales that can provide us with useful insights. Like boundaries, choosing scales is an art form and can be contentious. And like boundaries, controversies often exist because a situation is being examined from different scales without those involved realizing it. The controversy arises because thinking at different scales will lead to different understanding. The real issue is about which scales are relevant.

Context

The trout and beaver examples also illustrate the importance of context. The migration of beavers represents a change in the external environment of the HNA that ultimately would result in internal changes in the HNA. The openness of the HNA means that its legal property boundaries were permeable to beaver. It is as important to think through the external context of complex systems at appropriate scales as it is to think through its internal makeup. A mistake often made is to focus exclusively on what is inside the system boundaries to the exclusion of external influences. Given the frequency of negative surprises generated by a fixation by managers on boundaries, if the term boundary were not so well entrenched in the systems literature and if there were a clear alternative, one might be tempted to abandon it altogether.

One surprise we managed to avoid in the Huron Natural Area project involved the groundwater feeding the cold-water trout stream. The preservation of trout streams normally focuses on maintaining the riparian zone around the stream and the surface water flows from the stream's watershed, a normal strategy in such circumstances. However, as noted earlier, groundwater flows are the important source of fresh cold water for the stream. This prompted a study of the groundwater system at a scale beyond the local watershed of the stream. This study revealed a planned sand and gravel extraction operation a couple of kilometers away from the stream. This operation could have seriously modified the groundwater system, probably diminishing the cold-water seepage into the stream sufficiently to eliminate the trout stream regime. Two years of study were carried out to understand the dynamics of the aquifer. If this had not been done, there was a real danger that the trout fishery might have disappeared (a nasty surprise), and I suspect no one would have connected the disappearance of trout to the construction of the sand and gravel quarry, a change in the system's context, which occurred kilometers away.

Type

The holarchy described in fig. 2.1 is from the perspective of surface water. The last example was from the perspective of groundwater, a perspective that leads to a different systems description of the Huron Natural Area. For example, the aquifer boundaries and the subwatershed boundaries are not the same. This introduces another issue to be considered when developing a systems description, type of description. Consider a house. There are many possible descriptions of a house. The homeowner might describe it in terms of the number of rooms or the layout. An electrical engineer would describe the electrical system, the wiring, plugs, lights, fuses, etc., and their interconnection. A mechanical engineer would describe water flow through plumbing systems. The same physical objects described from different perspectives lead to different system descriptions.

Looking at one biophysical entity, one can frame it in terms of communities, landscapes, energy and material flows, population dynamics, etc.—different windows on the same world. So, in addition to considering the question of appropriate scales of description, the question of appropriate types of perspectives is equally important and for the same reasons. The earlier example of the impact assessment of the effects of a proposed pipeline on caribou herds is about type. The assessment only looked at the situation from a reproduction perspective and not from a food supply perspective. The example about the Ucayali region also illustrates issues of type. The perspective of cattle farming was used to examine the situation and not the perspective of fishing. In both cases the issue of type of perspective was not thought through.

Figure 2.2 describes the different types of biophysical[6] perspectives, disciplinary analysis, and systems descriptions that were required for the Huron Natural Area. The analysis and systems descriptions of the Huron Natural Area made by

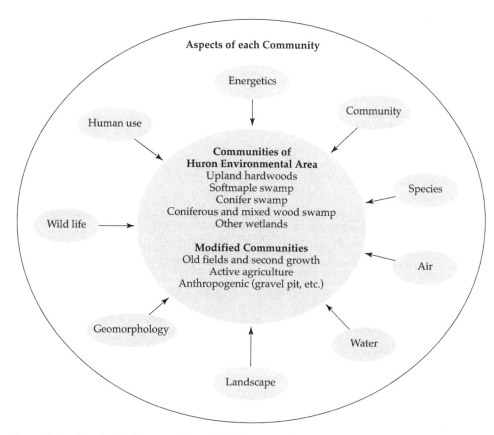

Figure 2.2   Biophysical perspectives, disciplinary analyses, and systems descriptions of the Huron Natural Area.

a forester are very different from those made by a hydrologist, and both are different from the description by a recreational trail designer. Each of these disciplinary analysis and systems descriptions are correct, but none are complete. Understanding complex problem situations requires different perspectives, different forms of disciplinary analysis, and different types of systems description, a polyocular description. All these descriptions need to be synthesized together into an overall description of the situation. This synthesis for the Huron Natural Area is represented in fig. 2.3. This notion of synthesis is only introduced here; it is discussed in more detail in later chapters.

This brings us to the last systems principle that will be discussed in this chapter. Complex problematic situations require multiple perspectives, perspectives that come from observations made at different scales and using different types of criteria. Understanding, albeit not complete, comes from synthesizing these multiple perspectives. Each perspective sees one facet of the situation. It is, in part, the multifaceted nature of the problem situation that makes it complex.

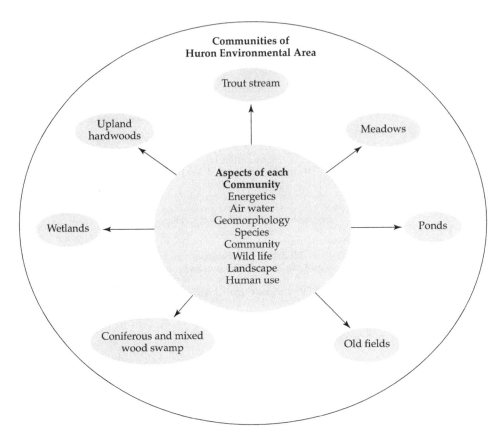

Figure 2.3   Synthesis diagram of the Huron Natural Area.

Hierarchy Theory

Allen and others have developed a theory for the consideration of multiple perspectives, which they refer to as hierarchy theory (Allen and Starr 1982; Allen and Hoekstra 1992; Ahl and Allen 1996). Hierarchy theory provides us with many powerful insights into how to frame situations, some of which are discussed by Allen in chapter 3. The theory is about holarchies, but it also applies to nonnested hierarchies such as organizational hierarchies. Once the scales of observation, types of perspectives, and boundaries have been established, it is then a question of generating systemic description[7] of the appropriate holons and holarchy.

It should be noted that a holon refers to a system considered at one scale and from one perspective. A holarchy (or holonocracy) is a nesting of holons, each at a different scale of observation, but all seen from the same perspective. A hierarchical description is a set of holarchic descriptions, each holarchy reflecting a relevant type of perspective. A specific entity, like the Huron Natural Area, can be part of several nested holons (e.g., the groundwater holarchy, the terrestrial community holarchy, and the regional trail system holarchy).

The application of hierarchy theory in practice is not commonplace, in spite of the work of Allen and others, and despite the work of the projects discussed later in this book, all of which have demonstrated its power as a tool. The impediment to using the wholesale incorporation of hierarchy theory is, again, that it is a different way of thinking about situations. Application of hierarchy theory requires practitioners who can function in a highly interdisciplinary environment. Traditional science education does not prepare people for this, as has been recognized by the 10-year outlook report of the National Science Foundation of the United States in 2003 (Pfirman and the ACE-ERE 2003).

One noteworthy challenge facing us is the so-called cross-scale problem. The issue is how to understand how things occurring at one scale are manifested at other scales. Giampietro and Mayumi (Giampietro 2003) have made some headway in developing a theory and practice of cross-scale interactions related to agroecosystems, and some of the work of the Resilience Alliance addresses this with regard to fast- and slow-moving variables in ecosystems (Gunderson and Holling 2002). A related issue, in many ways more intractable, is about how observations made from different perspectives can be used to inform each other. It is insufficient simply to invoke an analysis that may well be sufficient for academic scholarship. Out in the world of real-time problems and solutions, there are some significant additional challenges facing those who would undertake to produce a system synthesis, which is necessary for management.

System Identification and Dynamics

The discussion of systems has so far been about identifying what the system to be considered is. This phase of studying a system is known as the system identification phase. It is about establishing how we will look at the problematic situation, the scale and type of observations, what will and will not be included, and the relationships between the components that make up the system, that is, the system's structure. The relationships can be divided into two types of relationship in a hierarchy: horizontal and vertical. Horizontal relationships are between holons at the same scale. Vertical relationships are between holons at different scales, i.e., holarchic nesting and system-environment relationships. The system identification phase ends up with a rich picture of the situation, a conceptual roadmap of the interrelationships.

Having identified the system, that is, having built a mental map to frame the situation, the next phase is to describe how the system changes over time, that is, its dynamics. Much of the discussion in the literature has been about relationships that are framed as physical flows (i.e., food, water) between components. This type description is an input/output description, as it looks at the inputs into a system, their flow through a system, and the resulting outputs. There has been much written on dynamical descriptions of input/output systems. Ulanowicz, Hannon, and Patten have written extensively about the dynamics of ecological flows in input/output systems (Ulanowicz 1997; Hannon and Ruth 2001). Most of this thinking has its roots in the foundational work of Odum (1994).

State Description

There is a second and equally important type of system description, the state description. Consider the heating system in a house. It can be described in terms of flows of energy from an external source into a furnace, where the energy is burned and converted into heat and waste, the waste going out the exhaust, and the heat going into the house and ultimately out the walls into the environment. The description will tell us about the energy consumed, heat and waste produced, the inputs and outputs, and how this happens. For most of us this is not especially relevant. What is important is if we are warm or not, that is, the temperature of the house. The temperature of the house is a stable variable, and a description of the house in terms of temperature is called a state description. State descriptions and input/output descriptions are very different characterizations of the same situation. Telling me how much fuel a house consumes tells me little about how comfortable I am in the house (and vice versa).

State descriptions are fairly new and have their origins in thermodynamics. Dynamical discussions in terms of state descriptions focus on changes in state variables. In many situations, it is the state of the situation that we first see and are interested in. What is the state of the economy, we ask, or the health of a community? When describing a system in terms of states, a system description is about influences between the elements of a system that cause states to change. These descriptions take the form of influence diagrams (see Neudoerffer, chapter 15; Gitau et al., chapter 12). Influence diagrams describe which things influence other things in the system and what the influence is. They take the form of statements like, "the state of entity A influences the state of entity B by . . ." (e.g., the availability of lichen as food influences the health of caribou). Influence diagrams are maps of causality in a system. Checkland's (1981) "rich pictures" are a type of influence diagram (see chapters 9 and 11).

When we talk about the state of a system and how that state results from a set of influences, we are talking about a systems organization. Many academics spend much time trying to generate influence diagrams that explain the states of a system. As we shall see in chapter 3, self-organization is of particular importance when we try to understand the biophysical world. But first, some closing remarks about multiple perspectives, systems thinking, and post-normal science are in order.

Multiple Perspectives, Systems Thinking, and Post-Normal Science

To reiterate, complex socioecological situations can be understood only from multiple, nonequivalent perspectives. There exist multiple "windows" to investigate different aspects of the situation. A truly complex situation cannot be adequately "captured" or represented from any single perspective. The sheer diversity of elements and interactions requires a plurality of perspectives. Indeed, degree of complexity has been equated with the number of nonequivalent descriptions of a system (Casti 1994). These nonequivalent descriptions will result in nonequivalent evaluations and hence different strategies for the future.

The need to incorporate multiple perspectives, when dealing with sustainability, has important methodological implications. The plurality of different legitimate perspectives and the inability of one particular view to capture the whole necessitates a variety of forms of inquiry, inclusion of, and dialogue with, persons representing different interests and different worldviews (Waltner-Toews and Wall 1997). Multiple modes of investigation and analysis and multiple sources of evidence are necessary to understand the system.

This assertion, often repeated in recent environmental, health, and sustainable development literature, leaves unanswered the question as to how various perspectives are selected and legitimized. Greater multi- and interdisciplinarity in the research into and management of complex socioecological situations is necessary but not sufficient. Left to themselves, researchers will inquire into those aspects of the system that they, for their own personal, disciplinary, or political reasons, deem to be important. This is not a very sure foundation for learning our way into a sustainable future. Situations to be managed develop out of particular historical conditions, with no single preferred future state. The goals selected for management are necessarily value-laden. Because of the interaction between historicity and values, it is crucial that the values, concerns, and knowledge of local stakeholders and actors be central to any inquiry.

Systems thinking is a heuristic tool for dealing with and integrating different perspectives. It is a tool for framing the investigations of complex socioecological situations. It helps us to generate a mental map of the situation with particular emphasis on relationships and interconnections, appropriate scales of observation and analysis, and types of analysis and descriptions. It provides us with an approach for forging meaning in a situation, "sense making," if you will.

Through the process of identifying the system of interest, developing systems diagrams, and generating a system description, people involved in a complex socioecological situation are brought together, often for the first time. They learn about thinking from other perspectives and how others think about the situation. They gain a common language for discussion and a broader appreciation of the system. They start thinking outside their usual frame of reference, and this often unlocks their creative energies. In our experience this phase of developing a system's understanding of a situation is often sufficient to catalyze significant action to resolve the situation.

The starting point of any systems identification exercise has to be the identification of the issues and questions that motivate the exercise in the first place. Our systemic description of a student residence cafeteria would have been very different if the root issue had been minimizing cost, maximizing profit, or customer satisfaction as versus minimizing waste production. In the Ucayali example, if the environmental impact of land use in the region is of interest, most attention would be focused on pastures and cattle as they comprise the largest percentage of land that is intervened. However, if agricultural income and household wealth were emphasized, research would examine the different annual cropping systems critical to the livelihood of both subsistence and commercial colonists. If food security and nutritional status were the issues of concern, it would soon be discovered that fish are the most important food source (12 times more protein comes from fish than from beef). Therefore any attempt to alleviate the high levels of malnourishment must address the role of aquatic resources in household consumption. The issues not only define the systems of interest but also the relevant elements, actors, and interconnections in those systems.

The issues orient the system identification phase. They provide the rational for choosing boundaries, scales, and types of descriptions. Without a clear statement of issues, a systems investigation is rudderless. In fact, there is often a recursive process of trying to identify appropriate boundaries, scale, types, etc., and revisiting what the issues really are.

In most situations, there are multiple issues. For example, in the Huron Natural Area, people had different concerns. Some people were interested in preserving the natural landscape, some in maintaining a trout fishery, some in a recreational site for city dwellers, and some in a natural education site for school children.

This immediately begs the question: whose concerns and whose values are the ones that will be deemed legitimate and guide the systems identification exercise? Furthermore, how do we identify and involve these people? Who gets to decide? These questions cannot be avoided and, in fact, are highlighted by a systems investigation. The arena of discourse here lies in the domain of politics and ethics. This is the reason why a systems investigation must go hand in hand with a participatory process and why so much of this book is about participatory process. Participatory approaches are the means of bringing various people to the process, of identifying their values, their issues, and their concerns. Systems thinking is the means by which participants can gain understanding of the ecological, particularly the biophysical, aspects of the situation. It provides a framework for exploring the ecological realities of the situation and as such underpins an ecosystem approach.

Systems thinking is not about the pursuit of objective "truth" but rather the pursuit of many "truths" as defined by the actors and investigators. This is why I do not use the term "system science" in association with this activity. The term system science was adopted to give legitimacy to a systems posture. This was done because "hard scientists" saw systems thinking as "soft" because of its observer dependence. Soft approaches do, indeed, lack objectivity, which is a serious weakness in the view for hard scientists, and it is a weakness from which hard science explicitly wishes to distance itself. In the world of normal science, there is little use for a scheme that cannot say which singular answer is correct!

These objections of normal science analysis are based on a fair characterization of systems thinking, but they are still misplaced. In fact, lack of objectivity and a refusal to assert just one answer is precisely why systems thinking is not normal science but rather post-normal science (Funtowicz and Ravetz 1993). In the domain where systems thinking applies, normal science also cannot come up with unequivocal objectivity, but what if the important issues lie in the middle of that arena? Systems thinking has the courage to continue to pursue happy outcomes anyway, and it simply accepts that the only tools left deny unequivocal objectivity. Our characterization of reality depends on how you look at it! There is a certain irony, as Ravetz and Funtowicz have pointed out, that systems thinking can help us deal with the "hard" (as in difficult) questions that normal science cannot resolve. Systems thinking can help us sort out the implications of looking at a situation in different ways and synthesizing these different perspectives into a sense of the whole. Normal science comes into play only when the whole has been defined, and so it is not well equipped to make decisions about what is the whole. Systems thinking acknowledges that there is not a true and correct perspective, just different perspectives that provide us with more or less insights. Rosen (1991), Allen et al. (2001), Giampietro (2003), and Funtowitz and Ravetz (1993), to name a few, have discussed the pursuit of a different sort of truth that has fundamental implications for the agenda of science.

There is more to the story of complexity and systems thinking than the issues brought to the fore by a generating systems description. There are also the issues of uncertainty and multiple possibilities that characterize self-organization, a subject we shall return to in chapter 4, after a more in-depth look by Allen (chapter 3) at the profound implications of considering scale and type.

Notes

[1]Organization should not be confused with order. Organization is about the configuration of the system so as to carry out a task or function. Order characterizes the way in which a collection of things is assembled. These are not the same thing. Something that is ordered may not be organized (a diamond); furthermore, something that is organized is not necessarily ordered (the Internet).

[2]There is ambiguity in how the term structure is used in the system's literature and in this chapter. Structure is about how the system is put together. At the component scale, structure is about how the components are interconnected and the relationship between the components. (e.g., the electrical wiring that interconnects the electrical components in electrical systems like a computer chip or the food web in a forest.) At the system level, structure is about what we see as a static whole because of how the system is put together (e.g., a computer chip or all the trees in a forest). Discussion of these notions in terms of systems can be found in Ahl and Allen (1996), Flood and Carson (1993), Nicolis and Prigogine (1977, 1989), and Prigogine and Stengers (1984).

[3]It should be noted that there are sophisticated techniques for analyzing flow networks that help us to understand the nature of the relationships in system. See works on network topology, graph theory, input/output analysis, and network thermodynamics.

[4]See pp.144–147 in Weinberg (1975).

[5]This boundary reflects the legislated responsibility of the agency. They are only responsible for what happens inside the power plant, and so this is where they draw their boundaries.

[6]Different kinds of sociocultural descriptions are needed as well (e.g., as a recreation area, as education resource, as part of a zoning/planning process).

[7]Sometimes systems description is used to refer to the description of a single holon, from one perspective at one scale. Other times, it is used to refer to the hierarchical description at all scales, from all perspectives. The single-scale, single-perspective description should really be referred to as a holon description, but, unfortunately, the ambiguity exists in the literature and in this book.

References

Ahl, V. and T. F. H. Allen. 1996. *Hierarchy Theory: A Vision, Vocabulary and Epistemology.* New York: Columbia University Press.

Allen, T. F. H. and T. Hoekstra. 1992. *Toward a Unified Ecology.* New York: Columbia University Press.

Allen, T. F. H. and T. B. Starr. 1982. *Hierarchy: Perspectives for Ecological Complexity.* Chicago, Ill.: University of Chicago Press.

Allen, T. F. H., J. A. Tainter, J. C. Pires and T. W. Hoekstra. 2001. Dragnet ecology— "Just the Facts Ma'am": The privilege of science in a postmodern world. *Bioscience* 51:475–485.

Casti, J. L. 1994. *Complexification*. New York: HarperCollins.

Checkland, P. 1981. *Systems Thinking, Systems Practice*. New York: Wiley.

Cobean, R., K. Karpan, M. Massey, T. L. Riley and S. Santarossa. 1995. "Indoor Pest Management at the University of Waterloo" (Sept. 9, 2004), at www.adm.uwaterloo. ca/infowast/watgreen/projects/disclaimer/780.html.

Flood, R. L. and E. R. Carson. 1993. *Dealing with Complexity* (2nd ed.). New York: Plenum Press.

Funtowicz, S. O. and J. R. Ravetz 1993. Science for the post-normal age. *Futures* 25(7): 739–755.

Giampietro, M. 2003. *Multi-scale Integrated Analysis of Agroecosystems*. Boca Raton, Fla.: CRC Press.

Gunderson, L. H. and C. S. Holling (eds.). 2002. *Panarchy: Understanding Transformations in Human and Natural Systems*. Washington, D. C.: Island Press.

Hannon, B. and M. Ruth. 2001. *Dynamic Modeling* (2nd ed.). New York: Springer-Verlag.

Koestler, A. 1967. *The Ghost in the Machine*. New York: MacMillan.

Koestler, A. 1978. *Janus: A Summing Up*. London, UK: Hutchinson.

Lister, N. M. and J. J. Kay. 2000. "Celebrating Diversity: Adaptive Planning and Biodiversity Conservation." In: S. Bocking (ed.) *Biodiversity In Canada: Ecology, Ideas and Action*. Toronto: Broadview Press. pp 189-218.

Maturana, H. R. and F. J. Varela. 1987. *The Tree of Knowledge: The Biological Roots of Human Understanding*. London, UK: Shambhala.

Nicolis, G. and I. Prigogine. 1977. *Self-Organization in Non-Equilibrium Systems*. New York: Wiley.

Nicolis, G. and I. Prigogine. 1989. *Exploring Complexity*. New York: W. H. Freeman.

O'Connor, D. 2002. "Walkerton Commission of Inquiry Reports" (Sept. 7, 2004), at www.attorneygeneral.jus.gov.on.ca/english/about/pubs/walkerton/.

Odum, H. T. 1994. *Ecological and General Systems: An Introduction to Systems Ecology*. Boulder: University Press of Colorado.

Pfirman, S. and the ACE-ERE. 2003. Complex environmental systems: Synthesis for Earth, life, and society in the 21st century, A report summarizing a 10 year outlook in environmental research and education. Washington, D. C.: National Science Foundation.

Prigogine, I. and I. Stengers. 1984. *Order Out of Chaos: Man's New Dialogue with Nature*. Toronto, Canada: Bantam Books.

Rosen, R. 1991. *Life Itself: A Comprehensive Inquiry into the Nature, Origin and Fabrication of Life*. New York: Columbia University Press.

Tobias, T. and J. J. Kay. 1994. The bush harvest in the northern village of Pinehouse. *Arctic* 47:207–221.

Ulanowicz, R. 1997. *Ecology, The Ascendent Perspective*. New York: Columbia University Press.

Waltner-Toews, D. and E. Wall. 1997. Emergent perplexity: in search of post-normal questions for community and agroecosystem health. *Social Science and Medicine* 45:1741–1749.

Weinberg, Gerald M. 1975. *An Introduction to General Systems Thinking*. New York: Wiley.

3

Scale and Type
A Requirement for Addressing Complexity with Dynamical Quality

T. F. H. Allen

It is humbling for students of complexity to realize that their main issue turns merely on whether or not the observer has a paradigm. One might have hoped for something grander to circumscribe a life's work on complexity. Kuhn (1962) defines a paradigm as a normative framework that is characterized by shared tools, vocabulary, protocols, and questions. In a simple system, that normative framework defines the entities of interest. There is a variety of case studies in this book, and the different directions from which authors have approached the subject (health, planning, and governance). The upshot is: what is of interest is equivocal in complex systems. The agenda and methods of complexity science are completely different from the necessarily incremental nature of normal science (Kuhn 1962). Whereas normal science takes a prescribed framework for granted, complexity demands a protocol that seeks new frameworks because the ones at hand have proven inadequate. Students of complexity challenge accepted contexts when they seek new frameworks that can bring order and predictability to heretofore unruly systems. The other chapters in this book make a natural whole because of the unity of that protocol for finding new frameworks. The absence of a paradigm notwithstanding, there are protocols and intellectual devices for dealing with complexity, and this chapter seeks to lay out some of those known to the author.

The Role of Paradigms in Complexity Science

Rosen (2000) suggests that complexity cannot be modeled. In both the lay vernacular and the specialist literature, complexity is characterized as systems being hierarchical and organized to deal with elaborate structures. In the light of Rosen (2000) it appears that common parlance is mistaken. Organized, elaborate constraint is not the central characteristic of complex systems but rather the story one had to tell to make the system simple, so that it could indeed be modeled. There is no conflict between complexity science and normal science because the former sets up the situation so that the latter can perform well. Paradigms assert patterns that make complexity simple and tractable. One cannot model complexity, but one can approach it using analogy and narrative. A paradigm is a narrative, which is then improved by the models that are used by the normal scientists within the arena of discourse (Allen et al. 2005).

Dealing with complexity amounts to searching for a paradigm, and a paradigm is a narrative that highlights what matters. This is merely a different way of saying what was said in chapter 2, when Kay spoke of deciding on the boundaries of the system and its meaning when those boundaries have been set. One facet of complexity is that there are no accepted definitions that distinguish between significant change in a system and incidental dynamics within that system. Another facet is the lack of a threshold of change that distinguishes continuity from discrete change. Without a paradigm, change that amounts to the emergence of structure is indistinguishable from the change of state of a recognized structure, namely behavior. What should be quantified, as opposed to what deserves a qualitative treatment, has not been unequivocally decided in a complex system. The observer of complex systems lacks a normative framework and therefore can neither bound the system nor define its parts until such a framework is asserted. A paradigm decides on the framework making all the necessary distinctions: (1) between behavior and emergence, (2) between structure and dynamics, (3) between significant change and incidental dynamics, and (4) between the continuous and the discrete. Dealing with complexity has the long-term goal of finding simple expressions of the system by making decisions that remove the above dilemmas. Of course, normal scientists who work only inside their respective paradigms suspect systems thinking and the notion of complexity because it solves a problem they never face. At the conclusion of the investigation of complexity, success is achieved by imposing a workable paradigm, whereupon normal scientists can then make their contribution.

In the first stage of any investigation of some genuinely new situation, the researcher will be interested but will have not yet have decided on how to look at the phenomenon in a formal way. Normal science research is some version of tidying up some established problem. I do not mean to suggest that normal science is trivial or unimportant. Indeed, it has a long track record of great success, arising from the lifelong struggles of some great minds. However, because informality necessarily precedes the structured attention of the formal approach, complexity is the default setting at the outset, if the problem situation is genuinely new. Formalizing a significantly new investigation is a process of choosing from alternatives. Those choices impose limits, which may build to create a large new paradigm.

A paradigm becomes recognizable when small changes, wrought by choosing from alternatives, accumulate to make a distinctive difference. The final step that causes a paradigm to emerge need not be much different from previous small choices, except that it makes an observer notice that there is a new intellectual framework. Paradigm shifts need not be as large as when relativity replaced aspects of Newtonian mechanics. In fact, paradigm shifts could be quite small, resulting from a slight change in perspective. The antipathy between normal science working inside a paradigm and the science of complexity looking for paradigms is unfortunate and unnecessary. It is a political distinction imposed on what is simply a matter of scale. If only normal scientists could recognize that theirs is the same method as that in complexity science but a more local version that gives the crucial underpinning of the whole enterprise! Taming complexity is a continuous process wherein the tools of normal science have their role to play in chipping away at complex issues to show the sticking points.

The paradigms that define simple systems do not pertain to any material simplicity, whatever that might mean. Rather, paradigms put the study of a system in a frame of reference that recognizes constraints that impose certain limits. Those limits lead to observing only certain changes and calling only some of them events. Other changes are overlooked or actively excluded. In experimental science the system is physically and literally put under constraints so that simplicity prevails. Simplicity comes from a process of orderly exclusion of some behaviors so as to bring the regularity of other behaviors to the fore. Information outside the paradigm is suppressed by calling it noise or error, but that is how normal science keeps its focus and gains its leverage. All we post-normal and complexity scientists are doing here is suggesting ways to relax suppression of information at the critical times when the normal science approach gets stuck.

Paradigms and Reality

The conventional realist posture of modern science relies on prediction as the benchmark. As opposed to a postmodern position, a modernist view asserts that models approach reality because, whatever else is happening, prediction improves with scientific progress. However, prediction and approximation to ultimate reality are not isomorphic. In fact, one cannot tell the difference between increasing verity of a scientific model and the effects of tightening constraints. The scientist imposes constraints until the system has few options but to meet the prediction. Scientists regularly impose constraints in scientific conception or experimental protocol. There is presumably an infinity of even feasible models that do not approach reality, so finding or even approaching which one is right is unlikely without prior knowledge. Therefore the short odds are on the constraints as causes of improved prediction, with approaching reality as the distinct long shot explanation. As the paradigm develops, its constraints leave the material system fewer options for behavior, and that is the most probable basis of improved prediction. Understanding constraints makes systems simple and predictable, something that is valuable in itself, without any need to seek an external reference such as ultimate reality.

Choosing one paradigmatic framework necessarily excludes, for the time being, all others. Excluding other frameworks imposes constraints by eliminating options. Simplicity comes from an explicit juxtaposition of parts so as to expose a relationship that is significant, but significance is subjective. The act of recognizing significance imposes constraints that make the system simpler as the paradigm emerges. A paradigm amounts to a web of significance. The shared vocabulary, methods, and ideas that Kuhn (1962) uses to characterize paradigms follow from translating significance into action.

Complexity is invoked so that complex systems may be made simple. Simple systems are tractable. Tractability comes at the cost of closing options. One cannot solve all problems at once, so the loss of options in order to achieve the tractability of simple systems costs only a little. Sacrificing some generality has its price (Rosen 2000), but loss of some generality to achieve some tractability is not such a bad deal. That explains the willingness of normal scientists to live within their paradigms. Normal science deals with systems that are not complex, even if they are still complicated. In general, reductionism narrows the focus until only emergence is left and meaning is excluded. This last exclusion comes at a higher cost.

Systems defined here as simple can still be very complicated. Complication refers to the number of parts or connections. Complication introduces degrees of freedom, and so complicated systems often have complicated behavior. Organization strips away degrees of freedom by imposing asymmetric relationships. Pattee (1972) recognized system complexification as arising through the imposition of constraints, a process he called self-simplifcation. This is the same paradox that emerges in organizations and is the hallmark of a complex system, placing complexity at odds with complicatedness (Allen et al. 2001).

It is important not to dismiss normal science as being somehow easy, for the very complications that characterize normal science can be extremely challenging. For example, there are the details of temperature of the medium for a biochemical reaction and the exigency of scrupulous cleanliness in handling material for DNA amplification. However, follow the protocol and the system behaves predictably, as the same DNA sequence occurs in replicates. Such precision is hard to achieve.

Quality Versus Reality as a Benchmark

Normal science is generally a high-quality activity. The quality embodied in the meticulous attention to detail in normal science is what Funtowicz and Ravetz (1992) identify as structural quality. That structural quality requires the scientist to perform extended activities with great care. The care must be sufficient so that, if necessary, the performance can be repeated and have it all work out the same every time. To achieve repeatability, the situation must be exquisitely controlled. In other endeavors, structural quality might involve making perfect dovetail joints in French furniture or painting the hairs on a head in a Pre-Raphaelite painting. All structural quality is difficult to achieve, and the practitioners of normal science are justified in the pride that they take.

In contrast to the structural quality that prevails in normal science, dealing with complexity pertains more to dynamical quality. Maintaining structural quality is complicated. Dynamical quality arises by addressing the context of structural quality. Dynamical quality, while accepting the importance of getting the complicated details right, challenges the framework in which the details are cast.

Without dynamical quality, the paradigm tires, and the enterprise descends into pointless compulsion. Yes, the predictions almost always come true, but one must ask: what is the point? Science in that condition amounts to data or experiments for their own sakes. Periodic changes in meaning are crucial, and those are found in addressing complexity by confronting the conventional definitions. Sometimes normal scientists do challenge with dynamical quality, and this activity dominates when there is healthy debate, when disciplines are creative and outward looking. At that point, normal science is indulging in complexity science, but not in a self-conscious way. While dynamical quality is the hallmark of complexity science, it is still to be found in normal science, and the more of that we see in normal science, the better science will serve.

The challenge to prevailing frameworks and their conventions occurs in science at various levels. Indeed, the constant challenge that is routine in science at many levels is one source of the privilege that science enjoys and deserves. Even in a postmodern world, "science still deserves to be privileged, because it is the still best game in town" (Allen et al. 2001:476). Challenges at the lowest level amount to the simple rejection of a null hypothesis. At a higher level, there is recognition of new notions, such as photorespiration, which deals with confounding relationships between constructive photosynthesis and destructive respiration. At the highest level, the challenges of dynamical quality in science have led to large paradigm shifts such as the replacement of Newtonian mechanics with relativity.

All that being said, the emphasis on dynamical quality of post-normal and complexity science is particularly needed at this time. Humanity is faced presently with radical changes in its circumstances, such as global climate change and the limits to the economic viability of carbon based fuel. In such situations, science must invest in dynamical quality to a greater extent than is business as usual. Not all major work in science comes from dynamical quality and accepting radical ideas; it depends on the times. For instance, Newton was conservative for his time, in that there were many outlandish theories about gravity. He stood his ground and solved the problem. Times are different now. Dynamical quality is particularly needed in our contemporary setting.

Rudimentary as are the features that distinguish complexity from simplicity, there is a nuanced scaling of complexity that defies intuition. In fact, most facets of complexity in systems are addressed operationally as issues of scale and type. Scale pertains to issues of size of span, as well as the size of the finest distinction made in a given study. The notion of type, as a system characteristic, turns on what is recognized as being in the foreground, as opposed to everything else. The challenge embodied in dynamical quality comes from either changing the scale at which the system is addressed or making a qualitative change to recognize that some new type of system is in the foreground at a given scale. Complexity turns on issues of scale and type, and the body of this chapter addresses complexity in those terms.

Scale

The distinction between scale and type deserves further elaboration. It is tempting to allot scale to material effects only and to assign type to observer decisions only, but this is a mistaken assignation. It is important to realize that both scale and type reflect aspects of observation that are both internal and external to observer decisions.

If the observer chooses to look at a dung beetle instead of a water strider, then there are consequences that clearly arise external to that decision. The larger insect will be unable to skate across the surface of the water, resting on surface tension whether or not we wish that ability on the dung beetle. Note that there is something more significant to the change between insects than just a disconnected difference in size. The fact that they are both insects gives unity across which the comparison of system performance is assessed. That unity gives scaling criteria against which particular insects can be calibrated. The comparison between a dung beetle and a water strider is held together in many ways by their both being insects.

By contrast, a change between looking at a pond skater and a 5-cm-long 15-mm bolt could be argued as a meaningless change. As the 1970s British television comedians of Monty Python might say, "And now for something completely different." The bolt would be only something different, despite the fact that it, too, would break through the surface tension of water, exactly as would a dung beetle. The fact that both pond skaters and dung beetles are insects, and that they are recognized as such, gives a deeper meaning to the change in size and, in fact, defines it as an issue of scale. Dung beetles are under no selective pressure with regard to movement across water using surface tension. By contrast, that is exactly the design principle found in the evolved form of the pond strider. Buried in the difference between the sizes of insects is an implied discussion of selective pressures. A scale demands some sort of equivalence across the scale change. In the case of insects, the unity invokes a host of other biological phenomena, including evolution. If there were no link, as in the bolt and the pond strider, then the change would not be one of scale, but simply a disconnected substitution of one entity for another.

Now here's the rub. A physicist might indeed see the bolt/insect change as an issue of scale, but a biologist would not. Thus a given material situation may or may not be a matter of scale, depending on who is looking at it. Therefore scale pertains to the decision of the observer to link the two entities across the scale change. That means that scale pertains to more than a simple material effect, for it has values, choices, and emphases embedded in it.

Type

The easy mistake with regard to scale is to assert that it is purely a matter of material size; conversely, the easy mistake with regard to type is to imagine that type turns entirely on choices internal to the observer as to what is recognized as being in the foreground. Much as scale can only be properly characterized as the interaction of observer decisions and material externalities to those decisions, system type also requires an interaction between the observer and aspects of observation that are beyond observer decisions. That region "beyond observer decisions" is technically called the domain of the "other." The other is still part of observation, but it is not a possession of the observer. You choose to study lions, and choose to study spatial positioning of the animals, but you do not get to choose where they go. Where they go belongs to the other, because you cannot choose it. Assignment of responsibility of system type to the observer alone is inadequate. There is some subtlety in the role of experience in the notion of type, and so some dissection of the concept of experience is in order to understand the implications of the notion of system type.

The reference to that which is external to observer decisions need not degenerate into naive realism. The other does not have to be a crude reality. There are aspects of observation that appear to be above and beyond the decisions of the observer. Observation requires that there be an observer and something more that is other. The other need not be assigned to a reality that is independent of observation, for it arises here precisely held in the context of observation. What many would be happy to call material externality need not be more than the other in an observation. Even so, the view expressed here can tolerate a more straightforward realism, although the author would not embrace it eagerly. We cannot prove false a belief that there is a reality independent of observation. There is no proof one way or another as to whether or not the other in observation is closely related to a more or less autonomous externality. The spirit of this chapter seeks to occupy that large commonsensical middle ground between the extreme positions of antagonists in the science wars (Kleinman 1998). There is plenty of room between rude objectivism of the naive realist and the solipsism of extreme relativist subjectivism (Putman 1987).

The other that is embodied in observation does pertain to the notion of system type. Type defines some aspects of experience as being the foreground, and concomitantly type relegates all else to the background. Type is a matter of definition, and definitions are the entire responsibility of the one who makes them. In that case, whence comes the externality involved in notions of type? Linking definitions to observations invokes that which is external to the observer decisions in observation. Definitions encompass expectations as to what experience might actually meet the definition, so as to give an example of the thing defined. Of course, one does not expect to experience a living creature that meets the definition of a mythical creature such as a unicorn. So, there appears to be a definition of type that fails to invoke an externality through observation. Not so! First, one does experience images of unicorns, as in the Royal Crest of the British Monarchy. Furthermore, the definition of a unicorn invokes a horse-like animal, and horses are indeed observable as living creatures (Ahl and Allen 1996). The author employs here constructivist philosophy (Sismondo 1996), wherein experience is seen as heavily influenced by prior experience. Furthermore, any immediate prior experience was itself influenced by experience yet farther back in the past. Experience indicates what to expect, and expectations powerfully influence what is seen subsequently. Thus a cascade of prior experiences each has constructed the observer. That construction is mediated by an exchange between the observer and something external to the observer within the observer-observation complex. Prior experience through observation involves the other; the other that is a requirement of observation. Because definitions are made in the context of experience, the other is present in the definition. Material externality as manifested through the other in observation is thus linked to system type. The choice of system type, as it defines what is in the foreground, is linked to something external. Both scale and type demand both decisions of the observer and material externalities inside the observer-observation complex.

Relating Scale and Type Through Levels of Analysis

In biology, including ecology, there is confusion about levels as observed through the filter of a scale as opposed to levels that arise by definition. Ahl and Allen (1996) make much of the distinction between what they call levels of observation as opposed to levels of organization. A level is simply a class, with the special proviso that the level has an asymmetric relationship to some other class. The point of tension here is between a level of observation as opposed to a level of organization, and the nature of the asymmetries that make the hierarchy is different between the two sorts of levels.

A level of observation arises from changes in scale such that a move upscale reveals a system that exists at a higher level. Move from a microscope to the naked eye, and one sees the whole elephant instead of its epidermis. Conversely, a move down scale reveals a level below. Without observation, there is no level here. A level of observation is not to be confused with a level of organization, which arises out of definitions, not out of changes in scale. The asymmetry that makes a level of organization different from a simple class is part of the very definition of the level in question. Levels of organization involve asymmetric links to other levels that arise from the definitions of the respective levels of organization.

The difference between an organism and the organism level of organization is first that the organism level is a class to which the organism belongs. Second, the difference turns on an organism meeting a definition that does not particularly include links to other entities that exist at higher or lower levels. By contrast, the population level is defined to have such links between organisms. The definition of what it takes to be an organism need not include anything about how organisms may aggregate to make populations or be disaggregated to make a collection of some lower level entity. Allen and Hoekstra (1992) have a particularly austere, but general definition of an organism. It turns on three criteria that are met to various degrees by various organisms. First, organisms have a physiological coherence, where happenings in one part affect and are affected by other parts. Second, organisms tend to be bounded in space so as to separate them from other organisms. Third, organisms generally have a genetic identity, such that each has its own genome that is different from that of most or all others. Of course, there are organisms that are either genetically or physiologically incoherent, such as lichens or Portuguese Man O' Wars. Nevertheless, one or more of the three criteria apply to a significant degree for all organisms. The organism level, as a level not just a class, is occupied by things that meet the three criteria to a significant degree, but it is a level precisely because there are additional facets of the organism level that relate it to other levels. In addition to bounded genetic identity and physiological coherence within organisms, the organism level is defined as having members that aggregate into populations. This creates an asymmetric juxtaposition to the population level. Also, these organisms may be made of cells, and this can be used to place the organism level above the cell or organelle level.

The organism level has no spatiotemporal scale in particular, and so has no scale-defined relationship to any other level. Of course, organisms are members of populations that, being aggregates of organisms, must be larger. However, many populations are smaller than many organisms. Observe the dust mite as an organism. Now move up to a higher spatially defined level of observation where there are many mites in a population. All is well and good so far, but move to a yet higher level of observation and we find a human organism that the mites treat as a home and landscape. Suddenly, the order of levels of observation no longer corresponds to the commensurate levels of organization.

Allen (1998a) dissected the landscape level as a concept. The notion of level is often used carelessly. It emerges that ecologists often mean just a landscape when they insist on calling it a landscape level. When it comes to ecosystems, ecologists do make a clearer distinction between a forest and a forest ecosystem, the latter being identified as particularly coherent in its function as a whole. Simple forests, without the ecosystem rider, refer more to collections of trees or places or sylvan geographic regions. While ecosystems are well defined, the ecosystem level like landscape level is less fully meaningful. The notion of level is commonly attached to ecological types in an informal way. Saying, "landscape level" as opposed to "landscape" is often hand waving that announces, "No, believe me, I am looking at content, context, and process here." Level becomes a label for more holistic thinking.

Given this imprecise usage surrounding the notion of landscape, the use of the term "landscape scale" should come as no surprise. Landscape level can be used appropriately, even if generally, but landscape scale is a misnomer. Landscape is a type not a scale. Landscape refers to things where the primary organizer is spatial arrangement. The spatial arrangement defines what is in the foreground. A landscape becomes the thing in the foreground when the organizing criterion for the thing of interest is a collection of forms spatially arranged. As a contrast, a collection of forms arranged, say, in the order of what eats what, would be a food chain. There is no particular landscape scale, much as there is no organism scale. Particular organisms have a certain size, but the notion of organism as a type has exemplars that range in size from bacteria to redwood trees. Being less tangibly bounded, landscapes are wont to be burdened with misplaced notions of scale. Regional scale is, by contrast, a perfectly valid term or concept because a region is clearly larger than a back yard, smaller than a continent, and about the size of a county or so. Even though a region may encompass several small countries, as in the Caribbean Region it has a well-understood spatial implication as an area that is rather bigger than one can see across. However, landscape scale is an unnatural hybrid level of organization and level of observation.

The confusion that arises in landscapes is a symptom of a more general misapprehension. Biologists are fond of large overarching notions, and biological scientists use them in the hope that the disparate collection of material that is biological can be put neatly in some orderly scheme. Despite explicit criticism through the 1990s (e.g., from Allen and Hoekstra 1992), biology and ecology textbooks persist in putting biology into a grand hierarchy that goes from quarks to the universe. In the middle is a mixed pile of biological material that is ordered on scalar principles, while consisting of a ranking of levels of organization. Such arrangements are not wrong in themselves, but neither are they the general scheme that a textbook author might hope they will be. For instance, a redwood tree is an organism that is a longer resident of its site than are the biomes that pertain in that region. Coastal redwoods grow in fairy rings because there was, in a previous millennium, a tree that fell down. The ring of trees consists of stump sprouts from the original. The trees in the ring are genetically identical and in some sense constitute a single organism. Redwood trees are clearly hedging bets against ice ages and phases of global warming. Climate changes sufficiently every three thousand years or so for sea levels to change dramatically. Large, old organisms are the parameters, whereas biomes are the variables. If some organisms are larger scale than their own biomes, then the textbook hierarchy of life is not a scale-based system. It can be hammered into an order reflecting scale only at the cost of almost all generality. The scale of the levels of organization must be represented by examples that are carefully chosen to coincide with the levels of observation.

The consequence of mixing scale and type is to put all of life in a straightjacket that reflects an arbitrarily small part of what happens in nature. There are many aspects of organism aggregation that do not pertain to populations. In fact, the very things that make an organism part of a community have very little to do with what makes that same organism a member of a population. Populations do not have to be of only a single species, as in the case of winter bird flocks consisting of several species. However, the parts of a population have to be similar enough so that they are simply additive in some way. A population is characterized as being sufficiently homogeneous within so that one can write equations. Often, even usually, this means that the population consists of one species. Not so for the organisms in a community. In fact, the notion of community is precisely one that contains heterogeneity. The aspects of an organism that make it a member of a population are marginal considerations in the functioning of a community. A community member, say, a tree in a forest, interacts most strongly with whatever organism is its neighbor; often that is a member of a different species. Whereas population members can usually breed with each other, for the most part community members cannot. Insisting on interposing the population level of organization between the organism and community levels of organization sidelines most of the organismal functioning of community members. The trouble with the conventional ecological hierarchy is that it mistakes type for scale, and so it misses most of what is happening.

Allen and Hoekstra (1992) organized their book by explicitly separating scale and type. As a result, they facilitated nesting process-oriented ecosystem systems inside larger ecosystems that are still process-directed. The level above is not, in this case, a different type of ecological system, say, a landscape. The hierarchy of ecosystems becomes a set of processes, each held in the context of slower, more far-reaching processes: ecosystem within ecosystems, within ecosystem. And none of this preempts taking a given spatiotemporal scale as a level of observation, say, a particular piece of forest, and considering it as both an ecosystem and a community or even a piece of landscape. Here one changes the type as one holds the scale constant. Furthermore, one can move down scale in a shift between levels of observation and look to see which type of system offers an explanation for the ecosystem level above. One might choose to distinguish a whole dead organism, a rotting log, held on the context of a forest community. Alternatively, the log makes a perfectly good ecosystem in its own right. Part of that ecosystem, the log, functions as a landscape for a population of mosses that create a bed for germinating hemlock seeds in upper level community succession. The type-centered conventional hierarchy insists on landscape above ecosystems. However, this convention misplaces the landscape type here on the top of the log in these levels of observation. The narrative of the mosses on the rotting log is valid, even though it places landscapes below dead organisms and ecosystems.

Scale, Type, and the New Challenges

When the environmental scientist studies a complex system, there are many levels that pertain. The notion of level is itself multifaceted, for the scientist uses a level of analysis that marries levels of organization with levels of observation. The distinction between scale and type of system is crucial to maintain intellectual and operational flexibility. Ignore the distinction between scale and type, and the scientist becomes lost in the tiers of definition and observation that make the system complex.

While type is often a polar opposite of scale, there is a union of scale and type in issues of dynamical quality. Both changes of scale and changes of type break paradigms. In fact, those paradigm changes that are not clear changes of scale are attributable to changes in type. It is possible to keep the scale constant and change of type. The same material situation pertains, but the observer is rejecting one way of looking at the system and is embracing another. The coherence arises here from the material system acting as an anchor as some new aspect of the system is moved into the foreground. Of course, a change in scale may well be the stimulus for a change of type. Look closely at the human skin and it becomes a landscape for mites. The upshot is that both type and scale, either together or alone, engender change that imposes dynamical quality.

Ecologists often assert a command of issues of scale. Part of the precision of structural quality comes from exquisite scaling so that the calibrations work. To that extent, there are many ecologists who do indeed use scaling with effect. However, if a high degree of structural quality is good, even more structural quality is not necessarily better. More compulsive control of the description of simple systems, in improvements of structural quality, may actually get in the way when one needs a change of scale that reparameterizes the system characterization. Becoming more deeply engrossed in structural quality at one level is often of no help, although it is too often the excuse for more research funds. There is no one framework for structural quality, given the questions that ecosystem ecologists wish to ask. There is an acute need to take scaling issues in ecology beyond the scaling that amounts to rudimentary system calibration, important as calibration may be.

Changes of scale and type to accommodate several levels of functioning are as important as getting the scale properly imposed at any one level. The dynamical quality that comes from changes in scale and type is often less welcome in mainstream ecology, but in a complex system, dynamical quality is an essential tool of the ecosystem scientist. With all this changing of level, criteria for aggregation change. Models can become held in contexts that may be at odds with the original meaning of some of the models. The power of models is their internal consistency. The limitation of models is also their need for internal consistency. As scales and types change, paradox and contradiction come into play. The antidote for such dilemmas is to employ narratives. Narratives are improved by being made consistent with models, but narratives can still reach over and above models to unite disparate situations. Narratives do not have to be consistent, and are not about verity (Zellmer et al., 2006). Rather, narratives are statements of what the narrator considers important in relationships of scale and type. In this volume, large complicated, and complex systems are put under scrutiny and their management is made tractable. It is no accident that there is a powerful narrative quality to this volume as a whole. Faced with climate change, a looming energy crises, and attendant ecological devastation from the move to renewable resources (Allen et al. 2001), ecology and resource science must embrace the dynamical quality that comes from the astute but bold use of scale and type in a narrative setting.

References

Ahl, V. and T. F. H. Allen. 1996. *Hierarchy Theory, A Vision Vocabulary and Epistemology*. New York: Columbia University Press.

Allen T. F. H. 1998a. The landscape level is dead: Persuading the family to take it off the respirator. In *Scale Issues in Ecology*, eds. D. Peterson and V. T. Parker, 35–54. New York: Columbia University Press.

Allen, T. F. H. and T. W. Hoekstra. 1992. *Toward a Unified Ecology*. New York: Columbia University Press.

Allen, T. F. H., J. A. Tainter, J. C. Pires and T. W. Hoekstra. 2001. Dragnet ecology, "Just the facts Ma'am": The privilege of science in a post-modern world. *Bioscience* 51:475–485.

Allen, T. F. H., A. J. Zellmer and C. Wuennenberg. 2005. The loss of narrative. In *Ecological Paradigms Lost: Routes to Theory Change, Theoretical Ecology Series*, eds. K. Cuddington and B. E. Beisner, 333–370. New York: Academic.

Funtowicz, S. O. and J. R. Ravetz. 1992. The good, the true and the postmodern. *Futures* 24:963–976

Kleinman, D. L. 1998. Beyond the science wars, contemplating the democratization of science. *Politics and the Life Sciences* 16:133–145.

Kuhn, T. S. 1962. *The Structure of Scientific Revolutions*. Chicago, Ill.: University of Chicago Press.

Pattee, H. H. 1972. The evolution of self-simplifying systems. In *The Relevance of General Systems Theory*, ed. E. Lazlo, 31–41. New York: Braziller,

Putnam, H. 1987. *The Many Faces of Realism*. La Salle, Ill.: Open Court Press.

Rosen, R. 2000. *Essays on Life Itself*. New York: Columbia University Press.

Sismondo, S. 1996. *Science Without Myth on Constructivism, Reality, and Social Knowledge*. New York: SUNY Press.

Zellmer, A. J., T. F. H. Allen and K. Kesseboehmer. 2006. The nature of ecological complexity: A protocol for building the narrative. *Ecological Complexity* 3: 171-182.

4

Self-Organizing, Holarchic, Open Systems (SOHOs)

James J. Kay and Michelle Boyle

In the preceding chapters of this section, we have introduced some basic concepts of systems thinking and explored in more detail some problems of framing the situation and the role of the observer. We introduced the idea of self-organization, primarily through empirical examples, as a way of understanding the limits of normal science in dealing with true complexity. In chapter 3, Allen explored the importance of scale and type in how we conceptualize and manage ourselves and our environments. We now turn to look at self-organization in more detail and the nature of Self-Organizing, Holarchic, Open systems (SOHOs).

> The law that entropy increases—the Second Law of Thermodynamics— holds, I think, the supreme position among the laws of Nature. If someone points out to you that your pet theory of the Universe is in disagreement with Maxwell's equations—then so much the worse for Maxwell's equations. If it is found to be contradicted by observation—well, these experimentalists do bungle things sometimes. But if your theory is found to be against the Second Law of Thermodynamics I can give you no hope; there is nothing for it but to collapse in deepest humiliation.
>
> —Sir Arthur Eddington (Eddington 1930)

Introduction

The most important scientific challenge facing humanity is to understand the co-evolution of the natural world and the human-constructed world that together form the biosphere of our planet. Only with this understanding can we begin to manage our affairs such that the biosphere is healthy and vibrant, both now and in the future. The key to understanding coevolution is to understand self-organization, something that systems theory has much to tell us about. Having set out in previous chapters a general framing of systems in terms of scale and type, in this chapter the insights of systems theory will be brought to bear on the coevolution of the natural and human-constructed worlds. The focus is on the evolution of biophysical versus social and cultural dimensions. However, it always must be kept in mind that these aspects are intimately linked. The purpose of this chapter is to provide a framework for research, to identify key questions and issues that need to be explored, and to present a conceptual model (which we refer to as the SOHO model) to frame investigation.

Self-organization has posed a conundrum for the physical sciences. At first blush, any self-organizing system seems to contradict the edict of increasing entropy associated with the second law of thermodynamics. Boltzman, Schrödinger, and others contemplated this puzzle (Boltzman 1886; Schrödinger 1944). Prigogine provided us with the first glimpse of a satisfactory resolution when he demonstrated a set of conditions under which a stable steady state could exist for open thermodynamic systems that are not in thermodynamic equilibrium (Prigogine and Wiame 1946). So our discussion of biophysical self-organization must take into account thermodynamics, but first we must clarify what constitutes a self-organizing system. Another aspect of this discussion, ecological integrity, will be explored in the context of implementing an ecosystem approach where it relates directly to themes and goal setting for sustainability (chapters 14, 17, and 19).

The Notion of Self-Organization

We begin by noting that the essence of a system is a collection of things that we perceive to be a whole. They appear to have an identity and to do something either actively or passively, in that they fulfill some function in the outside world. Systems are made up of components that are interconnected in a particular way. In physical systems these relationships are about mass, energy, and information flow between the components. The particular way in which the components are interconnected constitutes the system structure. The components, configured together in a particular structure, constitute the system's organization, and it is this organization that allows the system to do something.

Of particular interest in this regard are open systems: systems that have an environment, or "context," that provides the source and sink for flows into and out of the system. Open systems are active. That is, they do something: they convert inputs into outputs. There are two complementary aspects to open system organization: functional organization describing processes occurring in the system and structural organization referring to how the system is configured. When we look at an open system, we see its structure but not the processes. We see the consequences of the processes, usually in the form of inputs and outputs of the system or in the ability of the system to maintain itself in a state that is different from that of its environment.

Open systems span a range of complexity. In the simple case, we have mechanical systems. In these systems, the *raison d'être* of the system is external. The components are brought in from outside and assembled according to some external blueprint. The processes within the system, as well as any changes in the system organization, happen according to externally defined rules. The causality of the system is linear; that is, there is a linear logic chain between inputs, their processing by the system, and the resulting outputs. Therefore inputs can be used to control outputs, and outputs will not change if inputs do not change. In short, the system's organization is due to external agency.

Complex self-organizing systems, in contrast, can change some combinations of their components, the behavior of components, or the way the components are interconnected (their structures). Consequently, different relationships and processes can develop, and the system can change its repertoire of behavior. In short, the system can change its organization through internal agency. This organizational change can be independent of external environmental change. External events may result in no change in the system, and system change can occur without change in the external world.

A central feature of self-organizing systems is that they can manifest coherent self-perpetuating behavior that is internally generated (Nicolis and Prigogine 1977, 1989; Prigogine and Stengers 1984). Such systems have a tendency to establish themselves in thermodynamically nonequilibrium steady states that are capable of persisting, even when the external environment changes. Each of these steady states represents an organizational mode: a particular configuration of components and processes that give rise to specific patterns of behavior into which the system is capable of locking itself. Giampietro refers to this event as mode locking, and he further notes that such organizational modes are "self-entailing" because they arise from internal processes. These organizational modes are not necessarily strictly homeostatic but, instead, may be confined to a limited domain in state space about a dynamically stable equilibrium point (Kay 1991a), described by some as an "attractor." As such, even though the behavior of the system may change, we still recognize it as the same entity. The ability of a self-organizing system to maintain its identity (i.e., specific organizational mode) can be attributed to the feedback loops in the system or as Maruyama refers to them, "morphogenetic causal loops" (Caley and Sawada 1994; Maruyama 1980). These feedback loops, especially those involved in autocatalytic (internally initiated) processes, give rise to coherent tendencies of the system that Ulanowicz refers to as the system's "propensities" (Ulanowicz 1997a). In summary, a complex self-organizing system is an intricate interplay between nested processes and elements (the organizational mode) perpetuated by morphogenetic casual loops that give rise to a set of propensities, which, in turn, are manifested as self-perpetuating behaviors that give the system its coherence as an entity. A description of all of these phenomena for a particular system is referred to as its "canon" (Kay and Regier 1999; Kay et al. 1999; Kay 2000; Ulanowicz 1997a).

In mechanical systems, linear causality and black box explanations may be adequate because behavior and output are determined by input. Explanation of self-organizing phenomena, however, must be in terms of internal system processes. Although these processes are open to the system's environment, they can

BOX 4.1   Systems Descriptions

The following are examples of how the terminology introduced in this section is used to describe a system and its self-organization. For more discussion of systems terminology, see Blauberg et al. (1977) and Flood and Carson (1993). For another example of these concepts used to describe an ecosystem, see Kay and Regier (1999)

1. Bénard cells: Given certain conditions and a specific temperature gradient across a fluid, a pattern spontaneously appears in the liquid that looks like the hexagonal "cells" of a beehive (Chandrasekhar 1961; Cross and Hohenberg 1993; Nicolis and Prigogine 1989; see www.etl.noaa.gov/about/eo/science/convection/RBCells.html, en.wikipedia.org/wiki/Rayleigh-Bénard_convection for illustrations).

Components
   • Molecules of fluid.
Interconnected relationships between components
   • How the molecules move relative to their neighbors.
   • Conduction: random directions and collisions.
   • Convection: move in same direction as neighbors, coherent behavior.
Environment/context
   • Fluid in a container with a heat source on the top, a cold sink on the bottom, and an external source of energy to maintain the temperatures of the source and sink. In other words, a gradient is imposed on the fluid.
Process
   • Energy transfer from hot source to cold sink. With Bénard cells present, this transfer is by convective process.
Structure
   • Bénard cells
System organization
   • Coherent fluid flow in the form of stable convective cells.
Functional organization, what the system does, the processes occurring in the system
   • Coherent energy transfer
Structural organization, how the system is configured
   • Hexagon pattern of cells (Bénard cells) that result from the coherent motion of the fluid.
Specific patterns of behavior/mode locking
   • Bénard cells

Morphogenetic casual loops/autocatalytic processes
   • Convective motion of the fluid.
Propensities
   • As gradient increases new structures emerge that transfer energy between the hot and cold source at a faster rate, which also increases the rate of entropy production and exergy destruction.
Canon
   • The emergence of stable convective cells that facilitate energy transfer between the source and sink.
Self-organizing
   • In the sense that the cells appear spontaneously without any external intervention.

(continued)

BOX 4.1   Systems Descriptions *(continued)*

2. Belousov-Zhabotinski reaction: Spatial patterns of homogenous color in a chemical reaction system. See Nicolis and Prigogine (1989:18) for a detailed description of the phenomena.

Components
- Chemical species.

Interconnected relationships between components
- Their spatial concentrations within the reaction vessel.

Environment/context
- Specific chemical species are added and removed from the reaction vessel.

Process
- Reaction of the different species with each other to give a particular spatial and temporal distribution of chemical composition.

Structure
- Stirred: periodic (fixed frequency) change in color (and therefore chemical species) throughout the reaction vessel (at any moment in time the contents, and color, of the whole reaction vessel is the same).
- Not mixed: Patterns of wave fronts of color (specific chemical compositions) move through the reaction vessel. One such pattern forms a checkerboard pattern in the vessel with the color of the squares changing with a fixed frequency.

System organization
- Coherent spatial and temporal distribution of chemical species and hence color in the reaction vessel.

Functional organization, what the system does, the processes occurring in the system
- Rates of reactions (which may be zero) between the constituent chemical species that give rise to coherent behavior.

Structural organization, how the system is configured
- The spatial and temporal distribution of color in the reaction vessel.

Specific patterns of behavior/mode locking
- The system gets locked into a specific spatial and temporal pattern as described in "structure."
- Morphogenetic casual loops/autocatalytic processes.
- Some of the reactions are autocatalytic, and others are part of stable sets of reactions that mutually entail each other.

Propensities
- To establish fixed reaction rates that give rise to coherent behavior in the form of coherent spatial and temporal distribution of chemical species, hence color in the reaction vessel.

Canon
- Behaves like a clock.

Self-organizing
- The structure (color patterns) emerges without external intervention.

3. A forest (energetics perspective only): This example is limited by our lack of thinking of biological systems in these terms and hence a lack of understanding and data about a forest as a self-organizing system.   *(continued)*

BOX 4.1   Systems Descriptions *(continued)*

Components
  • Trees and other species both above and below ground.
Interconnected relationships between components
  • Who eats who?
Environment/context
  • Incoming exergy, availability of nutrients and other material flows, information flows, and the physical environment.
Process
  • The utilization of exergy (and hence degradation of exergy) by the individuals of species and the flow of energy between individuals.
Structure
  • The food web (including both the grazing and detrital food webs).
System organization
  • A complicated and complex interrelationship between many species that allows the forest to maintain its integrity while increasing its utilization of exergy. For a discussion of ecological integrity from this perspective, see Kay and Regier (2000).
Functional organization, what the system does, the processes occurring in the system
  • Trees capture solar exergy and use it to pump large volumes of water.
  • A very small amount of solar exergy is converted to stored exergy (primary production) through photosynthesis. This stored exergy is consumed through herbivory, carnivory, and detritivory.
  • Material is brought into the system by several processes, particularly the water pumping of trees. It then cycles among the individuals in the forest, eventually to be broken down into raw nutrients in the soil for reuse by trees or leaches out of the forest. For detailed discussion, see Bormann and Likens (1979).
Structural organization, how the system is configured
  • An intricate interlocking canopy of leaves and roots. Primary producers are herbivores, carnivores, and detritivores. For an aquatic example, see Baird and Ulanowicz (1989).
Specific patterns of behavior/mode locking
  • The forest canopy is maintained in spite of the death of individual trees. Details depend on the specific type of forest.
  • An example of the patterns of behavior that can emerge is the Holling four-phase model for ecosystems (Holling 1992).

Morphogenetic casual loops/autocatalytic processes
  • Very dependent on the forest type. For a discussion for temperate deciduous forests, see Bormann and Likens (1979).
Propensities
  • Captures increasing resources (exergy and material), makes ever more effective use of the resources, builds more structure, and enhances survivability until the local carrying capacity is met. Then establishes some near steady state.
Canon
  • A forest is made up of a self-perpetuating canopy of trees.
Self-organizing
  • Forests emerge spontaneously on their own, if the context is present.

alter the type, quality, and quantity of inputs to the system. Furthermore, these processes can be changed by internal agency. Because these processes are made up of reciprocal or circular relationships in the form of feedback loops, explanation will necessarily be circular or nonlinear (in the sense that in feedback loops, the effect or output is also part of the cause or input). This characteristic, as documented by Maruyama (Caley and Sawada 1994:6), poses an immense challenge to our traditional scientific approach to explanation. Because the behaviors of self-organizing systems require explanation in terms of causal loops, reciprocal relationships, and the internal processes in the system, we perceive that self-organizing systems are behaving on their own. This is what gives rise to the sense of "self" that we have about these systems. It is the decoupling of changes in these systems from changes in their environment that distinguishes self-organizing systems from simpler types of organized systems.

Self-organizing systems may be emergent.[1] Emergence is when a self-organizing system exhibits a novel mode of organization that requires a different kind of explanation than was previously used to describe the system. Emergence should be distinguished from fast, dramatic reorganization of the system [in catastrophe theory terms, a flip to a new attractor (Kay 1991b)] that may be surprising but not novel. It is this possibility of emergence that allows self-organizing systems to evolve.

The examination and description of self-organization is challenging because of its complexity. Investigations must deal with the irreducible uncertainty that comes from the system's capability to modify itself (including the possibility of emergence), the existence of several distinct but possible organizational modes for the system (multiple attractors), and the hierarchical nature of such systems. We have discussed each of these above based on empirical observations and complex systems theory. Before we tackle the even more complex and challenging issues related to the governance and management of ecosystems for sustainability, it is useful at this point to examine the thermodynamics that underlies complex systems.

Thermodynamics, Exergy, and Self-Organization in Biophysical Systems

Schrödinger, Odum, and Prigogine have observed (at different times) that thermodynamics is the foundation for understanding self-organization and life (Odum and Pinkerton 1955; Odum 1988; Prigogine et al. 1972a,b; Prigogine and Wiame 1946; Schrödinger 1944). In particular, Prigogine developed a theory of nonequilibrium thermodynamics and self-organization in terms of dissipative structures. This theory explained how spontaneous coherent behavior and organization occurs in open systems. Central to understanding such phenomena is the realization that open systems are processing an enduring flow of high-quality energy. In these circumstances, coherent behavior appears in systems for varying periods of time. However, such behavior can change suddenly whenever the system reaches a catastrophe threshold and "flips" into a new coherent behavioral state (Nicolis and Prigogine 1977; Regier and Kay 2002). A "catastrophe threshold" is a point of discontinuity at which continuous change of some variable(s) generates a sudden discontinuous response. A simple example is the vortex that spontaneously appears in water from draining a bathtub or the dramatic appearance of tornadoes "from nowhere" (Kay 1991b).

Further discussion requires the introduction of the notion of quality of energy, or exergy. Over the past thirty years and especially in the last decade, exergy has emerged as a central concept in discussions on thermodynamics (Bejan 1997; Fraser and Kay 2000; Szargut et al. 1988; Wall 1986). Energy varies in its quality or capacity to do useful work; although all energy can be measured in joules, not all joules are equivalent. One can do less with a joule of crude oil than with a joule of household heating oil, and less with a joule of both of these than with a joule of electricity. The varying quality of each of these joules of energy is measured by exergy. During any chemical or physical process the quality or capacity of energy to perform work is irretrievably lost. Exergy is a measure of the maximum capacity of the available energy contained in a system to perform useful work as the system proceeds to equilibrium with its surroundings. For example, water at the top of a high cliff is a high-quality energy source (high exergy) because its potential energy can be used to perform work. We can use the falling water to turn a turbine and produce high-quality energy in the form of electricity. However, if the high-quality energy in the falling water is not run through a turbine and falls freely to the rocks below, it turns into low-quality dispersed heat energy. The exergy content of the water at the top is high, but the same water at the bottom, with the same energy content, has much less exergy. Exergy is a measure of the quality of energy.

In terms of exergy, the classical second law of thermodynamics can be stated as, during any macroscopic thermodynamic process, the quality or capacity of energy to perform work is irretrievably lost. Energy looses exergy during any real process. A conventional first law energy analysis does not account for differences in energy qualities. A first law efficiency only compares the total amount of energy put into a system to the total amount received out of the system. However, in the discipline of energy system analysis and engineering thermodynamics, both first law (energy, quantity) and second law (exergy, quality) analyses are necessary for the understanding and development of efficient and effective energy utilization systems (Gaggioli 1980, 1983; Hevert and Hevert 1980; Brodyansky et al. 1994; Edgerton 1982; Ford et al. 1975; Moran 1982).

Exergy is a function of the gradient between a system and its environment. It is a summation of the free energies in a situation. In effect, it measures how far a system is from thermodynamic equilibrium with its environment. Exergy is not a useful concept for discussing equilibrium situations, the domain of classical thermodynamics, because by definition its value is zero in such situations. However, it is a very powerful tool for nonequilibrium situations. The larger the value of the exergy, the more the situation is out of equilibrium.

Using the exergy concept, Kay and Schneider have extended Prigogine's work and reformulated the second law of thermodynamics for nonequilibrium situations (Kay 2000; Schneider and Kay 1993, 1994a,b; Fraser and Kay 2000). In exergy terms, the restated second law is, given that a system, exposed to a flow of exergy from outside, will be displaced from (thermodynamic) equilibrium, the response of the system will be to organize itself so as to degrade the exergy as thoroughly as circumstances permit, thus limiting the degree to which the system is moved from thermodynamic equilibrium. Moreover, the farther the system is moved from equilibrium, the larger the number of organizational (i.e., dissipative) opportunities that will become accessible to it, and, consequently, the more effective it will become at exergy degradation. This is the exergy degradation principle, the second law for nonequilibrium thermodynamic situations.

Exergy pumped into an open system moves it away from equilibrium, but nature resists this displacement. When the input of exergy and material pushes the system beyond a critical distance from equilibrium, the open system responds with the spontaneous emergence of new, reconfigured organized behavior that uses the exergy to build, organize, and maintain its new structure. This reduces the ability of the exergy to move the system farther away from equilibrium. As more exergy is pumped into a system, more organization emerges, in a stepwise way, to dissipate the exergy. Furthermore, these systems tend to get better and better at "grabbing" resources and utilizing them to build more structure, thus enhancing their dissipating capability. In principle, however, there is an upper limit to this organizational response. Beyond a critical distance from equilibrium, the organizational capacity of the system is overwhelmed and the system's behavior leaves the domain of self-organization and becomes chaotic. As noted by Ulanowicz, there is a "window of vitality," that is, a minimum and maximum level between which self-organization can occur (Ulanowicz 1997b).

A common example is the emergence of a vortex in bathtub water as it drains. The exergy is the potential energy of the water (due to the height of water in the bathtub), the raw material is the water, the dissipative process is water draining, and the dissipative structure is the vortex. The vortex will not form until a sufficient height of water is in the bathtub, and if the water is too deep, laminar flow occurs instead of a vortex.

In summary, the theory of nonequilibrium thermodynamics suggests that the self-organization process in open systems proceeds in a way that captures increasing resources (exergy and material), makes ever more effective use of the resources, builds more structure, and enhances survivability (Kay 1984; Kay and Schneider 1992; Kay 2000; Schneider and Kay 1994b). These seem to comprise the kernel of propensities of self-organization.

The manner in which these propensities manifest themselves as morphogenetic causal loops and dissipative processes (system organization) is a function of the given environment (context) in which the system is embedded, as well as the available materials, exergy, and "information." Which energy and mass transformation processes can occur is strongly governed by the physical environment. The rates of processes are a function of temperature, humidity, pressure, etc. Exergy provides the stimulus for organization. Depending on the form of exergy accessible, different processes will be possible and will be able to progress at different rates. Materials provide the raw matter for the processes and building blocks for structure. Information refers to the factors that act internally within the system to constrain its behavior. Information can also flow into the system from outside. It can catalyze certain processes and not others, thus promoting their emergence and persistence. Information therefore biases the direction of self-organization. It provides a record or template of successful organizational strategies that can be used in the future to quickly build structure. It should be noted that information also is important in terms of sources of exergy and materials.[2] The interplay of environmental conditions, exergy, materials, and information defines the context and constraints for the set of processes and structures that may emerge during self-organization. Generally speaking, which specific processes emerge from the potential set is uncertain. One of our major research challenges is to understand the relationship between the external context and self-organization, that is, the structure and processes that may develop.

In summary, self-organizing dissipative processes emerge in biophysical systems whenever sufficient exergy is available to support them. Once such processes are established, they manifest as structures. These structures provide a new context, nested within which new processes can emerge; these, in turn, beget new structures, nested within which, and so on. Thus emerges a self-organizing holarchic open system embedded in a physical environment: a nested constellation of self-organizing dissipative process/structures organized about particular sources of exergy, materials, and information that give rise to coherent self-perpetuating behaviors. This conception of self-organization is presented in fig. 4.1. Elsewhere, using Wicken's ideas, the interaction of exergy and materials to form a self-organizing living system is explored (Kay and Schneider 1992; Kay 1984; Wicken 1987). This work has been used as the basis for developing narrative descriptions of self-organization in ecosystems (Kay and Regier 1999; Kay 2000).

The World as a Self-Organizing Holarchical Open System: A Conceptual Model

Using the portrayal of self-organization above, we have developed an integrated SOHO system model that portrays ecological-societal systems as dissipative complex systems (Boyle et al. 2000; Corning and Kline 1998; Kay and Regier 1999; Kay et al. 1999; Regier and Kay 1996). This SOHO system model provides a conceptual basis for discussing ecological integrity and human sustainability (refer to figs. 4.2–4.4). It furnishes us with an integrated, nested ecosystem description of the relationship between natural and human systems and serves as a basis for understanding their coevolution.

In this model, the elements of the landscape (e.g., woodlots, wetlands, farms, neighborhoods) that make up the human and natural ecological systems are seen

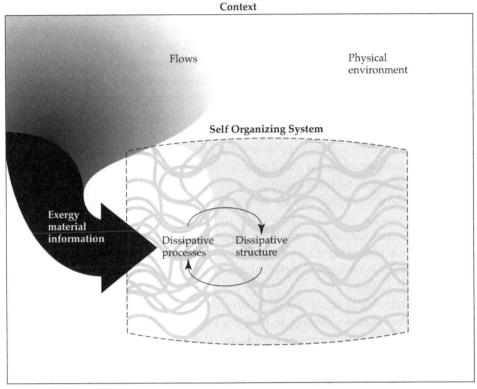

Figure 4.1   A conceptual model for self-organizing systems as dissipative structures. Self-organizing dissipative processes emerge whenever sufficient exergy is available to support them. Dissipative processes restructure the available raw materials in order to dissipate the exergy. Through catalysis, the information present enables and promotes some processes to the disadvantage of others. The physical environment will favor certain processes. The interplay of these factors defines the context for (i.e., constrains) the set of processes that may emerge. Once a dissipative process emerges and becomes established, it manifests itself as a structure. Copyright NESH (2004).

as self-organizing entities set in an environmental context. Self-organizing entities are understood through their constituent processes and structures and their relationships. (For example, in a woodlot, processes would be evapotranspiration and growth of biomass, structure would be the species that make up the woodlot, and a description of the relationship between these processes and structure would be Holling's four box model (Holling 1986; Gunderson and Holling 2002; fig. 4.5). The processes involve flows and transformations of exergy, material, and information. The structures are the objects (i.e., trees) we see on the landscape. The processes allow for the emergence and support of structures, which, in turn, allow for the emergence of new processes, and so on. The recursive relationship between process and structure is an important feature of this nested conceptual model.

Our conception of self-organization as a dissipative system was presented in fig. 4.1. It is a description of how a mass-energy transformation system emerges,

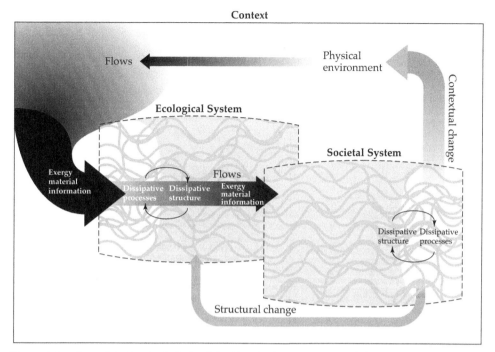

Figure 4.2   A model of a single holarchical level of ecosystems and their interactions. The ecological system forms the context for the societal system. The arrow across the bottom represents direct societal influence on the ecological system (i.e., structural change). The larger arrow across the top represents indirect societal influence on the ecological system (i.e., contextual change for the ecological system that cascades down to change the societal system). Copyright NESH (2004).

and it is the kernel of the SOHO system model. This basic description characterizes mass-energy transformation systems in terms of the exergy, materials, and information they consume and how these are used in dissipative processes. Thus it focuses on the consumption side of the production-consumption duality of systems. However, when more than one such element is connected to form a larger system, the production aspect of the system becomes clear. Each element not only consumes exergy, materials, and information but also produces exergy, materials, and information for the next element in the concatenation. Each element provides the context for another element. So horizontally, each element in the SOHO system model has two faces, its consumption face and its production face (see fig. 4.2).

The Coupled Ecological-Societal System

Natural ecological communities provide exergy, materials, and information required for human societies to sustain themselves. This is depicted in fig. 4.2. The human-constructed societal system depends on the flow of exergy, materials, and

Context

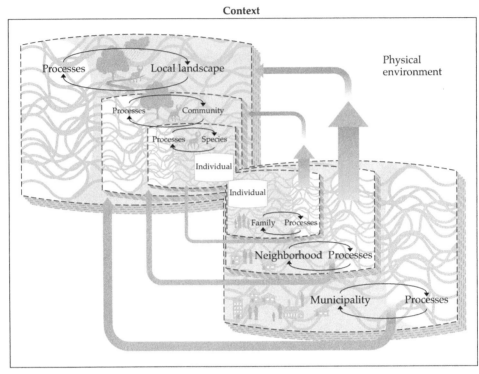

Figure 4.3   An example of components and interactions over three holarchical levels (beyond the individual). The "stacked deck" effect is a reminder that each level is made up of a conglomeration of defined systems. That is, many species together comprise an ecological community, and many communities together form the local landscape. It is the aggregation of these local landscapes that makes up the landscape mosaic of a region (such as a province or state). On the societal side, families and businesses comprise neighborhoods. Municipalities are made up of neighborhoods and finally, the province/state is politically divided into municipalities and counties. Note that this diagram demonstrates only one possible way of parsing the system. Copyright NESH (2004).

information from the natural ecological system to support its processes and structures. These flows, along with the biophysical environment provided by ecological systems, are the context for societal systems. This context constrains the possible societal processes and structures in a specific location. While fig. 4.2 illustrates a single ecological system providing the context for the societal system, the reality is that it is a suite of adjacent ecological systems (e.g., woodlots, fields, and wetlands adjacent to a farm) that provide this context. Alterations in these adjacent systems will alter the context and thus the possibilities for the system in question.

However, the societal system can also influence the ecological system in two ways. The first influence occurs when humans change the structure of the ecological system (e.g., cutting trees down in a woodlot, filling in wetlands, and all the human activities that involve adding, removing, or dismantling ecological structures on the landscape). Such actions alter the flows from the ecological systems and

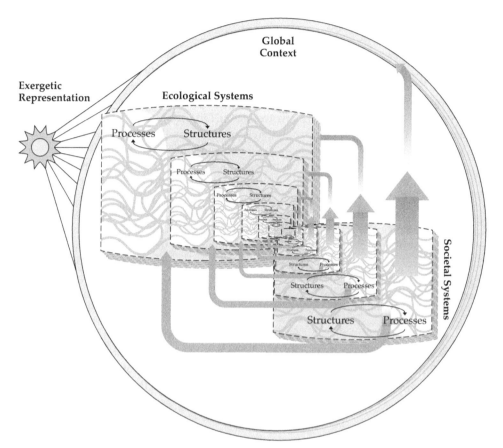

Figure 4.4   The Big Picture—Interrelationships and influences between ecological and societal systems in the biosphere. The arrow extending up to the circle represents the relatively new ability for humans to influence the biosphere and the global context. Copyright NESH (2004).

BOX 4.2   The Nature-Society Continuum

The term ecological-societal system is used here to refer to the intertwined natural and human-constructed ecosystems that together constitute our biosphere. Other authors have used social-ecological, socioecological, and ecosocial systems to mean more or less the same thing. None of the language is entirely satisfactory. Human-constructed ecosystem refers to the biophysical side of society's systems, meaning factories, cities, families, etc. The terms societal system or human system are interchangeable in this context. "Ecological system," here, refers to natural ecosystems or the natural world.

In some ways this is a poor choice of language as we really need to conceptually think of both the natural and human-constructed worlds from an ecological perspective. Splitting the world into "natural" and "societal" is itself an artificial dichotomy. We can differentiate between the two by defining a natural system as one whose local context, that is its physical environment, and flows of exergy, material and information into the system are not influenced by humans. Humans may be an important part of the system,

(continued)

BOX 4.2   The Nature-Society Continuum *(continued)*

---

as indigenous people are, but they are not part of the system's context. In contrast, a societal system is one whose local context is completely determined by humans (e.g., a factory). Between these two extremes is a continuum of systems whose organizational context is more or less influenced by humans. (An agricultural system lies somewhere in the middle of the continuum.) Notwithstanding this continuum, as discussed later in this chapter, this nature-society separation is an illusion in that all societal systems are nested within a larger nested system whose ultimate context is the natural world.

Historically, there were many examples of purely natural systems. However, this is no longer the case. Currently, all biophysical systems on the planet (except perhaps for deep sea vents) have some aspect of their context affected by humans. Changes in the ozone layer, global climate change, and the ubiquitous presence, even in the Arctic, of substances like DDT and PCB (which constitute information pollution) mean that there really are no "natural systems." Human influence on the biosphere is such that it is becoming increasingly artificial to separate the natural world from the human world. From the self-organizing open systems perspective, the two are becoming inextricably intertwined.

---

feed back to the societal systems. This is represented in fig. 4.2 by the lower arrow back from the societal system to the structure in the ecological system. The feedback to the societal system occurs because changes in the ecological structure change the context for the societal system.

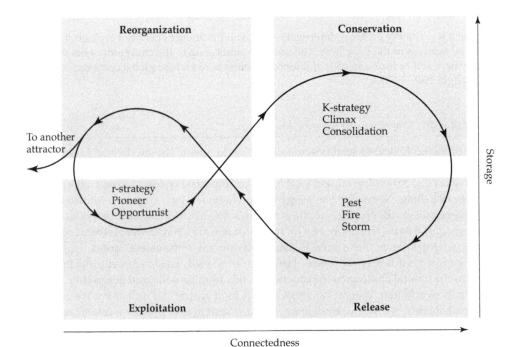

Figure 4.5   Holling's four-box model (adapted and redrawn).

The second influence occurs when the societal system alters the context for the ecological system. For example, the runoff into a wetland or stream may be altered by human activities on adjacent properties (depicted in fig. 4.2 by the upper arrow from the societal system to the context of the ecological system). This influence is qualitatively different from the structural influence just discussed. The resulting feedback loop has more steps and accordingly is more indirect. By changing the context of the ecological system, the societal system affects the ecological processes, and, in turn, the ecological structure and ultimately affects the societal system's own context. For example, modifying the runoff into a waterway can dramatically alter the character of the waterway, and hence the type of fish found in it, and therefore the sport fishery and associated economic system.

Referring to fig. 4.2, the relationship between societal systems and ecological systems is threefold:

1. Ecological systems provide the context for societal systems. That is, they provide the biophysical surroundings and flows of exergy, material, and information that are required by the self-organizing processes of the societal systems.
2. Societal systems can alter the structures in ecological systems (e.g., cutting down a woodlot, removing beaver from a watershed.) Changes in the ecological structure alter the context for the societal systems.
3. Societal systems can alter the context for the self-organizing processes of ecological systems (e.g., changing the drainage patterns into a wetland or paving a large area to create a heat island effect that changes the local microclimate for a woodlot.) Changes in ecological process can alter ecological structure and consequently the context for societal systems.

In our daily activities, most of us are not cognizant of the relationship that we and our complex human-constructed systems have with the rest of the biosphere nor do we think about the flows of exergy and materials that we consume. Furthermore, very few study information flows. Yet it is critical to understand human-ecological interactions. Flows between ecological and societal systems directly and indirectly set the stage for our bustling societal production and consumption system. The physical environment is a prerequisite for our economy and livelihoods. In recent years, however, insurance companies (as one example) are beginning to recognize the depth of our relationship with nature and the economic implications of global climate change.

When we are not in the concrete canyon of an urban center or safely nestled in our offices and homes, we see the structures that make up nature around us. We see trees and birds, woodlots and wetlands. However, when was the last time you gave a thought to how much cooling was being provided by the trees in your neighborhood pumping water or to the clean water that was filtered though a wetland? The vast majority of humanity in the developed world has never thought about the self-organizing processes in nature and how they ultimately provide the natural structures, physical environment and flows of exergy, materials, and information that allow our ecosystems (that is, our production and consumption processes) to flourish and evolve.

Given how unaware we are of our dependence on the natural world as context for societal self-organization, it follows that our effect on the self-organization of the natural world is rarely thought through (as apparent in some of the examples presented in chapter 2). When we remove structures from the landscape (cut down a woodlot, fill in a wetland), the immediate loss of structure is obvious. What is less obvious is how the loss of these structures alters self-organization in the natural world and, ultimately, our own context. In fact, in many, many situations we simply don't have the scientific understanding necessary for exploring the effects of loss of structure. So it is not surprising that we do not, in our minds, close the causal loop from our actions to change in the natural world that affects us. Because these feedbacks are happening at several scales at once, it is even more difficult to understand how changes and consequences combine and cascade through time and space. An example is the rarely contemplated relationship between driving a car and our food supply: the car emits greenhouse gases, contributing to global climate change, in turn causing crop failures.

Conceptually, the point of the SOHO model is to encourage people to think about the relationship between the natural and human-constructed world and how human activities can affect the organization of the natural world feeding back and cascading through to affect the context for society. The failure to think through this relationship has resulted in many unpleasant surprises that perhaps could have been avoided. If we wish to understand this dynamic, then we must develop a science that allows us to study the coevolution of the natural world and the human-constructed world in terms of SOHO systems.

The Nested Model

Our discussion surrounding fig. 4.2 applies to one scale of observation, but, as noted earlier, sustainability and integrity issues can only be understood in terms of "nested holons." Figure 4.3 illustrates this idea of nesting. On the ecological side, "local landscape" can be applied to a subwatershed, for instance. The hydrological cycle is an example of a process in the subwatershed. The structures that make up the subwatershed are the ecological communities (e.g., woodlots, wetlands, open fields, etc.). The communities, in turn, are made up of species. On the societal side, municipalities rest on the local landscape. These, in turn, are made up of neighborhoods, which are made up of families and businesses, and ultimately individuals.

In many cases, the local subwatershed defines the context for the local municipality. However, the municipality directly modifies the ecological communities in the subwatershed and thus its own context. Adjacent ecological communities also form part of the context for local neighborhoods. As we discussed in the previous section, however, the local neighborhood is capable of influencing ecological communities, through direct structural change (such as harvesting wood from a woodlot), or altering the context for an ecological community (e.g., changing drainage patterns into a wetland).

BOX 4.3   Holarchy, Holons, and Hierarchy

The language of hierarchy theory is still not "fixed," so a brief discussion of terms is in order. The systems we are discussing are nested and must be examined at different levels of scale. Each level is a system in its own right, made up of subsystems (the next level down), and has an environment (the next level up). After Koestler (1978), we refer to the level of focus as a holon. For Koestler, the nesting of all the holons (that is, all the levels) constitutes a holarchy. (See fig. 4.4.) Koestler did not use the term hierarchy because of its connotations of command and control from higher levels, which may be misleading in discussing linked social and ecological systems.

Not only is the consideration of the nesting of such systems important, but they must also be examined from different perspectives. For example, a factory could be studied in terms of the copper material flow through it, and in terms of the electrical flow through it. Each study would result in a very different system description. In this chapter, after Allen, we use the term hierarchy for a system description that is both holarchic (made up of nested holons) and viewed from different perspectives. Therefore a nested system description from a single perspective is a holarchic description, and one from multiple perspectives is a hierarchical description (Ahl and Allen 1996; Allen and Starr 1982).

Figure 4.3 and the examples are meant to be illustrative and not exhaustive, although they do demonstrate that such changes can cascade through the nested holons to ultimately affect individual families and businesses. A unique feature of this conceptual model is that it explicitly identifies these relationships and provides a framework to discuss them. Figure 4.4 shows the full conceptual model as a template to develop any particular conceptual model in which its specific environmental context and important levels, processes, structures, and influences/feedbacks are identified.

Some Insights from the SOHO Model

The full conceptual model (fig. 4.4) provides insight into the coupled ecological-societal system. It illustrates the dependence of human-constructed systems (societal systems) upon nature as their context. If the context were severely degraded or somehow removed, the self-organization of societal systems would be radically altered, if not halted all together. Of particular significance in this model is exergy flow (originating from the sun) from the top left corner to the bottom right corner, and as is dictated the second law, it is degraded at each step along the way. This illustration reflects the thermodynamic reality of the biosphere. Solar energy is the source of exergy for nature, and nature, in turn, is the source of exergy for society (and this source is rate limited). If society's source of exergy were turned off, self-organization in society would stop, just as the vortex in a bathtub stops when the water tap is turned off.

The model also highlights the feedbacks from society to nature. Of particular importance are the two outermost arrows. These feedback loops to the global scale have formed only very recently in human history (the past 50 years or so). Prior to this time, humanity did not have the capability of altering context on the oceanic or planetary scale. Now we do.

Now the societal-ecological system is an intertwined, coevolving system at all scales. This fact profoundly and fundamentally alters the relationship between humans and nature. Changes in self-organization in one affect the other, demanding new strategies to manage our relationship with nature. Historically, the contextual feedback loops from society to nature only allowed humans to alter the self-organization in nature at scales smaller than continental but still very large. So even if sizeable areas were severely degraded, it was only a portion of the biosphere, not the biosphere as whole, that was altered. If things went awry in a particular locale (even to the extent of a civilization collapsing), humans could rely on simply moving somewhere else. This strategy is no longer viable, so the cost of "getting it wrong" is much higher even than the collapse of a local civilization.

The model also provides a resolution of a long-standing conundrum about the relationship between nature and society. There have been two models of this relationship: a set of concentric nested circles and a Venn diagram of overlapping circles (ecological, economic, and social systems). Which version is correct has been debated passionately. The nested SOHO model would suggest that they are both right and that differences are those of perspective on the part of the observer (see the discussion by Allen on scale and type in chapter 3).

Consider the three-dimensional version of the SOHO model (fig. 4.6). The six views in fig. 4.7 are two- and three-dimensional orthogonal projections of this model. The top views show the ecological and social systems overlapping, as in the Venn diagram. This perspective highlights the coevolutionary nature of the relationship between the ecological and societal systems. The center views show the projection on the plane perpendicular to the direction of exergy flows, looking toward the sun. This perspective corresponds to the nested concentric circles and highlights the dependence of the societal system on nature for its context. In this sense, the societal system is nested within the natural ecological system. The right-hand views, looking from the side, show multilayered and intertwined connections between the two systems.

The degree to which the ecological and societal systems overlap reflects the extent to which processes and structure are shared by both systems and the extent to which processes and structures in one system are connected directly to those in another. So a modern urban city would have many processes and structures that are not directly coupled to natural systems and few that are shared. The degree of overlap would be relatively small. An agricultural community would have many more processes and structures that are directly coupled and reliant on natural processes and structures, as well as shared processes and structures. Hence the two systems would have a higher degree of overlap. A traditional indigenous community would make use of mostly natural processes and structures or processes and structures that are directly coupled to natural processes and structures. The overlap would be considerable.

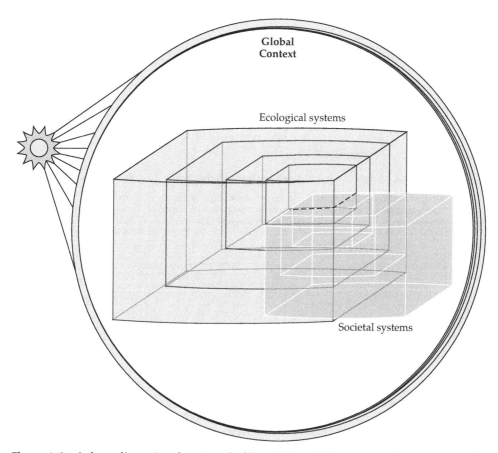

Figure 4.6   A three-dimensional portrayal of the relationship between socioeconomic and ecological systems. The variations show that much of the debate of the relationship between ecological and socioeconomic systems can be explained by using different observer-dependent perspectives on the same reality. Copyright NESH (2004).

However, this characterization of the urban versus agrarian versus indigenous cultures is not inherent to them. There is no reason why a modern community could not be organized so that there is much more overlap with nature. In fact, there are many good reasons to apply "greener" organizational strategies. In such a community, our dependence on nature would not be as hidden as it is in our current Western society. Perhaps the design of sustainable societies should strive to have as much overlap as possible with natural ecosystems, while preserving the ability of natural systems to self-organize.

Another aspect of the nature-society relationship demonstrated by the SOHO system model is the relative scale of different system perspectives. For example, if energy is the scale of the three-dimensional model, the natural system would be many orders of magnitude larger than the societal system, reflecting the total energy use in the two systems. If the scale is net primary production (in a photo-

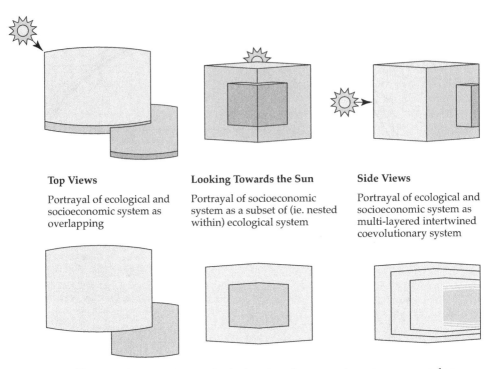

Top Views

Portrayal of ecological and socioeconomic system as overlapping

Looking Towards the Sun

Portrayal of socioeconomic system as a subset of (ie. nested within) ecological system

Side Views

Portrayal of ecological and socioeconomic system as multi-layered intertwined coevolutionary system

Figure 4.7   The top views portray ecological and socioeconomic systems as overlapping. A view looking "through" the figure with the sun behind shows the socioeconomic system as a subset of the ecological system. In this orthogonal view, society is in the shadow of the natural world and completely dependent on the natural world for its context. This view highlights the thermodynamic realities of the relationship between society and nature. The "side views" are from the bottom left corner. (The orthogonal projection onto a plane perpendicular to the page and including a line that runs from the bottom right corner to the top left corner.) In this orthogonal view we see two holarchicly nested systems, which are interconnected at every level. This view highlights the intertwined nature of the relationship between the two systems. Copyright NESH (2004).

synthetic sense), the natural system would only be double the size of the societal system. Repeating this exercise for a number of different variables (e.g., exergy, total material flow, information flow, etc.) may yield valuable insights into the relationship between these systems.

The SOHO model provides a basis for discussing the biophysical relationship between society and nature. It is based on a self-organizing systems approach, also referred to as an ecosystem approach (Kay et al. 1999; Kay and Regier 1999; Boyle et al. 2000). It provides a framework for identifying and discussing the important aspects of the relationship, that is,

- the structures and processes of self-organization and their interdependence;
- the contextual elements for these structures and processes (physical environment, forms of exergy, material, and information flow);

- the holarchic nesting of the structures and processes, and the recursive relationship between context, which begets process, which begets structure, which provides context, etc.;
- the feedback loops, particularly those coupling society and nature; and
- the thermodynamic realities of the relationship.

Thus it provides a basis for thinking through the interactions between these two systems and a better understanding of them as coupled self-organizing systems.

A sobering observation, however, is that we currently are not collecting data or even examining the relationships that the SOHO model suggests are important. In particular, many of the self-organization activities occur at the larger regional scale, a scale that is beyond that of local government but usually below that of state government. We have often found in our work that no heed is paid to the most basic system issues at this intermediate scale. Yet our experience is that it is precisely at this scale that many of the significant interactions and feedbacks between natural and societal systems occur. Furthermore, when information is being collected at any scale, it is not usually for the purpose of understanding self-organizing and systemic phenomena. Furthermore, virtually no one tracks the contextual elements, particularly exergy flows, of the ecological-societal relationship. We have argued elsewhere that the SOHO conceptual model should be the basis for monitoring for sustainability (Boyle et al. 2000; also chap. 16).

Applying the SOHO Model to Economics

One subset of important issues relates to how ecological and economic systems relate to each other. "Ecological economics" purports to address these linkages and provides an interesting example of how the SOHO model can be used. If ecological economics is the science of the relationship between economic systems and natural systems, then the SOHO model should be the conceptual foundation for the biophysical dimension. Bearing in mind that sustainability is about a wider array of social and cultural issues than we have discussed here, our premise is that sustainability is about maintaining the integrity of the combined ecological-societal system. Integrity is preserved when the system's self-organizing processes are preserved, something that happens naturally if we maintain the context for self-organization in ecological systems, which, in turn, will maintain the context for the continued well-being of the societal systems. If its promise of advancing the cause of sustainability is to come to fruition, ecological economics must focus on this task.

More specifically, production-consumption models of each element in the combined ecological-societal system must be constructed. These models will need to tell us about the relationship between the organizational state (attractor) of the system and its context. A description of the context, in turn, is comprised of four aspects: the flow of energy, the flow of material, the flow of information, and the physical environment. Given this description, one needs to link specific contextual states to organizational attractors. As noted earlier, this is a fundamental challenge because we are only beginning to explore the relationship between context and self-organization.

The greatest amount of progress has been made in discussing energy flow. If one considers only the energy aspect of the flows in a SOHO system description, then network thermodynamics, using graph theoretic techniques allows a complete system description (Mikulecky and Sauer 1988; Mikulecky 1985; Peusner 1970, 1986; Wong 1979; Wong and Chandrashekar 1982). At the core of this description are measures of the quantity and quality of the flow. Each component is described by the change in quality and quantity of flow between its inputs and outputs. The description effectively involves measuring the gradient drop across the component and the associated flow through the component. The components together, that is, the overall system, must conform to two rules known variously as Kirchkoff's laws: the cutset and circuit equation or the first and second laws of thermodynamics. In electrical systems the measure of quantity is current and the measure of quality is voltage. In hydrodynamic systems the measure of quantity is volume flow and the measure of quality is pressure. In general, for any energy flow system the quantity measure is flow of energy and the quality measure is exergy density (exergy per unit energy).

The details of how to measure and analyze these indicators are not important here and are left for the reader. What is important is that this body of work has demonstrated unequivocally that any description of a real physical flow system must be in terms of *both* a quality and quantity measure, if meaningful results are to arise from the analysis. Only with both types of measures can the first and second laws of thermodynamics be taken into account. Most energy analysis has traditionally looked at only energy flow (quantity), and it has been demonstrated in a number of works that this has lead to poor decisions, both at the micro (plant or building) level and at the macro (describing the economy) level. In spite of the power of exergy analysis, it has only begun to work its way into the engineering curriculum and textbooks in the past decade. In particular, very little macro level exergy analysis is done and this shortfall will need to be rectified if SOHO models are to be quantified.

Unfortunately, such a body of knowledge does not exist for the other necessary aspects of describing the context for self-organizing hierarchical open systems. Similar measures of the quality and quantity of material flow are needed. While quantity measures of material flow are self-evident, quality measures elude us. Yet these are critical if we are to evaluate the implications of such strategies as material recycling. Even less clear is the calculus of information. The central role of information in directing the emergence of self-organizing processes and structures has only recently been put forward, and a means of describing this role remains unclear.

These gaps in our knowledge present major challenges. There are profound and fundamental theoretical issues related to complexity and self-organization that we must resolve before we have a robust theoretical basis for discussing sustainability, ecological integrity, and ecological economics. These issues revolve around the question of system description and quality.

However, there is much to be gained even from considering only conceptual and qualitative SOHO descriptions. Take, for example, the premise that the marketplace can be used as a mechanism of mediating the relationship between nature and humanity; it is just a question of getting the price right. The market only works if there is someone to pay the price for something, someone whose decision will be influenced by cost. As the SOHO model shows, much of the self-organizing interaction between humanity and nature occurs at scales that are beyond that of individuals or even global corporations. The market cannot operate at these scales because there is no direct interaction involving individuals or corporations. Rather, the interaction occurs with emergent phenomena at much higher scales of system organization whose effect cascades down through the nested levels of the ecological-societal system to change the context for individuals or corporations. There seems no logical way to assign a cost or price to a higher-level emergent phenomenon so that the lower-level holons will alter their behavior to reflect the cost. How can one assign a cost to degradation of the ozone layer or global climate change when they are emergent properties of the collective ecological-societal systems operating across the planet?

Some have suggested using replacement costs as a surrogate for ecological services, but a little reflection suggests that this is nonsensical. For example, consider the water pumping due to evapotranspiration in forests that comprises part of the hydrological cycle. Some simple calculations suggest that evapotranspiration uses about 44,000 TW/year in comparison to the entire consumption of energy by humans in a year of 11 TW. Clearly, evapotranspiration occurs at a much larger energy scale than societal systems. Even if we wanted to replace evapotranspiration, we do not have the thermodynamic means to do so. Further, even if we could assign a genuine replacement cost, how could this cost be passed on to consumers? Charge people for each time it rains in their locale? What about the people who benefit downstream from the rain? As the SOHO model reveals, these are self-organizing phenomena occurring at scales that are not compatible with the scale at which the marketplace operates and at which it stimulates self-organization. The marketplace cannot be used as a mechanism for optimizing the coevolution of the natural world and human-constructed world because of the limited scale at which it operates; it can result only in suboptimization of ecological-societal systems. This conclusion is but one example of the insights that can be gained from using a SOHO system model in consideration of ecological economics.

Conclusion

The SOHO model provides us with a framework for exploring the multilayered relationship between the coevolving and coupled natural and human-constructed worlds that make up our biosphere. It helps us understand the realities of our relationship with nature, particularly our dependence on nature for our self-organizing context and the thermodynamic realities of this dependence. In this respect, it makes clear how vulnerable our systems are to large-scale changes in nature. The model furnishes us with a framework for discussing the biophysical aspects of sustainability. The importance of feedback loops, hierarchically nested processes and structures, self-organization, and emergence are drawn to our attention by the model. So, too, is the importance of the relationship between context and organizational state of the system. We need to develop tools that allow us to construct a SOHO description, particularly ways of describing the organizational state of structure and processes and their relationships and the quantity and quality of flows through the system.

The SOHO model alerts us to the aspects of the biosphere that we ought to be (and are not currently) monitoring. We do not, as a matter of course, think about self-organization and the coevolution of nature and society and our role in these dynamics. We certainly do not think through the implications of our actions in a hierarchical systems sense nor in a thermodynamic sense. Yet the systems considerations reflected in the SOHO model show us the importance of conducting these analyses. If we do not, we will be subject to the whims of events as they unfold. Given our lack of monitoring, we may not even realize that an event has occurred until its effects have cascaded through the hierarchical system to affect the context at our level. By this point, it is likely that the system will have been radically reorganized, many layers away from us, and it will be too late to do anything about it. Simply trying to mitigate local impacts will not work. This is the warning and the challenge that emerge from thinking through the coevolution of natural and societal systems from the perspective of self-organization theory and complexity.

Notes

[1]There are other behaviors of such systems, such as homeostasis, equifinality, and diversification (see Ahl and Allen 1996; Caley and Sawada 1994; Flood and Carson 1993; Jacobs 2000; Rosen 2000; von Bertalanffy 1968, 1975; Weinberg 1975).

[2]In fact, one could argue that it is information that shapes self-organization of the system by allowing it to access appropriate energy and materials.

References

Ahl, V. and T. F. H. Allen. 1996. *Hierarchy Theory: A Vision, Vocabulary and Epistemology.* New York: Columbia University Press.

Allen, T. F. H. and T. B. Starr. 1982. *Hierarchy: Perspectives for Ecological Complexity.* Chicago, Ill.: University of Chicago Press.

Baird, D. and R. E. Ulanowicz. 1989. The seasonal dynamics of the Chesapeake Bay ecosystem. *Ecological Monographs* 59:329–364.

Bejan, A. 1997. *Advanced Engineering Thermodynamics.* New York: Wiley.

Blauberg, J. X., V. N. Sadovsky, and E. G. Yudin, (1977). *Systems theory: Philosophical and methodological problems.* Moscow: Progress Publishers.

Boltzmann, L. 1974. The second law of thermodynamics. In *Ludwig Boltzmann: Theoretical Physics and Philosophical Problems,* ed. B. McGuinness, pp. 13–22. New York: D. Reidel.

Bormann, F. and G. Likens. 1979. *Pattern and Process in a Forested Ecosystem.* New York: Springer-Verlag.

Boyle, M., J. J. Kay and B. Pond. 2000. Monitoring in support of policy: An adaptive ecosystem approach. In *Encyclopedia of Global Environmental Change*, ed. R. Munn. pp.116–137. New York: Wiley.

Brodyansky, V., M. Sorin and P. LeGoff. 1994. *The Efficiency of Industrial Processes: Exergy Analysis and Optimization*. New York: Elsevier.

Caley, M. T. and D. Sawada. 1994. *Mindscapes: The Epistemology of Magoroh Maruyama*. Langhorne, Pa.: Gordon and Breach Science.

Chandrasekhar, S. 1961. *Hydrodynamics and Hydromagnetic Stability*. Oxford, UK: Oxford University Press.

Corning, P. and S. Kline. 1998. Thermodynamics, information and life revisited, Part 1: To be or entropy. *Systems Research and Behavioral Science* 15:273–295.

Cross, M. C. and P. C. Hohenberg. 1993. Pattern formation outside of equilibrium. *Reviews of Modern Physics* 65(3):851–1112.

Eddington, A. E. 1930. *The Nature of the Physical World*. New York: MacMillan.

Edgerton, R. H. 1982. *Available Energy and Environmental Economics*. Lexington, Mass.: D. C. Heath.

Flood, R. L. and E. R. Carson. 1993. *Dealing with Complexity: An Introduction to the Theory and Application of Systems Science* (2nd ed.). New York: Plenum.

Ford, K. W., G. I. Rochlin and R. H. Socolow. 1975. *Efficient Use of Energy*. New York: American Institute for Physics.

Fraser, R. and J. J. Kay. 2000. Exergy, solar radiation and terrestrial ecosystems. In *Thermal Remote Sensing in Land Surface Processes*. eds. D. Quattrochi and J. Luvall, pp. 283–360. Ann Arbor, Mich.: Ann Arbor Press.

Gaggioli, R. A. (ed.). 1980. *Thermodynamics: Second Law Analysis*. Washington, D. C.: American Chemical Society.

Gaggioli, R. A. (ed.). 1983. *Efficiency and Costing; Second Law Analysis of Processes*. Washington, D. C.: American Chemical Society.

Gunderson, L. H. and C. S. Holling (eds.). 2002. *Panarchy: Understanding Transformations in Human and Natural Systems*. Washington, D. C.: Island Press.

Hevert, H. and S. Hevert. 1980. Second law analysis: An alternative indicator of system efficiency. *Energy—The International Journal* 5(8–9):865–873.

Holling, C. S. 1986. The resilience of terrestrial ecosystems: Local surprise and global change. In *Sustainable Development in the Biosphere*, eds. W. M. Clark and R. E. Munn, pp. 292–320. New York: Cambridge University Press.

Holling, C. S. 1992. Cross-scale morphology, geometry, and dynamics of ecosystems. *Ecological Monographs* 62(4):447–502.

Jacobs, J. 2000. *The Nature of Economies*. Toronto, Canada: Random House.

Kay, J. J. 1984. Self-organization in living systems. PhD thesis. Waterloo, Ontario, Canada: Systems Design Engineering, University of Waterloo.

Kay, J. J. 1991a. *The Concept of "Ecological Integrity", Alternative Theories of Ecology, and Implications for Decision-Support Indicators*, pp. 22–58. Ottawa: Canadian Environmental Advisory Council.

Kay, J. J. 1991b. A non-equilibrium thermodynamic framework for discussing ecosystem integrity. *Environmental Management* 15(4):483–495.

Kay, J. J. 2000. Ecosystems as self-organizing holarchic open systems: Narratives and the second law of thermodynamics. In *Handbook of Ecosystem Theories and Management*, eds. S. E. Jørgensen and F. Müller, pp.135–160. Boca Raton, Fla.: CRC Press.

Kay, J. J. 2002. On complexity theory, exergy and industrial ecology: Some implications for construction ecology. In *Construction Ecology: Nature as the Basis for Green Buildings*, eds. C. Kibert et al., pp. 72–107. London, UK: Taylor & Francis.

Kay, J. J. and H. A. Regier. 1999. An ecosystem approach to Erie's ecology. In *The State of Lake Erie (SOLE)—Past, Present and Future, A Tribute to Drs. Joe Leach & Henry Regier*, eds. M. Munawar et al., pp. 511–533. Leiden, Netherlands: Backhuys Academic Publishers.

Kay, J. J. and H. A. Regier. 2000. Uncertainty, complexity, and ecological integrity: Insights from an ecosystem approach. In *Implementing Ecological Integrity: Restoring Regional and Global Environmental and Human Health*, eds. P. Crabbé et al., pp. 121–156. Dordrecht, Netherlands: Kluwer Academic.

Kay, J. J. and E. D. Schneider. 1992. Thermodynamics and measures of ecosystem integrity. In *Ecological Indicators, Proceedings of the International Symposium on Ecological Indicators*, eds. D. H. McKenzie et al., pp. 159–182. New York: Elsevier.

Kay, J. J., H. A. Regier, M. Boyle and G. Francis. 1999. An ecosystem approach for sustainability: Addressing the challenge of complexity. *Futures* 31(7):721–742.

Koestler, A. 1978. *Janus: A Summing Up*. London, UK: Hutchinson.

Maruyama, M. 1980. Mindscapes and science theories. *Current Anthropology* 21:589–599.

Mikulecky, D. C. 1985. Network thermodynamics in biology and ecology: An introduction. In *Ecosystem Theory for Biological Oceanography*, eds. R. E. Ulanowicz and T. Platt. Canadian Bulletin Fisheries and Aquatic Sciences 231:163–175.

Mikulecky, D. C. and F. A. Sauer. 1988. The role of the reference state in nonlinear kinetic models: Network thermodynamics leads to a linear and reciprocal coordinate system far from equilibrium. *Journal of Mathematical Chemistry* 2:171-196.

Moran, M. and H. Shapiro. 1993. *Fundamentals of Engineering Thermodynamics*. New York: Wiley.

Moran, M. J. 1982. *Availability Analysis: A Guide to Efficient Energy Use*. Saddle River, N. J.: Prentice - Hall.

NESH, 2004. *The Network for Ecosystem Sustainability and Health*. (http://www.nesh.ca) An on-line not-for-profit community organization working for health, agriculture and resource management.

Nicolis, G. and I. Prigogine. 1977. *Self-Organization in Non-Equilibrium Systems*. New York: Wiley.

Nicolis, G. and I. Prigogine. 1989. *Exploring Complexity*. New York: W. H. Freeman.

Odum, H. T. 1988. Self-organization, transformity, and information. *Science* 242:1132–1139.

Odum, H. T. and R. C. Pinkerton. 1955. Time's speed regulator. *American Scientist* 43: 321–343.

Peusner, L. 1970. *The Principles of Network Thermodynamics: Theory and Biophysical Applications*. Boston, Mass: Harvard University.

Peusner, L. 1986. *Studies in Network Thermodynamics*. New York: Elsevier Science.

Prigogine, I. and I. Stengers. 1984. *Order Out of Chaos: Man's New Dialogue with Nature*. Toronto, Canada: Bantam.

Prigogine, I. and J. M. Wiame. 1946. Biologie et thermodynamique des phenomenes irreversible. *Experientia* II:451–453.

Prigogine, I., G. Nicolis and A. Babloyantz. 1972a. Thermodynamics of evolution. *Physics Today* 23(11):23–28.

Prigogine, I., G. Nicolis and A. Babloyantz. 1972b. Thermodynamics of evolution. *Physics Today* 23(12):38–44.

Regier, H. A. and J. J. Kay. 1996. An heuristic model of transformations of the aquatic ecosystems of the Great Lakes-St. Lawrence River Basin. *Journal of Aquatic Ecosystem Health* 5:3–21.

Regier, H. A. and J. J. Kay. 2002. Phase shifts or flip-flops in complex systems. In *Social and Economic Dimensions on Global Environmental Change*, ed. P. Timmerman, pp. 422–429. New York: Wiley.

Rosen, R. 2000. Essays on Life Itself. New York: Columbia University Press.

Schneider, E. and J. J. Kay. 1993. Exergy degradation, thermodynamics, and the development of ecosystems. In *Energy, Systems, and Ecology*, vol. 1, *Proceedings of ENSEC '93 Conference*, ed. J. Szargut et al., pp. 33–42. Krakow, Poland.

Schneider, E. and J. J. Kay. 1994a. Life as a manifestation of the second law of thermody-namics. *Mathematical and Computer Modelling* 19(6–8):25–48.

Schneider, E. D. and J. J. Kay. 1994b. Complexity and thermodynamics: Towards a new ecology. *Futures* 24(6):626–647.

Schrödinger, E. 1944. *What is Life?* New York: Doubleday.

Szargut, J., D. R. Morris and F. R. Steward. 1988. *Exergy Analysis of Thermal, Chemical, and Metallurgical Processes.* New York: Hemisphere.

Ulanowicz, R. E. 1997a. Ecology, *The Ascendent Perspective.* New York: Columbia University Press.

Ulanowicz, R. E. 1997b. Limitations on the connectivity of ecosystem flow networks. In *Biological Models: Proceedings of the 1992 Summer School on Environmental Dynam-ics*, eds. A. Rinaldo and A. Marani, pp. 125–143. Venice: Istituto Veneto di Scienze, Lettere ed Arti.

von Bertalanffy, L.1968. General Systems Theory. New York: George Brazillier.

von Bertalanffy, L. 1975. Perspectives on General Systems Theory. New York: George Brazillier.

Wall, G. 1986. Exergy–A Useful Concept. Goteborg, Sweden: Physical Resource Theory Group, Chalmers University of Technology.

Weinberg, G. M. 1975. An Introduction to General Systems Thinking. New York: Wiley.

Wicken, J. S. 1987. Evolution, Thermodynamics, and Information: Extending the Darwin-ian Program. Oxford, UK: Oxford University Press.

Wong, F. C. 1979. System-theoretic models for the analysis of thermodynamic systems. Ph.D. thesis. Waterloo, Ontario, Canada: Systems Design Engineering, University of Waterloo.

Wong, F. C. and M. Chandrashekar. 1982. Thermodynamic systems analysis. Energy: The International Journal 7:539–566.

5

So, What Changes in a Complex World?

James J. Kay

In this first section of the book, we have focused on ecosystems primarily as bio-physical systems and have begun to explore the ambiguous role of the observer in characterizing those systems. Complexity is characterized by irreducible uncertainty, multiple attractors, and nested hierarchical scales viewed from multiple perspectives. We have established that complexity is qualitatively different from complicatedness, the terrain explored by normal science. In this brief summary chapter for this section, I wish to ask a question from the point of view of researchers and managers: so what?

In fact, viewing the biophysical world through the lenses of complexity changes a great many things for both researchers and practitioners involved in environmental management. In other words, there are huge social implications for this change in stance with regard to the natural world.

Fundamentally, our understanding of the world changes. Where once we saw clockwork mechanisms, we now see self-organization and nested hierarchies characterized by evolution and emergence, attractors, rapid change, and flips. In the old way of seeing, cause and effect can be neatly separated; with complexity, feedback loops dominate, and effects are also causes. A complicated system can still, in principle, be predictable; a complex one is irreducibly uncertain. What this adds up to is that, for the merely complicated, scientists can arrive at a right answer in principle. Under conditions of complexity, science can arrive at a set of answers that are not even probably correct but are only, at best, possibly correct. These fundamental changes in understanding have implications for practice. In the context of complexity, being correct must change its value.

Under conditions of complexity, for instance, our framing of the situation changes from one (correct) perspective to multiple perspectives and scales. Our explanations shift from linear cause-effect models to feedback loops and from quantitative, predictive models to narratives about self-organization, attractors, canons, and propensities. From searches for maxima and minima we shift to solutions that seek to balance competing perspectives and claims of truth.

Investigators into complexity do not seek prediction, control, right answers, or efficiency. These are not sensible goals under conditions of complexity. Rather, the investigators seek understanding, adaptability, and resilience. Scientific inquiry, more than ever, becomes an act of collaborative learning and knowledge integration. The role of the expert shifts from problem solving to an exploration of possibilities and from giving correct advice to sharing information about options and trade-offs. In fact, those who cling to being the old sort of expert lose their expertise.

Because there is no correct answer and no definitive perspective, decision making under conditions of complexity must be broadly participatory. Perhaps the greatest challenge for civic society is that the kinds of ongoing collaborative learning demanded by this new understanding will require new institutional arrangements and broad public participation. Management shifts from top-down command and control to collaboration and encouragement of self-organization about desired attractors. Equally important are discouraging activities that tend to push the system to undesirable states. We can no longer manage nature nor can we manage people because we ourselves are all part of this fundamental complexity. The best we can do is to anticipate what might happen and nurture adaptive interactions with the systems in which we are embedded.

Managing under conditions of complexity requires us to rethink what we mean by monitoring. This is no longer about detecting what, if anything, we have done wrong and correcting the mistakes. It is not clear what wrong and mistaken mean under these circumstances. Without a linear cause-effect model, it would also not be clear what could be corrected. We can, however, determine if we are moving in desirable directions and direct our decisions to alter course. This, in turn, requires a kind of continual, self-aware storytelling, learning from our experience and seeking to improve what we are doing.

We can no longer fall back on quantitative measures of correctness, on "just the fact," or on our rigorous ability to solve problems and avoid surprises. However, this does not mean that our storytelling is simply a sort of collective fantasy. The quality of our information can still be (vigorously) assessed, according to its fitness for the purposes we have chosen. Does our understanding take us where we would like to go? Quality control thus requires an understanding of the possibilities open to us, a consideration of different perspectives, and an ability to adapt and cope with surprises on the path we have chosen. As in all science, this quality control requires a community of peers challenging and questioning and probing each other's methods, perspectives, and facts. It is this mutual challenging and "correction" that makes this a dynamically high-quality science and not merely a kind of postmodern free-for-all. The key difference from normal science is that our community of peers is enlarged to include a much wider array of epistemological communities.

Given that all this changes, we are faced with some serious challenges as we create a new science for sustainability. How can we enable collaborative learning systems through infrastructure and policy? How can we build the capacity—theory, practice, and experience—to operate collaborative learning systems? How do we develop a serious understanding of the coevolution of ecological and social systems?

These are the kinds of questions that we will address in subsequent chapters. From the work reported here and the rapidly growing literature on participatory learning, resilience, and ecosocial evolution within which their work is situated, readers will see that these challenges are already being faced, met, and, in many cases, transcended. As they shall also see, there is a kind of convergence of the insights we get from examining thermodynamics and the study of complex ecosystems and those that emerge from the study of governance and social systems.

6

Bridging Science and Values
The Challenge of Biodiversity Conservation

Nina-Marie E. Lister

Each day, on my way to work in the heart of Canada's largest city, I cycle past a large construction site—another office tower and condominium development, a stone's throw from the city's bustling financial district. While waiting for the light to change, I noticed a group of construction workers taking a morning break. One of the laborers caught my attention. He was a big man with his shirtsleeves rolled up, revealing an artfully tattooed bicep. He was chuckling softly, bent to the ground, sandwich in one hand, coffee beside him on the ground. His other hand was outstretched, his calloused and nicotine-stained fingertips delicately offering tiny bits of bread to a sparrow. The sparrow adopted its cautious but cheeky stance, its dingy feathers fluffed, darting back and forth to retrieve each morsel as it was offered. I stood watching, drawn in by this display of tenderness—for that's exactly what it was. The man did not do this absentmindedly: he did not merely toss some scraps to the ground nor did he hide his act of kindness for a tiny creature from his coworkers. Rather, it was a deliberate and carefully orchestrated act in which he shared his meal with a bird: he knelt awkwardly, moving slowly so as not to frighten the animal, and he waited patiently for it to come to his fingers, and he was clearly amused by this, pleased that his gift was rewarded with close contact.

This is not an unfamiliar pattern in the city: a grandmother feeds the pigeons daily in a nearby park, the tourists feed peanuts to already fat squirrels, a kind neighbor nurtures every feral cat who stops by, and so on. Although it might be argued that urban biodiversity is approaching an oxymoron, still, people will go out of their way to engender a relationship with wild creatures, no matter if they are considered common pests by others. Indeed, the diversity of wild species in most North American cities is dominated by those that most closely mimic human strategies of success: like us, they are adaptable, omnivorous opportunists (e.g., rats, squirrels, raccoons, pigeons, cockroaches, and house sparrows among others considered as pests). The man I watched probably neither knew nor cared that the bird he was feeding was not part of the biological diversity native to this part of Canada. The house sparrow (*Passer domesticus*), rather like most Canadians, came from elsewhere; it is an immigrant to these shores, and finding abundant opportunity, it has proliferated. In fact, the house sparrow has been, by all accounts, invasive in that it breeds with native species, and their hybrid progeny now outnumber and displace, even threaten, native sparrows as well as other avian species. Nevertheless, this relative newcomer is now inextricably woven into the diversity of species dwelling in our increasingly altered landscapes.

So which species do we choose to protect, to cull, to revere, or to vilify? These questions are central to biodiversity conservation specifically and to an ecosystem approach more generally, for they cannot be answered by ecological science alone (see, e.g., work by Foster and Sandberg 2004). These are questions of ethics and related values—sociocultural and political—as much as they are matters of ecology (Clewell and Aronson 2006; Daugstad et al. 2006). Indeed, such questions are core to managing for biodiversity in the context of long-term ecological health and sustainability. Characterized by both scientific and sociocultural dimensions, biodiversity conservation is a key challenge of the ecosystem approach as set out in this book.

Biodiversity literally refers to the variety, distinctiveness, and abundance of life-forms and processes. It is within this context that biodiversity has traditionally been studied; e.g., ecologists' long intrigue with and debate over the role of biodiversity in ecosystems and biologists' systematic classification of taxa. Figuratively, however, "biodiversity" encumbers a multitude of different, often conflicting, and possibly even irreconcilable sociocultural values for species, and the related politics of management, i.e., the necessary trade-offs and choices for conservation and exploitation of various species. Following the *Global Convention on Biological Diversity* in 1992, biodiversity joined the ranks of "sustainable development," "ecosystem health," and other ecological imperatives. Furthermore, the emergence of post-normal, ecosystems-based science described in this book, with its hallmarks of uncertainty and complexity, reveals ecological realities that force consideration of multiple contexts, including varied and sometimes conflicting sociocultural values for biodiversity.

It is argued in this chapter that biodiversity conservation to be meaningful and effective over the long term necessitates an ecosystem approach as postulated and advocated by authors throughout this book. What are the implications of an ecosystem approach on the ways in which we study, consider, and undertake biodiversity conservation? This question is addressed, first, through identifying current perspectives of biodiversity; and, second, through tracing the emergence of a broader ecological context for biodiversity conservation. From this analysis the challenges and implications of biodiversity conservation in the context of the ecosystem approach are considered, with specific reference to an adaptive planning process.

Conventional and Emerging Perspectives on Biodiversity

A closer look at the layered nature of biodiversity reveals that a multifaceted approach to its conservation, including ecological function and sociocultural values is a strategic choice for both study and management. The issues of scale, type, measurement, and boundaries in observation pose obstacles in defining biodiversity. Furthermore, most definitions do not deal explicitly with the complexity and uncertainty inherent in both the ecosystem and biodiversity concepts. Accordingly, it is worth exploring richer, more illuminating perspectives from both scientific function and sociocultural values.

Biodiversity definitions in the ecological and conservation literatures fall within three general categories. The most prevalent focus is on ecological structure (form), followed by a focus on function (ecological processes), and then an implicitly value-oriented focus on wealth or richness (resources). There is an emerging, broader perspective of biodiversity as *information* (Lister 1998), influenced by systems thinking in general and complex systems science in particular. This notion resonates with Wood (2000:40), who observes that "biodiversity is not the property of any one entity. Rather it is an emergent property of collections of entities. More precisely, it is the differences among them."

Despite the complexities inherent in the biodiversity concept, the most common definitions of biodiversity focus primarily on its relationship to ecological structure and to a lesser degree on function. The United Nations' *Convention on Biological Diversity* (United Nations Environment Programme 1992) and Agenda 21 (United Nations Conference on Environment and Development 1992) as well as major national policies derive their definitions from two sources: McNeely et al. (1990) and the *Global Biodiversity Strategy* (World Resources Institute 1992). These documents define biodiversity structurally and functionally according to the ecological hierarchy elaborated by Allen and Hoekstra (1992) and O'Neill et al. (1986). The ecological hierarchy stipulates three generalized levels of observation and of organization: genes, species, and ecosystems or landscapes. In short, the traditional (or structural) perspective of biodiversity is that "[it] is the totality of genes, species and ecosystems in a region" (World Resources Institute et al. 1992:2), where genetic diversity is the variability of the genetic code within a species and among constituent populations (considered the "building block" of biodiversity, facilitating fitness, adaptation, and evolution), species diversity is the variety of genetically similar organisms present in a given spatial unit, and ecosystem or landscape diversity is the variety of large-scale (here meaning sized) ecosystems or distinct landscape patterns in a given region.

Table 6.1 depicts both the conventional and the emerging perspectives of biodiversity within a range of spatial levels and contexts according to the common ecological hierarchy. The most tangible level of observation and the most common (but implicit) focus of conservation efforts occurs at the species level. In contrast, landscape diversity is the most complex level (subsuming and constraining genes, species, populations, and communities) and therefore difficult to measure.

TABLE **6.1**   Perspectives on Biodiversity: *Range of Levels and Contexts*

Perspective	Level	Context	Measures / Tools
	Genetic	DNA: genes; alleles	Various; incompatible between levels; e.g.: *Conservation Biology*:
Conventional	Species	species, taxa	-species richness -Shannon-Weaver/ Simpson Indices
Structural and Functional	Population	populations of species	-minimum viable population -species-area curves -relative abundance
(Hierarchical; Observer-dependent)	Community	ecological communities (observer-defined)	*Landscape Ecology:* -patch dynamics -relative patchiness
	Landscape	landscapes; ecosystems (observer-defined)	-network analyses -correspondence analysis
Emerging: Cultural and informational	All levels (except genetic)	Human culture, ethology	Heuristic; qualitative notion
(Holarchical; system-wide)	All levels	Information: genetic, behavioral, cultural, self-organizational	

Source: Lister 1998.

The interrelated concepts of scale and type have been carefully disentangled by Allen (chapter 3). Allen's discussion is important to the notion of biodiversity principally because it facilitates an understanding of the complexity inherent in the concept. Although many ecologists use the term "scale" broadly to mean spatial area or "size of span," Allen (chapter 3) points out that conventional ecological hierarchy is more precisely concerned with levels of observation and levels of organization. From this reasoning comes the important recognition that biological structure, function, tools, and measurements are dependent on *both* the level of observation and level of organization. For example, to measure species richness (a structural perspective on species), one would use an index unique to that phenomenon; this measure could not in any useful way be compared to genetic diversity, relative abundance, or connectedness at the landscape level. Thus there is no single convenient method of "measuring" biodiversity (Secretariat of the Convention on Biological Diversity 2006). As Nowicki (1993:65) notes, "the condition of being different or having differences does not provide a good yardstick with which to measure diversity. It is more convenient to define diversity as a complexity of systems. . . . Diversity does not imply that all components of the system are complex; it only implies that the system itself is complex."

As Allen (chapter 3) suggests, the measurement or analysis of biodiversity is dependent on the observer of the system, his or her choice of what and where to observe, and the measurement tool associated with the chosen level. Specifically, decisions of level of observation, system type, scale (in Allen's terms, including both of "size of span" and resolution of focus), boundary determination, the measurement and scope of the data to include in research or policy, are all significant means by which biodiversity is rendered dependent on the observer (Crawford et al. 2005). Limited to either one level of the ecological hierarchy (e.g., species) or one perspective (e.g., structure), conventional definitions of biodiversity used in making conservation plans and management decisions can be problematic when they fail to recognize *explicitly* the level and observer dependency of the diversity concept.

Finally, even at the species level of diversity, for which rudimentary measures and absolute counts of individuals exist, there is a staggering degree of uncertainty. The United Nations' Environment Programme's *Global Biodiversity Assessment* (Heywood 1995) puts the number of currently known (described) species at 1.7 million, yet experts' estimates of the total number of Earth's species are commonly between 10 and 30 million (Jeffries 1997). However, Wilson (1992) concedes that a more probable estimate lies anywhere between 10 and 100 million species, and Soulé and Orians (2001) acknowledge that the total number of species may exceed 50 million. Even at the species level of diversity, about which we have the most complete knowledge, the total scope of diversity is unknown to within an order of magnitude.

The analogy of a library, in which biodiversity acts as an information reserve, is another perspective offered in conservation literature. Janzen (1988) suggests that genes and species are analogous to books containing vital information, and landscapes or large ecosystems are akin to a library housing the books. Similarly, Ehrlich and Wilson (1991:760) observe that "biodiversity is a precious 'genetic library' maintained by natural ecosystems." In this broad sense, biodiversity essentially encompasses the capacity for a living system to renew and reorganize itself as part of a dynamic life cycle. As depicted in table 6.1, information phenomena are system-wide, affecting all levels from genes to landscapes. This notion of an information perspective for biodiversity holds potential for conservation in particular and an ecosystem approach more generally by virtue of its permeation through the whole living system rather than a single selected level.

Similarly, a cultural perspective of biodiversity is an important consideration in the context of the ecosystem approach. In the anthropological sense, i.e., the beliefs, customs, practices, and unique way of life a community, culture is absent from most definitions of biodiversity. Through the separation of human diversity from that of other creatures, the conventional definition of biodiversity implicitly removes humans from nature. Considering a cultural perspective would go some distance to recognizing a vital linkage between humans and nature, where culture and nature are mutually intertwined, each influencing the other (Donner 2006; Knight et al. 2006; Pedroli et al. 2006). The World Resources Institute (World Resources Institute et al. 1992:5) notes that

> "this linkage has profoundly helped determine cultural values. Most of the world's religions teach respect for the diversity of life and concern for its conservation. Indeed the variety of life is the backdrop against which culture itself languishes or flourishes."

Furthermore, the depletion of natural biodiversity is generally accompanied by and is a partial result of the loss of cultural diversity (Shiva 1993; World Resources Institute et al. 1992). It is important to realize that the homogenization of nature and the creation of biological monocultures have been a consequence of the domination of white, Western (mono) culture over many indigenous cultures—a phenomenon Shiva (1993) has called "monocultures of the mind." Through further pluralizing biodiversity to include its cultural aspect, we recognize and validate the variety of humanity, and in so doing the notion of *difference*. In so doing, we legitimize a place for other cultural values for and knowledge of biodiversity in terms of the way we choose to use, exploit, or conserve it—and, by extension, the ecosystems in which we dwell (O'Brien 2006).

Despite the complex nature of biodiversity as a concept, major management strategies and policy reports continue to use a definition limited to its structural description according to three hierarchical levels—most commonly, the species level. Similarly, scientific papers often to fail to define either or both the ecological perspective or the social context in which they use the term, and the conclusions are often extended by implication to another level of the hierarchy—most commonly the landscape level (Lister 1998; Lister and Kay 2000). Such definitions do little to assist managers and decision makers in developing effective conservation plans. Consequently, a broader set of perspectives for biodiversity, explicitly incorporating the vital notions of scale and context (both ecological and sociocultural) and the qualities of uncertainty and complexity, is proposed for consideration within the framework of the ecosystem approach discussed in this book. Through these enrichments, the notion of biodiversity may prove more useful as a heuristic concept, effective in guiding conservation planning and management within the context of the ecosystem approach.

Ecological Contexts for Biodiversity Conservation

Given that there exists no single measure of biodiversity, establishing a reasonable estimate of loss or decline on a large scale is problematic, if not impossible. Localized estimates, while possible, are generally confined to the structural perspective of diversity at the species level, yet these are still tenuous because the total number of species on Earth remains unknown. It is generally agreed that species loss has been increasing steadily over the past 400 years—since humans began colonization—and that present rates are significantly greater than background extinction; in fact, they are unprecedented (Hanski and Ovaskainen 2002). This concern led to the *Global Convention on Biological Diversity* in 1992 and was reiterated in the 2005 *Millennium Ecosystem Assessment's Biodiversity Synthesis Report.* It has been estimated that fully one quarter of all known species on Earth could be irrevocably lost in the early decades of this century (Environment Canada 1995:12).

Norgaard (1987) attributes the human-induced loss of biodiversity to three macrophenomena: increasing human population, technological change (resource extraction, industrial pollution, agricultural technologies, etc.), and social organization based on a deterministic worldview that assumes control over resources is possible, largely through market forces. Over and above direct exploitation of resources, the primary cause of biodiversity loss (at all levels, but most noticeably at the landscape level) is habitat destruction as a result of population growth and technological change (Ehrlich 1988; Ehrlich and Wilson 1991; McNeely et al. 1990; Reid and Miller 1989; World Resources Institute et al. 1992; Destefano et al. 2005). In tropical moist forests, areas considered hyperdiverse or "hot spots" of (especially genetic, species, and community) biodiversity, habitat destruction is usually irreversible and occurring at the fastest rate. A secondary cause of biodiversity loss related to habitat destruction is fragmentation, the process of reducing contiguous natural landscape cover to isolated, disconnected patches (Wiegand 2006). Wood et al. (2000) observe that habitat fragmentation and alternation and the related effects of pollution, climate change, and overharvesting are more properly identified as *proximate causes* of biodiversity loss; the *root causes*, they agree, are socioeconomic. Ultimately, the root cause of biodiversity loss is a dominant worldview of growth and consumption, entrained in the market forces, policies, and institutional structures that, taken together, provide a profound disincentive for biodiversity conservation in particular and sustainable behavior in general.

Of course, biodiversity conservation is a normative endeavor. Like medicine, the science that is used to support biodiversity conservation is motivated by an ethic (Norton and Ulanowicz 1992), and it is analogous in the goal to "heal and cure" pathologies related to habitat destruction and alteration (Theberge et al. 2006). Consequently, in the conventional sense, there can be no strictly scientific "ecological" basis for conservation. Rather, there exists a range of value orientations resulting from ethics that underlie conservation motivation (Orr 2006; Callicott et al. 2003). Interestingly, the environmental management literature makes much of the "scientific values" for biodiversity. However, it is possible to consider this a "trespass of semantics" to some degree because the field of ecology itself is roughly polarized between reductionist and systems-based specializations.

Ecological Perspectives and Paradigms

During the development of ecology as a science (confined here to the twentieth century), the perspective of biodiversity and the resulting contexts for its conservation have been influenced by the dominant themes and associated paradigms in ecological thinking. A key factor in the way biodiversity is perceived has to do with a schism that once characterized ecology and is still reflected in various management strategies. This schism has as its genesis the fact that ecology developed along two divergent routes: population ecology and ecosystem ecology. Both routes share three main areas of investigation that are designed to answer questions of "what," "how," and "why": structural ecology (concerned with description, classification, and natural history), functional ecology, and evolutionary ecology. The development paths of ecology became schismatic arguably because population ecology remained largely fixed in a conventional scientific approach, while ecosystem ecology began in the last two decades to include a systems perspective.

The schism in ecology may be due in part to the fact that it was not considered a "real" (i.e., reductionist) science until the late 1960s (Lister 1998). With its acceptance into the fold of normal science, the volume of ecological research has increased significantly, with biodiversity research specifically increasing dramatically over the past fifteen years, most notably, during the mid-1990s. This rise in research attention to biodiversity was coupled with considerable media and political attention; both scientific and political awareness of the concept are clearly linked to the United Nations' *Earth Summit* in 1992 in Brazil, at which the *Global Convention on Biological Diversity* was signed and later ratified by (most) member nations (Jeffries 1997).

Given that the majority of research has been established in an era of growing environmental concern, ecology has been linked to environmentalism—in both the media and by scientists themselves—and therefore to normative science (MacIntosh 1976). Furthermore, there is a broad spectrum of explicitly ethical contexts and bases for biodiversity conservation. Although not discussed here, the ethical contexts for conservation are eloquently explored in a rich and growing literature (see, e.g., Cronon 1996; Callicot 2003; Norton 2003).

Biodiversity in Ecological Science

Two of the major ecological subdisciplines, population and ecosystem ecology, have for many years debated the role of biodiversity in ecosystems, and yet there is no rigorous theory of biodiversity (Solbrig 1991a,b). However, it is generally recognized that biodiversity has strong feedbacks to ecosystem structure and functional processes, although cause-effect relationships are not well understood and are rarely quantifiable. Both ecosystem and population ecologists, although frequently divided in approach, perceive two general classes of roles for biodiversity in ecosystems: ecosystem stability and ecosystem function.

Biodiversity and Ecosystem Stability

That biodiversity is connected to ecosystem stability is an old and dominant theme in ecolog. Stability, the conventional generalization that there is an inherent "balance" or equilibrium in nature, is linked to successional theory: as systems become more diverse during succession, it is believed that they become more stable. The maintenance of "ecosystem stability" is a pervasive theme in the literature and has been frequently advocated as a basis for conservation (see, e.g., Burton et al. 1992; Tilman and Downing 1994; Smith and Swanson 1994). However, the diversity-stability connection may be considered a flawed premise for three reasons: First, it is not a rigorously defensible theory; second, there is a wide range of contrasting interpretations for both terms; and third, post-normal science holds insights that have fundamentally challenged the view of living systems upon which it is based. For these reasons, the diversity-stability association alone constitutes a weak basis for the conservation of biodiversity.

MacIntosh (1976) writes that the direct association of diversity with stability came about with the postulation that stability is imparted by increasing the number of links in the ecological food web. Kay and Schneider (1994) note that this idea, put forth by MacArthur (1955, in Kay and Schneider 1994), was misinterpreted by Hutchinson (1959, in Kay and Schneider 1994), elaborated by Margalef (1963, in Kay and Schneider 1994), and led to the eventual codification of the diversity-stability hypothesis. Along with May (1974) and others who refuted the diversity-stability hypothesis on mathematical grounds, Goodman (1975) analyzed its foundation and showed that there is no robust basis for the hypothesis. In addition to these counterarguments, there is continuing field evidence that shows mature forests contain fewer species than transitional (seral) forests. Despite the evidence against it, however, the diversity-stability hypothesis has become "almost axiomatic to some biologists" (MacIntosh 1976:366) and is still cited in many ecology texts today. Yet, as Real and Brown (1991:188) surmise,

> "[t]oday we realize that the whole debate about diversity and stability is flawed to a large extent by the imprecision in definition these terms. . . . Depending on the definition one chooses for these terms, one can obtain diametrically opposed results."

But stability is also a fuzzy concept that, until recently, was rarely defined. Pimm (1984, 1991, 1993) has written extensively on the range of interpretations of stability, arising from five distinct meanings: strict (mathematical) stability, resilience, variability, persistence, and resistance. He notes that, in general, the meanings are related to the similarly abstract notion of "balance" in nature (Pimm 1991). Kay (1991a) observed that stability is generally meant to convey that ecosystems are "well-behaved," although a more formal definition is problematic. In the strict mathematical sense, stability is defined as a numerical function (a state point) having constant value to which the system tends and returns following disturbance (Kay 1991a). Although the classical definition harkens intuitively close to the elusive "balance of nature," Schneider and Kay (1994) noted that there is no clear answer as to which state function should be measured to determine stability; they emphasized that the choice of one function—whether net productivity or food web complexity—represents only one aspect of stability, not a perspective on the whole system. Clearly, the juxtaposition of two complex and abstract concepts, diversity and stability, poses a tenuous context for biodiversity conservation, especially given the political pressure put on ecologists to provide rigorous management rules in the context of conservation.

Nevertheless, the diversity-stability connection is still advocated as an ecological basis for conservation. Wilson (1992:12, emphasis added) paraphrases the essence of the still popular diversity-stability hypothesis:

> "Biological diversity is the *key to the maintenance of the world as we know it.* Life in a local site struck down by a passing storm springs back quickly: opportunistic species rush in to fill the spaces. They entrain the succession that circles back to *something resembling the original state of the environment.*"

The notion of the "balance of nature," a return to equilibrium, is implicit in Wilson's conclusion and reflects the underlying and still dominant view of ecosystems, at least where management of them is concerned: that there exists an *ideal* stable state, and its maintenance depends at least in part on some threshold of biodiversity. As Solbrig (1991b) observes, the notion of equilibrium in nature has persisted because the dominant perception has been that ecosystems could be described according to deterministic laws. Yet, in a strictly thermodynamic sense, living systems cannot be at equilibrium: an organism at thermodynamic equilibrium with its surroundings is dead. Even in the purely physical sense of balance (where all forces acting are equal), equilibrium is not a useful substitute for stability, as living systems are characterized by fluctuations such as weather, populations, biomass, etc. As noted by Schneider and Kay (1994), these ecosystem functions are now recognized as dynamic; i.e., they are not stable themselves and thus cannot be used to measure even strict stability.

Post-normal science, and complex systems science in particular, have provided a different and broader perspective of living systems, fundamentally different from the conventional perspective, as Kay details in the first part of this book (see also, e.g., Manson 2001). Notably, in the context of the diversity-stability debate, living systems are shown to have multiple steady states, leading to what Holling (1986, 1992), and others have termed a "shifting steady state mosaic." Thus the idea of a single optimum and homeostatic state is replaced by the reality of multiple steady states. The fact that ecosystems may "shift" or diverge from any one of a number of steady operating points is a critical revelation of the recent paradigm of ecosystems as complex systems. Furthermore, Holling (1986, 1992) has shown that divergence from a given operating point or state by means of natural catastrophe, such as fire, pest outbreak, or human-induced perturbation, is a *normal* and usually *cyclic* event. As such, the diversity-stability hypothesis alone is insufficient as a basis for biodiversity conservation.

Biodiversity and Ecosystem Function

Another key class of roles for biodiversity may be found in ecosystem function. The essential processes of living systems (nutrient cycling, carbon and water cycling, productivity, etc.) are certainly dependent to some degree on the diversity of genes, species, populations, communities, landscapes, and information whose structures and composition perform these functions. Furthermore, the *diversity of functions themselves* is undoubtedly critical to the maintenance of ecosystems, and ultimately, the Earth's life-support system (Odum 1993). Here, I consider only the role of (structural) biodiversity in ecosystem function, focusing, first, on the species level, and second, with evidence from ecosystems science to the macro community/landscape or ecosystem level. The role of functional diversity in overall ecosystem function has not yet been explicitly considered, as this demands yet another level of complexity that has so far been beyond the realm of current ecology. Indeed, the acceptance of the perspective of ecosystems as complex systems may mean that there are limits to what is "knowable" and therefore useful from a management perspective.

It is a dominant proposition in the literature (Ehrlich 1988; Naeem et al. 1994; Schulze and Mooney 1993; Wilson 1985; Mooney et al. 1996; Jeffries 1997) that the loss of structural and functional biodiversity (largely through habitat destruction and fragmentation) impairs ecological systems and their ability to continue self-maintenance. Although diversity-function feedbacks are generally poorly understood and rarely quantified (Solbrig 1991a), the premise that ecosystem function is dependent on biodiversity is advocated as a basis for conservation in key policy reports (McNeely et al. 1990; Heywood 1995; Reid and Miller 1989; World Resources Institute et al. 1992). Concern for the maintenance of unimpaired ecosystem function lies in the assumption that it is ultimately integral, at the biospheric level at least, to the provision of human life support. Accordingly, there is a growing body of literature that attempts to link structural diversity to ecosystem processes. Such research has been generally confined to groups of organisms whose specific functions are known, such as nitrogen-fixing bacteria, although broader studies have been undertaken in the works of Schulze and Mooney (1993) and Naaem et al. (1994).

Of the plethora of possible diversity-function links, this analysis is confined to a conventional premise that connects species to ecosystem functions. Paine (1966) introduced the idea that certain species, termed "keystone species," are largely responsible for "community integrity" (here meaning stability) and have an effect on the survival of other species. The notion of keystone species has since become a major platform for the connection of biodiversity to ecosystem function. Bond (1993:237) argues that keystone species "should be conserved because they have a disproportionate effect on the persistence of all other species" through their actions, which may be directly or indirectly tied to ecosystem functions. Westman (1990) observes that certain species, particularly among decomposer microorganisms and litter invertebrates, may be directly tied to ecosystem function and that these "critical link species" are rarely considered for conservation action. Similarly, Lambeck (1997) has argued that certain ecologically demanding species, particularly those with large area requirements, may necessarily encapsulate the needs of many coexisting but less demanding species. Yet it remains unclear whether, or to what degree, such "focal species" (or "umbrella" species) confer conservation benefits to other species from dissimilar taxa (Roberge and Angelstam 2004). The discussion over keystone, critical link, and umbrella species remains an ongoing debate among ecologists in the determination of which species play vital roles in maintaining ecosystem functions and thus which should be targeted for priority conservation action and management.

Erhlich and Erhlich (1981) contended that all species play a small but significant role in ecosystem function, and if conservation emphasized only those species considered keystone species, it would be a serious mistake. The opposing view is that most species are redundant and only a small set of keystone species and processes is critical to ecosystem structure and function (see, e.g., Holling et al. 1995). However, both these perspectives leave conservation managers with a fundamental conundrum: the threshold of functional collapse will likely be crossed before it is known how many or which species can be considered expendable. (This point is reiterated by Kay and Boyle (chapter 4) in their discussion of the importance of SOHO analyses.)

The ongoing debate over the role of species in ecosystem function is characteristic of the search for simple rules to describe complex systems from which conservation and management policy can be derived. However, it may be that asking "which species play which role" is asking the wrong question. Rather than being an issue of "either keystone/or redundant," it is far more likely that species play *both roles*, although at different times in different ecosystem states or contexts. Thus it is not a matter of "either/or" but "when" (Lister 1998). This contention is based in part on insights from complex systems science (see, e.g., Allen, chapter 3; Allen and Hoekstra 1992; Jørgensen 1992; Kay and Schneider 1994) that reveal that the use of simplistic "rules" established at one level of the hierarchy or, in one specific ecological context, cannot be transferred meaningfully to another level or context to make generalized statements about the role or function of biodiversity. Thus there can be no universal "magic number" or threshold for any level of diversity let alone the whole system. Although it is reasonable that *some* level of biodiversity is critical to proper system function, it would almost certainly be unique to each discernible system at each observer-defined level. Recent research by Naeem et al. (1994) indicates that the theoretically possible identification and conservation of select keystone species would be limited to a specific context, which if altered through pest outbreak or climate change, would leave nothing as "back up." Indeed, there is now widespread recognition in the literature (see, e.g., Allen et al. 2001; Noss 1990; Pimm 1991; Solbrig 1991a; Pedroli et al. 2006) that ecological research must be carried out on multiple spatial and temporal scales, and at multiple levels of observation and organization, using scale-specific (e.g., population ecology) and systemic (e.g., ecosystem ecology) approaches if a deeper understanding of ecosystem function and diversity is to be achieved and an operational context for biodiversity conservation derived.

Biodiversity and Information

The third major class of ecological roles for biodiversity lies in information. It has been recognized that there is an informational role for biodiversity in terms of genetic structure and function as well as emergent properties. Genes are often considered as a "library" of information (Ehrlich and Ehrlich 1992; Janzen 1988) that facilitates evolution through the combined forces of natural selection and adaptation. More recently, discourses in evolutionary theory include a role for genetic biodiversity beyond variation and mutation, considering, for example, cooperation and other emergent phenomena associated with a complex systems under-standing of contemporary ecology (see, e.g., Nowak 2006).

Kay (1991b) and others have shown that living systems are open and self-organizing; i.e., they are capable of self-renewal or regeneration following disturbance. This perspective, arising from Prigogine's (in Kay 1991b) work on dissipative systems, is relatively recent. As part of systems-based science, self-organization theory allows for the spontaneous creation of order from disorder. In living systems the result of self-organization is life itself. Kauffman (1991) observes that self-organization provides a new basis for evolution beyond the conventional forces of natural selection and adaptation. However, he is quick to emphasize that Darwinism (natural selection) is not wrong *per se* only that it is *incomplete* (Kauffman 1994). Similarly, Wesson (1991) calls for a modernization of Darwinism and looks to complex systems theory for a synthesis. Wesson (1991:36) suggests that "self-organization is the essence of the origin of life and its complexification, that is, evolution." Given this, the gene is not the "driver" of evolution in an information context but an enabler, along with other larger-scale emergent phenomena such as cooperation. In this perspective, Kay (Kay and Schneider 1994) and others have observed that genes *constrain* the process of self-organization. That is, in an information sense, the role of the gene is to bound the structuring process in living systems, "remembering" fitness and facilitating adaptability within the ongoing process of evolution.

A hallmark of self-organization (and thus complex systems) is the emergence of new structures, properties, functions, and behaviors. Solbrig (1991b) defines emergent properties as those qualities of ecosystems *that are not present in the constituent parts* (genes, species, populations, or communities) yet are a result of interaction between system components and the self-organization process. (Examples are trophic and niche structure, seral stages, and system complexity itself.) Funtowicz and Ravetz (1993) emphasize that emergent complexity is a defining feature of living systems, rendering them literally "more than the sum of their parts." In this respect, biodiversity can be seen as emergent information. Specifically, the creation of new information (and diversity) occurs at all ecosystem scales and contexts during self-organization and the ensuing processes of ecosystem self-renewal/regeneration. This idea might be extended throughout a living system, with a role for biodiversity in an informational sense at each scale.

If biodiversity acts as information at scales beyond the genetic, what might be its primary role in the life cycle and functions of living systems? Considerable work (Holling 1986, 1992; Holling et al. 1995) in complex systems science has shown that ecosystems follow cyclic (rather than linear) paths of development, regularly punctuated by sudden, unpredictable, and rapid episodes of change to a variety of other possible states. Disturbance or perturbation from a seemingly steady state (by means of fire, storm, pest outbreaks, etc.) is now known to be a normal and integral part of living systems occurring on a more or less regular basis (Marzluff 2005). Therefore the ability of species, populations, communities, and whole systems to recover, reorganize, and adapt in the face of regular change is critical to survival (Berkes and Seixas 2005).

It is in this context that biodiversity, in an informational sense, is vital to system function. To illustrate, let us return briefly to the earlier discussion of species as keystone or redundant. It was emphasized that it was not a matter of "either/or" but a question of *when* a species acts in which role. Given the propensity for systems to move between multiple operating states, it is reasonable to expect that keystone species in one ecosystem state may be redundant in another. Folke (personal communication) has used the term "insurance species" rather than "redundant" to convey the notion that species are likely to perform different functions under differing circumstances and therefore in different contexts. As Holling et al. (1995:67) note, "the [keystone] role of species may only become apparent every now and then under particular conditions that trigger their key [organizing and] structuring function." Furthermore, our understanding of species-function links is weak even in present, let alone future, ecosystem states. Holling et al. (1995) suggest that the dominant conviction that only a few species are keystone (or critical to the structuring processes that result in ecosystem function) may be based on the limited choice of variables that researchers have modeled. Because we do not know whether our conclusions about the role of species in ecosystem function are a consequence of the models or of actual ecosystem dynamics, the notion of "insurance" species, or biodiversity as "information" in a broad sense, becomes an important consideration for conservation management decisions.

In keeping with hierarchy theory, it is not useful to reduce information to a single level of observation or organization. Although theoretically it may be possible to measure biodiversity in an information context, it would require simplification of system phenomena to the point where the emergent complexity is no longer apparent, i.e., to the point where the system can be reasonably predicted and controlled. In doing so, Schneider and Kay (1994:19) observe that the "very phenomena that ecology seeks to understand" would be lost. Furthermore, the issue of context is essential to any discussion of biodiversity conservation in an information sense. Because biodiversity information acts in the biophysical environment for which it is adapted, it must be conserved *in situ* if the information is to have meaning in a regenerative sense (Robbins et al. 2006).

Figure 6.1 depicts the dynamic cycle of ecosystem development, modified from Holling et al. (1995). As systems evolve, they do so discontinuously, creating "lumpy" geometry and distribution of elements. Following inevitable and sudden disturbance, the system reorganizes to "renew" or regenerate itself to a similar or perhaps new state. It is at this stage, immediately following disturbance, that information is most critical: the volume and type of information available in the diversity of a system's structures and functions (e.g., species) will determine its ability and direction of regeneration through reorganization. In this way, the informational aspect of biodiversity serves to facilitate the essential, life-giving process of self-organization. Clearly, given the pattern of ecosystem development supported by complex systems science, it is not a matter of redundant diversity but of what might be "investment" diversity that should be the basis for precautionary conservation, and by extension, a key aspect of an ecosystem approach. Despite the uncertainty about "how much" diversity is enough, it may be more helpful to consider biodiversity as

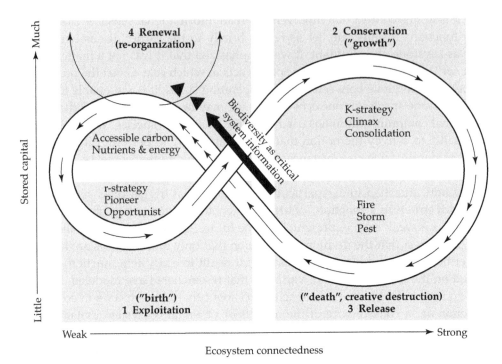

Figure 6.1   Ecosystem dynamics: Holling's modified figure eight.

an investment for the future: i.e., biodiversity protection is essential because ecosystem change is an *inevitability*, not a contingency (Lister 1998).

Biodiversity considered as ecosystem information provides us with broad, systems perspective in which the diversity of life is characterized by complex *interactions* and *relationships* rather than a collection of individuals. Thus, as a library of historical and emergent information, biodiversity provides not only a multiplicity of evolutionary and adaptive pathways for future development of life on Earth but also the *essential regenerative capacity for all living systems* (Lister 1998). This observation is echoed by Wood (2000:49), who notes that "biodiversity is a necessary precondition for adaptive evolution and the self-regulatory nature of ecological systems." The emerging notion of biodiversity as an investment in system adaptability holds significant potential as a heuristic for both the ecosystem approach in general and conservation management in particular. Specifically, the adaptive and regenerative capacity of biodiversity can illuminate trade-offs necessary in conservation planning specifically and ecosystem-based management more generally.

The Challenges of Complexity

A complex systems-based perspective of science accepts uncertainty, complexity, and diversity as natural phenomena. As Kay has explained through chapters 1, 2 and 5, in the absence of certainty, scientific inquiry must confine itself to illuminating choices and trade-offs. Decision makers must consider these choices and trade-offs in conservation goal selection—the process of which is value-driven, such as which species to conserve and which to forsake or exploit. However, current conservation decision making relies largely on the traditional scientific context of biodiversity to *determine management choices*, rarely considering other values (e.g., social-cultural, political, and economic values) explicitly. In fact, the problem of biodiversity loss itself can be partially attributed to sector-based (as opposed to issue-based) decision making (Dale 2001). This has resulted in the fragmentation and manipulation of natural systems, which has, in turn, ultimately resulted in the homogenization of landscapes and diversity. Therefore a key challenge for effective biodiversity conservation is to reform decision making from control-oriented, predictive, and interventionist *management of the environment* to adaptive, flexible, and participatory *management of human activities*.

Control-oriented management of the environment often reduces diversity—a paradox identified by Ehrenfeld (1991). In moving toward adaptive rather than control-oriented management, conservation decision makers must focus on ecosystem processes. "Because ecosystems are self-organizing and creative. . . . management must have as a central goal the protection of the system's creativity" (Norton 1992:37). This means that conservation of biodiversity (as insurance for system regeneration) is essential despite the uncertainty about how much is enough. Insights from complex systems science assure us that the unexpected will occur; with respect to biodiversity conservation, decision makers are clearly challenged to invest in the future: they must plan *adaptively* for inevitable ecosystem change.

Systems thinking, and its corollary, post-normal science (discussed by Kay in Part I of this book) are predicated on the meaningful and explicit inclusion of an extended peer community. We can interpret this to mean, in a pragmatic context, a variety of voices and values into the planning process, particularly in the goal-setting stage of any conservation plan. However, the ability to do this requires enlightened conservation managers who are open to a broad set of informed scientific and ethical perspectives on biodiversity. In essence, conservation planning must become a participatory and cooperative endeavor (Knight et al. 2006; Pinto-Correia et al. 2006). The need for both reductionist and systemic approaches to research is emphasized by Jørgensen (1992), Allen et al. (2001), Pedroli et al. (2006), and several authors in this book. Because conventional science alone cannot adequately describe ecosystems or biodiversity, much less determine "good" management, a cooperative and interdisciplinary approach to research is essential (Lamont 2006; Marzluff 2005). Of course, any effort at cooperation—whether between institutions, individuals, disciplines, agencies, or sectors—necessarily encumbers recognition of different values, and some of these may be ultimately, or even fundamentally irreconcilable. Indeed, the other side of cooperation is conflict (Pinto-Correia et al. 2006). As a result, many have argued that what is needed goes considerably beyond "disgruntled cooperation" and, instead, must embrace collaborative learning, shared power, and collective responsibility in decision making (Decker et al. 2005; Noble et al., chapter 9, this volume; Šunde, chapter 20, this volume).

However, collaboration is neither easy nor simple: it is often characterized by difference among perspectives. When new voices are invited to the decision-making table, they must be free to speak, and by extension, they must be heard. Even the tacit acknowledgement of "new voices" is a step forward in the recognition of diversity and difference and the associated range of values for biodiversity in all its contexts. As such, conversation values must be made explicit if management goals are to reflect the desired ecosystems states (and hence biodiversity) to which decision making is ultimately targeted (Knight et al. 2006). Notably, this entails legitimizing values of biodiversity beyond conventional anthropocentric ethics. Given that complex systems science demands recognition of humans as part of a mutually constraining living system, rather than externally dominant or controlling, there emerges the need to validate a new class of ethics. Manuel-Navarrete, Dolderman, and Kay offer a spectrum of related discourses (chapter 19, this volume). Various philosophers and ethicists have articulated the need for ethical holism, that is, the extension of moral consideration to whole living systems (e.g., Šunde, chapter 20; Des Jardins 1993; Callicott 2003; Farina and Belgrano 2006). A broad and explicit range of ethical considerations should necessarily be built into a framework for meaningful biodiversity conservation. In the end, ecological science can merely illuminate options. The identification of conservation choices, in a democratic political context, will be determined by value-driven management goals. This is the intersection of science and values, and it is in their murky confluence that biodiversity must be navigated (Daugstad et al. 2006).

Building an Ecosystem Approach: Toward Adaptive Design

The implications for decision making discussed here effectively form the basis for an adaptive decision-making framework for conservation and, by extension, for an ecosystem approach in general. Systems thinking, as the authors in this book have made clear, challenges decision makers to become less concerned with prediction and control and to move toward more organic, adaptive, and flexible management (Lister and Kay 2000). In the absence of certainty and predictability the implication for decision making is that greater participation in the process is necessary—decisions must be discussed, debated, negotiated, and ultimately *learned* rather than predetermined by rational choice (Noble et al., chapter 9; Pinto-Correia et al. 2006).

How Then Should a Decision-Making Framework for Biodiversity Conservation Evolve?

Figure 6.2 depicts a necessary evolution in decision making if our planning institutions are to deal adequately with complexity and uncertainty, particularly in an ecosystem context. On the left, a generalized sector-based and top-down structure is shown, in which a hierarchy of experts supplies decision support information in relative isolation, i.e., separated by discipline (see, e.g., Dale 2001). On the right,

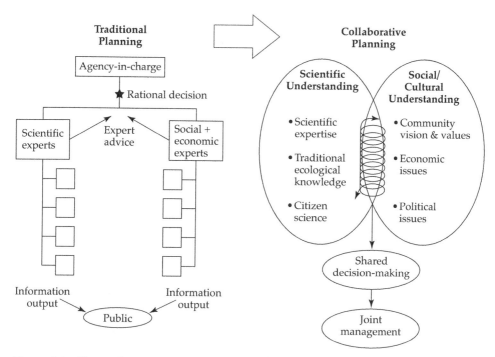

Figure 6.2   The evolution of planning under complexity and uncertainty.

a more flexible and organic model is offered, in which many forms of knowledge and values—including the conventional, disciplinary sciences as well as citizen science, traditional ecological knowledge (TEK), and economic, political, and cultural values—are *integrated* into a collaborative and shared decision-making process. A collaborative model ideally incorporates both top-down (expert) and bottom-up (citizen or grassroots) approaches to management, as managers, planners, and field personnel must work iteratively to share knowledge. Citizen and community participants should possess or establish a mutually acceptable degree of ecological literacy in order to support effectively the decision-making context (Lister and Kay 2000; Folke et al. 2005).

While the adaptive management concept appeared in the ecological literature in the late 1970s and has been evolving slowly ever since, it is only within the last few years that it has begun to move from theory to practice. Recently, scholars and practitioners in the management sciences and interdisciplinary environmental studies have begun to synthesize the notions of complexity and uncertainty in ecosystems with adaptive organizational management (e.g., Gunderson and Holling 2002; Folke et al. 2005; Lumb and Healie 2006). In this context, biodiversity conservation is a domain of theory and practice well-suited for this synthesis. In particular, biodiversity conservation as discussed here highlights the ill fit between open, dynamic ecosystems and the closed, brittle, decision-making structures through which we attempt to plan and manage them. In this way, adaptive management is not a new approach to conservation as much as it is fundamentally refocused to be *flexible, integrative,* and *resilient.* In short, it is a process for decision making that more closely models the living systems it is intended to shape.

However, as the literature and practice of adaptive planning and management matures, it is important to distinguish between types and intents of adaptive management (see, e.g., Gunderson and Holling 2002). What is proposed here is *conscious and intentionally* adaptive planning and management that together, rely on explicit, proactive, learning-based adaptation *within* an ecological context. This is in sharp contrast to passive adaptive management suggested by some of the scientific (e.g., climate change) literature in the context of unconscious, implicit, spontaneous, or reactive adaptation to externally imposed change (Lister and Kay 2000; Prato 2003). In this latter sense, humans are not necessarily seen as within or part of an ecological context. Furthermore, in the proactive, learning-based context, adaptive management constitutes sensitive management that is responsive to change, e.g., responding to new ecological information in a timely way, before critical and irreversible thresholds are crossed (Dale 2001; Lamont 2006; Umemoto and Suryanata 2006). In this way, adaptive management is "to learn to manage *by change* rather than merely reacting to it" (Gunderson et al. 1995:xi).

Figure 6.3 depicts an adaptive planning process, emphasizing the cyclic and continuous, rather than discrete, nature of planning and management. The adaptive context is one where learning is a collaborative and conscious activity, derived from empirically monitored or experientially acquired information, which, in turn, is transformed into knowledge through adapted behavior in the next planning cycle (Lister 1998).

The adaptive planning model shown here, in which individual phases are identified, is a variation on the complimentary adaptive methodology for ecosystem sustainability and health (AMESH) model discussed by Waltner-Toews and Kay in this book; it is the result of an applied research project in community-based conservation (Kay, chapter 2, this volume; Lister and Kay 2000). Although the conservation planning process can begin reasonably at several points, because the process is necessarily iterative, the "ideal" process begins with visioning, a social process of stakeholder and value analysis identifying what is desirable. The process proceeds interactively and iteratively with collaborators, through setting planning goals, objectives, and targets, planning criteria, and interactive workshops or a design charette, in synthesis with a scientific process of ecological analysis to determine what is possible. The process results in several outputs of synthesized information based on social and ecological analyses, e.g., conceptual models of how the local ecosystem works and relationships to the broader social system context, an open-ended master plan, and a management strategy. From this, the conservation plan should be refined, implemented, and monitored for ecosystem performance indicators, from which learned feedback is used to begin another planning cycle (Lister 1998; Lister and Kay 2000). This conceptual approach is based on proactive learning that is interactive and iterative rather than prescriptive and linear. That is, each element feeds information into another and the process shifts in a cyclic

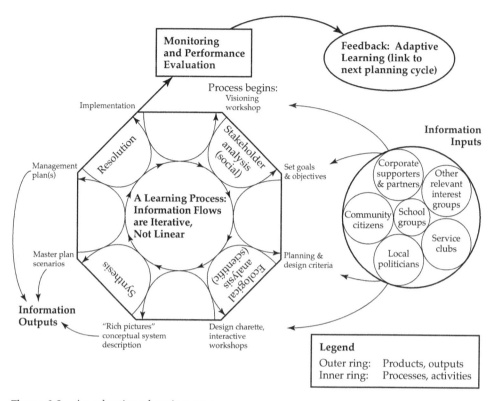

Figure 6.3   An adaptive planning process.

pattern as feedbacks result in directional adjustment. The approach is a holarchy: no single element can lead the other, yet all impose constraints on one another.

Given our preference for neat, simplified, and quick fixes, it is reasonable to question whether there exists the political will to implement such an open approach to conservation decision making. In the context of sustainability as discussed in this book, the precautionary principle should apply; every effort should be made to adjust to the realities of uncertainty and complexity before conservation managers are forced, and inadequately prepared, to do so. The adaptive planning approach outlined in fig. 6.3 is conceptual and generalized. It must be shaped according to each unique conservation context and issue. In the absence of a single "best model" provided by conclusive ecological science, the reasonable approach to conservation planning is to search for a set of contextualized common elements that work in a majority of similar cases with a diversity of committed, engaged collaborators.

Reflections

With insights from a paradigm of ecosystems as open and complex, a broader role for biodiversity in ecosystem dynamics has emerged. It is important to emphasize that this is not to say that a "new science" has replaced an outdated science but rather that normal science alone is insufficient; the recent perspective of ecosystems requires this and more—a *broader set of tools*. Clearly, the standard approach to ecosystem management through high-quality normal science is well suited to closed, mechanical systems and to smaller-scale, well-understood environmental problems with a few, controlled variables (Allen et al. 2001). The post-normal, systems-based perspective of biodiversity discussed here challenges conservation planners and managers to move away from prediction and control-oriented strategies and move toward more adaptive management of human activities. On the basis of proactive learning and empowerment of a range of different voices and values for biodiversity, this perspective demands an iterative process of change, between and among individuals and institutions. For practical purposes this means that we must look to build into our decision-making processes the adaptive capacity for change, flexibility, and resiliency; collaboration with a variety of interest groups and their divergent voices; and an appreciation of ecological context, place, and history. Biodiversity conservation must embrace art and science, politics, and culture. In doing so, we engage in an act of creation, and this is ultimately a process of *design*. Indeed, an ecosystem approach to biodiversity conservation demands that we expect uncertainty, embrace complexity, and celebrate diversity.

References

Allen, T. F. H. and T. W. Hoekstra. 1992. *Toward a Unified Ecology.* New York: Columbia University Press.

Allen, T. F. H., J. A. Tainter, J. C. Pires and T. W. Hoekstra. 2001. Dragnet ecology, "Just the facts Ma'am": The privilege of science in a post-modern world. *Bioscience* 51:475–485

Berkes, F. and C. S. Seixas. 2005. Building resilience in lagoon social—Ecological systems: A local-level perspective. *Ecosystems* 8(8):967–974.

Bond, W. J. 1993. Keystone species. In *Biodiversity and Ecosystem Function*, vol. 99, eds. E. D. Schulze and H. A. Mooney, pp. 237–292. New York: Springer-Verlag.

Burton, P. J., A. C. Balisky, L. P. Coward, S. G. Cumming and D. D. Kneeshaw. 1992. The value of managing for biodiversity. *The Forestry Chronicle* 68(2):225–237.

Callicott, J. B. 2003. The implications of the "shifting paradigm" in ecology for paradigm shifts in the philosophy of conservation. In *Reconstructing Conservation: Finding Common Ground*, eds. B. A. Minteer and R. E. Manning, pp. 239–262. Washington, D.C.: Island Press.

Clewell, A. F. and J. Aronson. 2006. Motivations for the restoration of ecosystems. *Conservation Biology* 20(2):420–428.

Crawford, T. W., J. P. Messina, S. M. Manson and D. O'Sullivan. 2005. Complexity science, complex systems, and land-use research. *Environment and Planning B* 32(6):792–798.

Cronon, W. (ed.) 1996. Uncommon Ground: Rethinking the Human Place in Nature. New York: Norton.

Dale, A. 2001. *At the Edge: Sustainable Development in the 21st Century*. Vancouver, Canada: University of British Columbia Press.

Daugstad, K., H. Svarstad and O. Vistad. 2006. A case of conflicts in conservation: Two trenches or a three-dimensional complexity? *Landscape Research* 31(1):1–19.

Dawkins, R. 1986. *The Blind Watchmaker*. New York: Norton.

Dawkins, R. 1989. *The Selfish Gene*, rev. ed. New York: Oxford University Press.

Decker, D., D. Raik, L. Carpenter, J. Organ and T. Schusler, 2005. Collaboration for community-based wildlife management. *Urban Ecosystems* 8(2):227–236.

Des Jardins, J. R. 1993. Environmental Ethics: An Introduction to Environmental Philosophy. Belmont, Calif.: Wadsworth.

Destefano, S., R. Deblinger and C. Miller. 2005. Suburban wildlife: Lessons, challenges, and opportunities. *Urban Ecosystems* 8(2):1083–8155.

Donner, D. 2006. Determining What's Important about Landscapes. *Conservation Biology* 20(2):592–594.

Ehrenfeld, D. 1991. The management of diversity: A conservation paradox. In *Ecology, Economics, Ethics: The Broken Circle*, eds. F. H. Bormann and S. R. Kellert, pp. 26–39. New Haven, Conn.: Yale University Press.

Ehrlich, P. R. 1988. The loss of biodiversity: Causes and consequences. In *Biodiversity*, ed. E. O. Wilson, pp. 21–27. Washington, D.C.: National Academy Press.

Ehrlich, P. R. and A. H. Ehrlich. 1981. Extinction: The Causes and Consequences of the Disappearance of Species. New York: Random House.

Ehrlich, P. R. and A. H. Ehrlich. 1992. The value of biodiversity. *Ambio* 21(3):219–226.

Ehrlich, P. R. and E. O. Wilson. 1991. Biodiversity studies: Science and policy. *Science* 253:758–762.

Environment Canada. 1995. *Canadian Biodiversity Strategy*. Ottawa, Canada: Queen's Printer.

Farina, A. and A. Belgrano. 2006. The Eco-field Hypothesis: Toward a cognitive landscape. *Landscape Ecology* 21(1):5–17.

Folke, C., T. Hahn, P. Olsson and J. Norberg. 2005. Adaptive governance of social-ecological systems. *Annual Review of Environment and Resources* 30:441–473.

Foster, J. and L. A. Sandberg. 2004. Friends or foe? Invasive species and the public greenspace of Toronto. *The Geographical Review* 94(2):178–198.

Funtowicz, S. O. and J. R. Ravetz. 1993. Science for the post-normal age. *Futures* 25(7): 739–755.

Goodman, D. 1975. The theory of diversity-stability relationships in ecology. *Quarterly Review of Biology* 50(3):237–266.

Gunderson, L. H. and C. S. Holling (eds.). 2002. *Panarchy: Understanding Transformations in Human and Natural Systems*. Washington, D.C.: Island Press.

Gunderson, L. H., C. S. Holling and S. S. Light (eds.). 1995. *Barriers and Bridges to the Renewal of Ecosystems and Institutions*. New York: Columbia University Press.

Hanski, I. and O. Ovaskainen. 2002. Extinction debt at extinction threshold. *Conservation Biology* 16:666–673.

Heywood, V. (ed.). 1995. *Global Biodiversity Assessment*. Cambridge, UK: Cambridge University Press for the United Nations' Environment Programme.

Holling, C. S. 1986. The resilience of terrestrial ecosystems: Local surprise and global change. In *Sustainable Development of the Biosphere*, eds. W. C. Clark and R. E. Munn, pp. 292–320. Cambridge, UK: Cambridge University Press.

Holling, C. S. 1992. Cross-scale morphology, geometry and dynamics of ecosystems. *Ecological Monographs* 62(4):447–502.

Holling, C. S., D. W. Schindler, B. W. Walker and J. Roughgarden. 1995. Biodiversity in the functioning of ecosystems: An ecological synthesis. In *Biodiversity Loss: Economic and Ecological Issues*, eds. C. Perrings et al., pp. 44–83. Cambridge, UK: Cambridge University Press.

Janzen, D. H. 1988. Tropical dry forests: The most endangered tropical ecosystem. In *Biodiversity*, ed. E. O. Wilson, pp. 130–137. Washington, D.C.: National Academy Press.

Jeffries, M. J. 1997. *Biodiversity and Conservation*. London, UK: Routledge.

Jørgensen, S. E. 1992. Integration of Ecosystem Theories: A Pattern, vol. 1. Ecology and Environment Series, Dordrecht: Kluwer Academic.

Kauffman, S. A. 1991. Antichaos and adaptation. *Scientific American* August: 78–84.

Kauffman, S. A. 1994. The Origins of Order: Self-Organization and Selection in Evolution. New York: Oxford University Press.

Kay, J. J. 1991a. The concept of ecological integrity, alternative theories of ecology, and implications for decision-support indicators. In *Economic, Ecological and Decision Theories: Indicators of Ecologically Sustainable Development*, eds. P. A. Victor et al., pp. 23–58. Ottawa: Canadian Environmental Advisory Council.

Kay, J. J. 1991b. A non-equilibrium thermodynamic framework for discussing ecosystem integrity. *Environmental Management* 15(4):483–495.

Kay, J. J. and E. Schneider. 1994. Embracing complexity: The challenge of the ecosystem approach. *Alternatives* 20(3):32–38.

Knight, A., R. M. Cowling and B. M. Campbell. 2006. An operational model for implementing conservation action. *Conservation Biology* 20(2):408–419.

Lambeck, R. J. 1997. Focal species: A multi-species umbrella for nature conservation. *Conservation Biology* 11:849–856.

Lamont, A. 2006. Policy characterization of ecosystem management. *Environmental Monitoring and Assessment* 113(1–3):5–18.

Lister, N. M. 1998. A systems approach to biodiversity conservation planning. *Environmental Monitoring and Assessment* 49(2/3):123–155.

Lister, N. M. and J. J. Kay. 2000. Celebrating diversity: Adaptive planning and biodiversity conservation. In *Biodiversity in Canada: Ecology, Ideas and Action*, ed. S. Bocking, pp. 189–218. Toronto, Canada: Broadview Press.

Lumb, A. and R. Healie. 2006. Canada's Ecosystem Initiatives. *Environmental Monitoring and Assessment* 113(1–3):1–3.

MacIntosh, R. P. 1976. Ecology since 1900. In *Issues and Ideas in America*, ed. B. J. Taylor and T. J. White, pp. 353–372. Norman: University of Oklahoma Press.

Manson, S. M. 2001. Simplifying complexity: A review of complexity theory. *Geoforum* 32(3):405–414.

Marzluff, J. 2005. Island biogeography for an urbanizing world: how extinction and colonization may determine biological diversity in human-dominated landscapes. *Urban Ecosystems* 8(2):157–177.

May, R. M. 1974. Biological populations with non-overlapping generations: Stable points, stable cycles and chaos. *Science* 186:645–647.

McNeely, J. A, K. R. Miller, W. V. Reid, R. A. Mittermeier and T. B. Werner. 1990. *Conserving the World's Biological Diversity*. New York: Wiley.

Mooney, H., et al. 1996. Functional Roles of Biodiversity: A Global Perspective, 480 pp. New York: Wiley.

Naeem, S., L. J. Thompson, S. P. Lawler, J. H. Lawton and R. M. Woodfin. 1994. Declining biodiversity can alter the performance of ecosystems. *Nature* 368:734–736.

Norgaard, R. B. 1987. Economics as mechanics and the demise of biological diversity. *Ecological Modelling* 38:107–121.

Norton, B.G. 1992. A new paradigm for environmental management. In *Ecosystem Health: New Goals for Environmental Management*, eds. R. Costanza et al., pp. 23–41. Washington, D. C.: Island Press.

Norton, B. G. 2003. Conservation: Moral crusade or environmental public policy? In *Reconstructing Conservation: Finding Common Ground*, eds. B. A. Minteer and R. E. Manning, pp. 187–206. Washington, D. C.: Island Press.

Norton, B. G. and R. E. Ulanowicz. 1992. Scale and biodiversity policy: A hierarchical approach. *Ambio* 21(3):244–249.

Noss, R. F. 1990. Indicators for monitoring biodiversity: A hierarchical approach. *Conservation Biology* 4(4):355–364.

Nowak, M., et al. 2006. Five rules for the evolution of cooperation. *Science* 314:1560–1563.

Nowicki, A. 1993. Diversity. *Trumpeter* 10(2):65–68.

O'Brien, W. 2006. Exotic invasions, nativism, and ecological restoration: On the persistence of a contentious debate. *Ethics, Place and Environment* 9(1):63–77.

Odum, E. P. 1993. *Ecology and Our Endangered Life-Support Systems* (2nd ed.). Sunderland, Mass.: Sinauer.

O'Neill, R. V., D. L. DeAngelis, J. B. Waide and T. F. H. Allen. 1986. *A Hierarchical Concept of Ecosystems*, vol. 23, *Monographs in Population Biology*. Princeton, N. J.: Princeton University Press.

Orr, D. 2006. Framing sustainability. *Conservation Biology* 20(2):265–268.

Paine, R. T. 1966. Food web complexity and species diversity. *The American Naturalist* 100:65–75.

Pedroli, B., T. Pinto-Correia and P. Cornish. 2006. Landscape: What's in it? Trends in European landscape science and priority themes for concerted research. *Landscape Ecology* 21(3):421–430.

Pimm, S. L. 1984. The complexity and stability of ecosystems. *Nature* 307:321–326.

Pimm, S. L. 1991. The Balance of Nature? Ecological Issues in the Conservation of Species and Communities. Chicago, Ill.: University of Chicago Press.

Pimm, S. L. 1993. Biodiversity and the balance of nature. In *Biodiversity and Ecosystem Function*, eds. E. D. Schulze and H. A. Mooney, pp. 347-360. New York: Springer-Verlag.

Pinto-Correia, T., R. Gustavsson and J. Pirnat. 2006. Bridging the gap between centrally defined policies and local decisions towards more sensitive and creative rural landscape management. *Landscape Ecology* 21(3):333-346.

Prato, T. 2003. Adaptive management of large rivers with special reference to the Missouri River. *Journal of the American Water Resources Association* 39:935–946.

Real, L. A. and J. H. Brown (eds.). 1991. *Foundations of Ecology*. Chicago, Ill.: University of Chicago Press.

Reid, W. C. and K. R. Miller. 1989. *Keeping Options Alive: The Scientific Basis for Conserving Biodiversity*. Washington, D. C.: World Resources Institute.

Robbins, P., K. McSweeney, T. Waite and J. Rice. 2006. Even conservation rules are made to be broken: Implications for biodiversity. *Environmental Management* 37(2):162–169.

Roberge, J-M. and P. Angelstam. 2004. Usefulness of the umbrella species concept as a conservation tool. *Conservation Biology* 18(1):76–85.

Schneider, E. D. and J. J. Kay. 1994. Complexity and thermodynamics: Towards a new ecology. *Futures* 24(6):626–647.

Schulze, E. D. and H. A. Mooney (eds.). 1993. *Biodiversity and Ecosystem Function*, vol. 99, *Ecological Studies*. New York: Springer-Verlag.

Secretariat of the Convention on Biological Diversity. 2006. *Global Biodiversity Outlook 2*. Montreal, Canada.

Shiva, V. 1993. *Monocultures of the Mind: Understanding the Threats to Biological and Cultural Diversity*. David Hopper Lecture on Development. Guelph, Ontario, Canada: University of Guelph.

Smith, F. and G. Swanson. 1994. *Biological Diversity and Ecosystem Stability: The Ecological Justification for Biodiversity Conservation*. Occasional Paper 10. Norwich, Norfolk, England: Centre for Social and Economic Research on the Global Environment, University of East Anglia.

Solbrig, O. T. 1991a. Biodiversity: Scientific Issues and Collaborative Research Proposals. MAB Digest 9. Paris: UNESCO.

Solbrig, O. T. 1991b. The origin and function of biodiversity. *Environment* 33(5):17–20/34–38.

Soulé, M. E. and G. H. Orians (eds.). 2001. *Conservation Biology: Research Priorities for the Next Decade*. Washington, D.C.: Island Press.

Theberge, J. B., M. T. Theberge, J. A. Vucetich and P. C. Paquet. 2006. Pitfalls of applying adaptive management to a wolf population in Algonquin Provincial Park, Ontario. *Environmental Management* 37(4):451–460.

Tilman, D. and J. A. Downing. 1994. Biodiversity and stability in grasslands. *Nature* 367:363–365.

Umemoto, K. and K. Suryanata. 2006. Technology, culture, and environmental uncertainty: Considering social contracts in adaptive management. *Journal of Planning Education and Research* 25(3):264–274.

United Nations Conference on Environment and Development. 1992. *Conservation of Biodiversity*, chap.15, Agenda 21. New York: United Nations.

United Nations Environment Programme. 1992. *Convention on Biological Diversity*. New York: United Nations.

Wesson, R. 1991. *Beyond Natural Selection*. Cambridge: Massachusetts Institute of Technology Press.

Westman, W. E. 1990. Managing for biodiversity: Unresolved science and policy questions. *BioScience* 40(1):26–33.

Wiegand, W. 2006. The shrinking world: Ecological consequences of habitat loss from a population and landscape perspective. *Conservation Biology* 20(2):590–592.

Wilson, E. O. 1985. The biological diversity crisis. *BioScience* 35(11):700–706.

Wilson, E. O. 1992. *The Diversity of Life*. Cambridge, Mass.: Belknap.

Wood, A., P. Steadman-Edwards and J. Mang (eds.). 2000. *The Root Causes of Biodiversity Loss*. London, UK: Earthscan.

Wood, P. M. 2000. Biodiversity and Democracy: Rethinking Society and Nature. Vancouver, Canada: UBC Press.

World Resources Institute, World Conservation Union, and the United Nations Environment Programme. 1992. *Global Biodiversity Strategy: Guidelines for Action to Save, Study and Use Earth's Biotic Wealth Sustainably and Equitably*. Washington, D.C.: World Resources Institute.

7

The Cultural Basis for an Ecosystem Approach
Sharing Across Systems of Knowledge

Fikret Berkes and Iain Davidson-Hunt

As this book and a series of others make it clear, the science of ecology is in the midst of two conceptual shifts that are central to a new ecosystem approach. One is the shift reflecting some of the more complex questions being asked and the recognition that simplifications that work well to answer "all other things being equal" questions in laboratory experiments are not successful in answering questions that embrace multiple interactions in space and time. The second shift reflects the inclusion of people inside ecosystems; this, in turn, involves a recognition that people have evolved as species inside the systems they are attempting to describe and that human population numbers and technological activities are now so extensive that there are no ecosystems that are unaffected by human activity. In the discourse on sustainability the simplification of reality into separate social and ecological systems, rather than conflating them into one social-ecological system that can be viewed from different perspectives, requires an untenable editing of the evidence (Gunderson et al. 1995; Berkes and Folke 1998; Levin 2000; Gunderson and Holling 2002). These two shifts are related, as they both pertain to the understanding of social-ecological systems as complex systems with humans as an integral part.

Several authors in this book argue for the necessity of multiple perspectives in managing for sustainability. What is implicit in many discussions about incorporation of such perspectives and their translation into governance, however, is that a common worldview (usually some version of scientific positivism) is assumed among participants, even those from different disciplinary or social positions (social activists, natural scientists, economists). But what if the collaborators see not only different things in the world, but see different worlds? We wish to extend this discussion further and examine more closely the topic of systems of environmental knowledge and its implications for ecosystem-based management.

The broader objective of this chapter is to address issues related to culture as a necessary and integrating context for action in an ecosystem approach. For example, how does one promote the collaboration between an indigenous elder (or a shaman) and an ecosystem scientist? How does one combine these different systems of knowledge? Or can they be combined at all (Reid et al. 2006)? These questions contribute to the discussion of post-normal science in dealing with complex adaptive systems. Post-normal science expands the domain of science to include scientists and the production of science itself, as well as transcending and incorporating a diversity of discipline-based paradigms. This chapter contributes to the discussion by exploring the implications of other knowledge systems, specifically traditional ecological knowledge, on the question of including humans in the ecosystem.

In this discussion, we use the term "social-ecological system" to refer to the integrated concept of humans and nature (Berkes and Folke 1998). Humans-in-ecosystem (Davidson-Hunt and Berkes 2003), or the "dwelling perspective" in the evocative terminology of the anthropologist Tim Ingold (2000) refers to the practical engagement of humans with others of the dwelt-in-environment. This practical engagement, building knowledge and ecological relationships, is the basis for putting humans back into the ecosystem. It involves the "skills, sensitivities and orientations that have developed through long experience of conducting one's life in a particular environment" (Ingold 2000:25).

We use "indigenous knowledge" (IK) as a generic term to refer to knowledge systems of indigenous peoples and "IK-holder" to refer to elders and other knowledgeable people, who, in the post-normal sensibility, expand the peer group and contribute multiple legitimate perspectives. We define "traditional ecological knowledge" (TEK) more specifically as "a cumulative body of knowledge, practice and belief, evolving by adaptive processes and handed down through generations by cultural transmission, about the relationship of living beings (including humans) with one another and with their environment" (Berkes 2008:7). Although many aspects of TEK are similar to scientific knowledge, there are also certain fundamental differences. TEK is a knowledge-practice-belief complex; it tends to be experiential knowledge closely related to a way of life. It is multigenerational, and it is passed on orally by *cultural* transmission rather than by book learning. Many traditional knowledge systems consider humans as part of an interacting web of life (Callicott 1994; Berkes 2008), a major distinction from the conventional Age of Enlightenment view of humans in the cosmos.

There is a long history of discussion in the environmental literature regarding humans in the ecosystem. Glacken (1967) observed that the concept of an external environment analytically separate from humans could be traced to post-Enlightenment philosophy in Europe. Bateson (1972) pointed out the importance of Cartesian dualism of mind versus matter, and hence humans versus the environment, in shaping the Western worldview regarding human-environment relationships. This dualism is the basis of our current ontology. The notion of "man's dominion over nature" is not only related to the basic tenets of positivism, rationalism, and reductionism in Western science, but it is also a key impediment to the task of putting humans back into the ecosystem (Merchant 1980).

If the positivistic tradition is a problem for human-ecosystem integration, a key question to ask is whether there are other traditions in Western science that fare better. A second question is whether there are systems of knowledge other than Western science that provide models worth examining. The scope of this chapter does not include the first question. Let us assume, however, that holistic traditions in Western science, including ecology, systems science, and quantum physics, are consistent with the inclusion of humans in the ecosystem (Capra 1982). We turn our attention, instead, to the second question as the focus of the chapter.

The philosopher Paul Feyerabend (1987) is often credited with the provocative idea that Western science should perhaps be considered as merely *one* of many systems of knowledge. Given the dominant position of Western science in the world, it is probably fair to say it cannot really be considered just another system of knowledge. Nevertheless, there are different ways of knowing, and there are diverse worldviews that conceptualize the relationship of people to the environment quite differently. The claim that there is only one science and one way of knowing has been characterized by Scott (1996) as "science for the West; myth for the rest." The importance and validity of multiple ways of knowing and a pluralistic science are starting points for the argument in this chapter.

We start with a consideration of ecosystem-like concepts in non-Western cultures and a selection of these concepts to illustrate the diversity of ways in which humans are treated as a part of the ecosystem in different cultures. We proceed to deal with indigenous knowledge research as it has been changing in the recent decades and how IK-holders are becoming partners in ecosystem research and management instead of being treated as *objects* of study. Such participatory research calls for the collaboration of Western scientists and IK-holders to share across systems of knowledge.

To illustrate participatory research, this chapter provides two examples from our recent work, one undertaken in cooperation with Iskatewizaagegan #39 Independent First Nation of northwestern Ontario and the other with the Inuvialuit of Sachs Harbour in the Canadian western Arctic. The concluding section explores the implications of expanding the knowledge base for ecosystem-based management to include not only the traditional ecological knowledge of non-Western peoples but also, more generally, the local ecological knowledge of a diversity of people who dwell in natural ecosystems.

Ecosystems and Shamans

Is it possible for ecological scientists and shamans to even begin a dialogue on ecosystem-based management? What is the evidence that there may be promising common grounds? Reichel-Dolmatoff (1976) was one of the earliest scholars to write that the activities and the role of shamans could be compared to that of ecosystem managers. In a study of the Tukano people of the Colombian northwest Amazon, Reichel-Dolmatoff emphasized the role of the shaman not only as a healer of individuals but also as an "ecosystem doctor," applying his treatment to the disturbed parts of the ecosystem. This the shaman did, for example, by establishing and helping enforce rules to avoid overhunting, the depletion of certain plant resources, and unchecked population increase. As the spiritual and social leader of the group, the shaman helped regulate human behavior in response to environmental feedbacks such as the local depletion of particular medicinal plants (Reichel-Dolmatoff 1976).

Reichel-Dolmatoff's analysis is clearly in the mode of an ecosystem and complexity approach, dealing with the two-way interactions of the Tukano and their tropical forest environment. As detailed by the researcher, Tukano cosmology is consistent with the ecosystem approach. The individual person considers himself or herself as part of a complex network of interactions that include people as well as the entire universe. These relationships are encoded in myths, and rituals help remember ecosystem management rules. The ecosystem discourse of the Tukano shaman uses highly symbolic language, and it is doubtful that it would be recognizable to the ecosystem scientist. Can the shaman and the scientist find a terminology and a set of concepts in common?

Take, for example, the concept of ecosystem. According to the conventional translations of American Indian languages, there is no word for "ecosystem." We have argued, however, that many of the words that usually get translated as "land" (*ashkii* in Cree, *aski* in Anishinaabe/Ojibwa, *ndeh* in Dene) carry a meaning that goes well beyond land as physical landscape (Berkes et al. 1998). Many of these terms encompass the living environment, including humans; in that sense, they are similar to Leopold's (1949) use of the term "land" as in land ethic. For example, the term used by the Dene groups of the western subarctic of North America, *ndeh*, is closer in meaning to ecosystem than to land because it conveys a sense of relations of living and nonliving things. It differs, however, from the scientific concept of ecosystem in that *ndeh* is based on the idea that everything in the environment has life and spirit (Berkes et al. 1998).

The landscape in which the Dene and Cree dwell is fully alive not only with fauna and flora but also with spirits—a sharp contrast to some of the older, mechanistic ecosystem concepts in which ecosystems were considered as little more than thermodynamic machines with cogs and wheels driven by solar energy! Newer conceptualizations of ecosystems are moving away from the positivistic and the mechanistic to embrace, instead, an ecology with a heart—a sacred ecology (Anderson 1996; Bradshaw and Bekoff 2001; Berkes 2008). The similarities of some of these conceptualizations to traditional ecological knowledge are not accidental. Ecological knowledge traditions of some non-Western cultures have been providing inspiration for the understanding of human-environment relationships.

In addition to the Tukano study, several other cases have taken a systems approach to the study of human-nature relations in traditional cultures (table 7.1). Some of these studies have used descriptive models, while others have used mathematical approaches. For example, the Peruvian llama herders' case in table 7.1 used simulation models to study herd dynamics, and the Balinese water temples

TABLE 7.1   Examples of Systems Approaches to the Study of Human-Nature Relationships in Non-Western Cultures

Country/Region and People	How the System Works	Reference
Amazon forest and the Tukano of Colombia	Tukano shamans manage both human health and ecosystem health through the rules they enforce	Reichel-Dolmatoff (1976)
Llama herders of the highlands, Ayacucho, Peru	A system of reciprocity and gift-giving regulates herd size of llamas and periodically reestablishes human-nature relations	Flannery et al. (1989)
New Guinea highland horticulturalists	A ritual of pig slaughter and tribal warfare, which occurs periodically, regulates resource management	Rappaport (1984)
Irrigated rice agriculture systems in Bali, Indonesia	Hindu priests manage a system of "water temples" that regulates use of irrigation water by villagers	Lansing (1993)

study used optimization models. In an extension of the Balinese study, Lansing et al. (1998) used a modification of the "Daisyworld model" to include feedbacks between the environment and the organism (rice irrigators). Modeling the selection of rice cropping patterns as a process of system-dependent selection (in which selection resulting from feedbacks constantly modifies subsequent selection), Lansing et al. (1998) were able to generate solutions that accurately predicted the observed patterns of rice production. Are the investigators perhaps "reading" too much into these systems? Or is there, in fact, evidence of ecosystem thinking in some traditional cultures?

Starting with this question, Berkes et al. (1998) reviewed evidence for ecosystem thinking in a number of ancient societies, spanning a range of cultures and geographical areas. In particular, the basic idea of watershed management, as a bounded ecological unit, may be found in a number of societies (table 7.2). There is good documentation from the sixteenth century onward of watershed-ecosystem thinking among the Chinese, Japanese, ancient Greeks, the Ottoman Turks, and the Swiss. Of all the ecosystem-like concepts in traditional societies, those in the Asia-Pacific region seem to be the most fully developed. Examples include the Fijian *vanua*, the Yap *Tabinau*, and the ancient Hawaiian *ahupua'a*. These were wedge-shaped watershed units controlled by local chiefs, stretching from the top of the cone of the volcanic mountain to the edge of the reef.

In some ecosystem-like concepts, people are considered to be an integral part of the system. Several of the Pacific Island cases and the Pacific Northwest examples have this feature (table 7.2). In the Solomon Islands, for example, *puava*

TABLE 7.2   Examples of the Applications of Ecosystem-like Concepts Among Non-Western Cultures

Country/ Region and People	System	Reference
Aborigines of Australia	Land is a living record of *dreamtime* events; *dreamtime* connects people to their ancestors, providing the bond between land, people and totemic beings	Wilkins (1993)
Raramuri of northern Mexico	The concept of *iwigara,* kincentric ecology, refers to the interconnectedness of life, both physical and spiritual	Salmón (2000)
Solomon Islands, Oceania	A *puava* is a defined, named territory, including all land and lagoon areas and resources associated with a descent (or kinship) group	Hviding (1996)
Amerindians of the Pacific Northwest of North America	Salmon rivers and associated hunting and gathering areas are managed as watershed units often associated with kinship-based groups or houses	Williams and Hunn (1982)

is a defined, named territory consisting of land and sea. It includes all resource areas associated with a descent group (lineage). *Puava* refers to an intimate association of a group of people with land, reef, and lagoon, and the entire livelihood produced by that territory, a "personal ecosystem" of a group of people (Hviding 1996). There are other systems in which specific groups of people are considered connected to their land but apparently without invoking a watershed ecosystem. The *dreamtime* concept of the Australian aborigines and the *iwigara* concept of the Raramuri of Mexico are examples (table 7.2).

Sharing Across Knowledge Systems through Participatory Research

If there are ecosystem-like concepts and ecosystem thinking in some traditional societies, there must also be elements of understanding of ecosystem-based management. If so, there are possibilities of ecosystem research collaboration between scientists and local knowledge holders. In turning to the question of research collaboration, it should be pointed out that the partnership of local stewardship groups with scientists is not a very unusual phenomenon. Such research collaboration for ecosystem management has been documented, for example, from Sweden (Olsson and Folke 2001) and Minnesota (Blann et al. 2003). Where there is a common knowledge base between scientists and ecosystem dwellers, meeting the challenge of research and management collaboration is relatively easy. However, the challenge is greater if the ecological understanding of the scientists and dwellers is based on different systems of knowledge.

Such is the challenge of participatory research involving IK holders. The diverse methodologies and practices used in IK or TEK research reflect the diverse origins of the discipline in ethnoscience and human ecology. They also reflect the diversity of purposes for carrying out IK research. These have ranged, historically, from research carried out for purposes of colonial domination, to research on local agricultural varieties, and research on species and natural chemicals important for medicinal or other commercial purposes (Posey and Dutfield 1996).

Over the years, ethnoscience has shifted from the descriptive to the analytical and from a focus on classification to an interest in understanding systems. There has been a shift of IK research focus from ecosystem components to ecosystem structure and function. Interests in folk taxonomy that used to dominate IK research, especially in terrestrial tropical ecosystems, have been replaced in part with ecological inquiries—indigenous perspectives about human-nature relationships and local understandings of ecosystem processes. These studies have already had an impact on the discipline of ecology and, to some extent, on professional ecologists by providing insights on ecosystem processes in such diverse geographical regions as the Brazilian Amazon, coastal ecosystems of Oceania, and the Canadian Arctic (Ford and Martinez 2000; Berkes 2008).

The practice of IK research itself has also been changing over the years. Clément (1998) identified three stages of research. He calls them pre-classical, classical, and post-classical and argues that these stages are helpful in understanding the changes that have occurred in IK research in the field of ethnobotany. The pre-classical and classical stages share a common interest in documenting "the economic use, vernacular nomenclature, and systematic classifications" of plants as well as broader interests such as the ". . . knowledge of resources and how to manage them." (Clément 1998:163) The pre-classical stage differed from the classical in its focus on producing "etic" representations that communicated the researcher's understanding of what people knew about a given subject. The classical stage emerged in the 1950s with the work of Conklin (1957) and moved to a focus on "emic" representations that communicated the researcher's understanding of how people knew what they knew about a given subject. The benefit of these stages of IK research was that indigenous knowledge and ways of knowing were given legitimacy by scientists.

The pre-classical stage showed that IK-holders knew a lot of information about the local environment. The classical stage demonstrated that ways of knowing included an awareness of complex social and ecological relationships and processes. However, at the same time that this research was documenting sophisticated knowledge and ways of knowing, it was concurrently eroding the authority of IK holders and the legitimacy of contemporary IK. For example, shamans' knowledge of medicinal plants was extracted from specific ecosystems, transferred to the public domain and then reinserted into specific ecosystems controlled by corporations through private property instruments. As Latour (1986) has caustically observed, the goal of such research was to transform knowledge with specific meaning and utility to "immutable mobiles" with universal relevance and application. The result of this approach to "sharing" knowledge has led to acrimonious discussions of the relationship between property rights and knowledge amongst IK holders and scientists (Posey and Dutfield 1996).

One problem that has emerged is that IK and indigenous ways of knowing tend to become "traditionalized" through the research process (Ingold 2000). Customs and practices of IK holders become "traditions" through a process of codification by research and legitimization by state resource management agencies that have obtained the authority to manage indigenous territories. The IK holder becomes a source of information on the environment; the insertion of this knowledge into management processes legitimizes the management system, without necessarily providing for effective participation. Indigenous ways of knowing, the adaptive and dynamic aspect of IK, become marginalized from the contemporary production of knowledge for management. The irony is that sharing knowledge can lead to the stripping of the IK holder for his/her authority over knowledge and the marginalization of the indigenous ways of knowing the ecosystem. Such experiences with sharing knowledge under conditions of inequitable power relations between indigenous people, scientists, and state management agencies have led to a new stage that Clément (1998:163) has called post-classical.

The post-classical stage emerged in the 1990s and was characterized by "marked cooperation between Western scientific researchers and Native peoples." The post-classical (or post-colonial) stage is where the ideas of post-normal science may resonate with the calls for cooperation, accountability, and transparency in the sharing of knowledge between IK holders and scientists (Davidson-Hunt and O'Flaherty 2007). This shift toward participatory research has been reflected in such documents as "The Declaration of Belem," which was one of the first sets of self-regulating guidelines prepared by ethnobiologists (Martin 1995). In fact, basic ethnobiological texts such as those by Balick and Cox (1996), Cotton (1996), and Martin (1995) have included discussions of how to create local benefits through research projects, cooperative research projects, intellectual property rights, and research ethics. Some indigenous political organizations have responded by drafting guidelines on the topics of research and the process by which research would be approved within a specified territory (Martin 1995).

International and national laws have been drafted that begin to discuss the rights held by indigenous peoples regarding intellectual property, cultural heritage, and natural resources. These instruments attempt to correct the power imbalances among IK holders, scientists, and state management agencies. However, there is also a theoretical shift in the post-classical stage. The IK holder, IK, and indigenous ways of knowing underwent a theoretical metamorphosis from being *objects* of research and management, to becoming *subjects* of research and management processes. Perhaps it is at this point where post-colonial research and post-normal science theory intersect that meaningful dialogue between IK holders and scientific researchers may contribute to ecosystem management (Davidson-Hunt and O'Flaherty 2007).

To summarize, ecosystem management has had the challenge of overcoming historic models of research and management in which humans were not considered part of the ecosystem. If an ecosystem is not an object to be understood or managed by external agents but a place of dwelling for a group of people who use it, then what are the implications of this for research and management? Observations of an ecosystem are shaped by cultural filters and organized according to the worldview of the observers. The resulting cumulative body of knowledge-practice-belief is grounded in a particular ecosystem. It is held by a group of people who not only dwell in that biophysical space but also shape it and are shaped by it in turn. A culturally sensitive use of the ecosystem approach seeks to understand TEK in these terms. Hence post-classical or post-colonial research emerges out of the shared experience of dwelling within an ecosystem or of choosing to know an ecosystem. Authority of knowledge and legitimacy of ways of knowing emerge out of the process of dwelling (Davidson-Hunt 2003).

Sharing Knowledge: Two Practical Cases from Canada

A shared goal for ecosystem scientists and IK holders could be the health and well-being of an ecosystem, leading to a consensus-based approach to determine research objectives, methods, conclusions, and outputs. Methods, or the ways by which the perception of the environment is built, can be a creative mix of scientific methodologies and indigenous ways of knowing. While the scientist may focus on producing artifacts (papers, books, models, reports, visual materials, etc.), IK holders may be more concerned with embodying what they have learned into their artifice (way of knowing).

This type of research model is an iterative, long-term process by which research goals, projects, and methodologies, along with new ecosystem management institutions, emerge as mutual understanding, respect, and trust builds between IK holders and scientists. The roles of IK holders and scientists transcend that of information provider and recorder and instead become based upon two-way flows of information and upon mutual learning. Given the history of conventional people-free ecosystem science, this alternative vision of ecosystem research and management is worth exploring. We use two cases to sketch out research approaches and processes that can work to build ecosystem knowledge cooperatively through mutual sharing and learning.

In these two cases, we seek to build models of participatory research that can allow sharing between IK and Western science in ways that do justice to both, while at the same time producing good ecology and social science. We analyze the main features of the partnership research model that we have used, and we attempt to derive some lessons from them. Both examples are based on recent research projects carried out with Canadian indigenous groups. One is on the use of non-timber forest products and the other in the area of climate change.

The first case is based on the project, "Combining scientific and first nations' knowledge for the management of non-timber forest products," carried out with the Anishinaabe people of Iskatewizaagegan #39 Independent First Nation (IIFN) (Davidson-Hunt 2003). IIFN is located on Shoal Lake on the southern border of northwestern Ontario and southeastern Manitoba. The project started with joint discussions between members of the Shoal Lake Resource Institute (SLRI), an organization of IIFN, and the researchers. The conversation began at a conference on Non-Timber Forest Products (Davidson-Hunt et al. 2001) and continued as consensus was sought on the potential themes for a research project. It proceeded through a series of meetings with both political leaders and elders of IIFN, culminating in a negotiated research protocol between the Natural Resources Institute, SLRI, and IIFN. This agreement established a research team composed of university personnel and community researchers. An advisory team made up of university researchers and IIFN political representatives and elders was also established. The research and advisory teams held a ceremony (feast) in order to establish proper social relationships among members of the two parties and to ensure proper ethical attitudes toward the plants to be studied before the research began.

Information was collected through elder interviews and participant observation. Biological data were obtained through sample plots in which non-timber forest products (e.g., berries, medicinal plants, and other culturally significant products such as birch bark) were collected and voucher samples taken. Elders took the initiative to take researchers to sites that they deemed important to teach them what they considered important, rather than researchers taking the elders to plot surveys (as originally planned). Results were recorded by audiotape and videotape to be deposited, along with photo records of plants, with SLRI. Results of the research were presented to research participants (elders) for verification and comment. Interim technical reports of the project were presented to the advisory team for approval and then distributed to government, research communities, and other First Nation communities through a workshop. Comments received from the advisory team and workshop participants were used to determine the activities undertaken in the second year of the research project. Figure 7.1 provides a summary of the structure of the partnership.

The second case is based on the project Inuit Observations of Climate Change, which was carried out with the Inuvialuit people of Sachs Harbour, in the Canadian Western Arctic. It was a project initiated by the people of Sachs Harbour who wanted their perceptions of climate change recorded and disseminated to the world. Led by the International Institute for Sustainable Development (IISD), the project started with a planning workshop that at the very start of the project asked the people of Sachs Harbour their objectives and what they considered important for the project to focus on. The priority issues, research questions, plans for video documentation, and the overall process for the project were all defined jointly by the project personnel and the community.

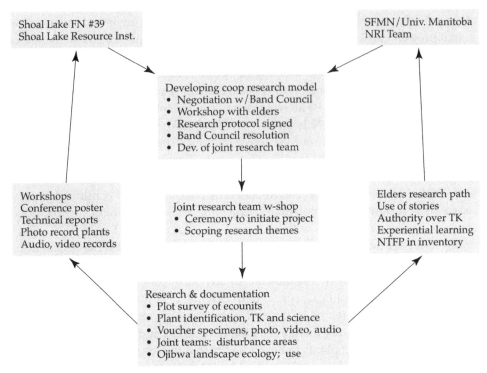

Figure 7.1   Structure of partnership with the Anishinaabe people of Iskatewizaagegan #39 Independent First Nation

Information was collected using a variety of interlinking methodologies: brainstorming workshops, focus groups, video interviews, individual semidirective interviews, and participant observation. The early versions of the video, along with technical trip reports and newsletters prepared for the community, were used for feedback and for the verification of results. In several of the key areas for climate change, the project invited southern scientific experts to Sachs Harbour to provide opportunity for one-on-one interaction between scientists and local experts on such topics as changes in sea-ice conditions. Repeat visits were made to the community around a yearly cycle, focusing on activities appropriate for that season. Continuity was provided by two members of the team who took part in all the trips and stayed for longer periods (Ford 2000; Riedlinger and Berkes 2001). Figure 7.2 shows the structure of the partnership arrangement.

The two research projects are different in organizational details, as well as being different in geographic setting and subject matter. However, they share a number of features that characterize a particular kind of partnership project. First, the project creates a "table" of equal partners, a forum in which different objectives can be discussed and the agendas of the parties laid on the table and made transparent. The actual objectives used, the research approaches, and the rules of conduct are all determined jointly by the community and the researcher. The actual research process has both a science and an IK component, and there is provision

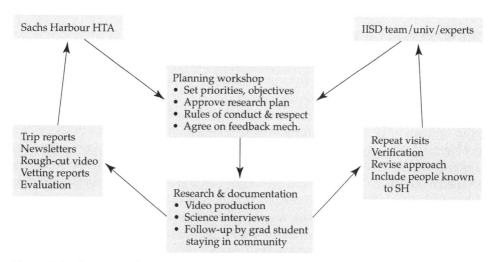

Figure 7.2   Structure of partnership with the Inuvialuit people of Sachs Harbour, in the Canadian Western Arctic.

for the two systems of knowledge to inform one another and to learn from one another. From the research process, there is feedback to the community in the form of provisional results; there is feedback to the research team in the form of revised approaches and verification.

The results of the project are shared as previously agreed upon, and the overall results are deposited with the community using media such as videos, poster boards, and photographic archives. Regarding the published work, the community retains control by indicating when certain kinds of knowledge, such as burial grounds, sacred areas, and other culturally sensitive information, should be left out. Community people receive credit for their knowledge and retain control over it. Legitimacy and authority of knowledge are not restricted to the researcher and the institution of the researcher but include the local people and their institutions as well.

Such a partnership approach satisfies the requirements for recognizing intellectual property rights of knowledge holders and helps the community retain the potential benefits of recording the information. As well, the approach has scholarly significance in at least two ways. First, it deals positively with the human ecological dilemma posed by Ingold (2000:3); that is, "human beings must simultaneously be constituted both as organisms within systems of ecological relations, and as persons within systems of social relations." In other words, the approach helps address both social systems and ecological systems, and bridges the divide. Second, by posing a fundamental challenge to the "expert-knows-best "convention of positivistic science, it shows how research can be created through processes of co-production in which scholars and stakeholders interact to define important questions, relevant evidence, and convincing forms of argument (Kates et al. 2001).

Implications for the Ecosystem Approach

The ecosystem approach as described in this volume is a primary tool to ask questions that include humans in the notion of an ecosystem. Much of the old environmental literature dealt with humans as merely "stressors." A more useful approach for the ecosystem-based management of the twenty-first century is to deal with humans as part of a complex adaptive system that includes both social and ecological subsystems (Berkes and Folke 1998). For practical purposes, all social systems depend on ecological life-support processes, and all ecological systems are shaped or affected by human activities. Thus cultural and social processes provide a necessary integrating context for action in the use of the ecosystem approach.

One of these action items for the ecosystem scientist is to expand the knowledge base to include a wider range of information sources than available from conventional science. Traditional ecological and local knowledge are relevant in this regard and provide some surprising insights that are consistent with current ecological thinking (Berkes et al. 2000). However, it would be a mistake to think that only traditional societies can provide information and insights for the ecosystem approach. Any group of people who dwell in an ecosystem have potential contributions to make.

A case in point is the local fishing association in the contemporary Swedish rural community Arvika studied by Olsson and Folke (2001). A mix of science and local knowledge is used to manage the crayfish and to deal with a series of environmental problems (acidification, fungal disease, overharvesting) that impact the crayfish resource. The characteristics of management at Arvika (monitoring at a variety of scales, use of flexible approaches, intervention based on ecological feedbacks) are precisely those that are needed to sustain ecological systems. The Arvika case, however, is based on one way of knowing, that is, Western science. In dealing with indigenous groups or other ethnoculturally diverse groups with their own ways of knowing and relating to the environment, the task of the ecosystem manager becomes more complicated and no doubt more difficult.

Part of the difficulty faced by the ecosystem scientist and manager lies in breaking out of the paradigm of "expert-knows-best" science, and part of it lies in the unfamiliarity of scientists and managers with different knowledge systems and different forms of cultural organization and institutions. Not many ecosystem managers have given thought to the dynamic processes by which organizations, institutions, and knowledge change. Science education curricula (including that of ecology) are not known for their strengths in epistemology and cross-cultural studies. Most ecosystem managers who deal with indigenous or non-Western cultures receive no training on these matters; they learn their craft on the job. It is only in recent years that traditional ecological knowledge is being discussed and experiences analyzed for learning, the beginnings of a new subfield (Berkes 2008).

As IK research has come to be taken more seriously in professional circles [witness the TEK special issue in *Ecological Applications* edited by Ford and Martinez

(2000)], many indigenous groups have also realized that "knowledge is power" and have become more guarded about their knowledge and more cautious about the risk of being stripped of the authority over the knowledge that they hold. Hence it has become imperative to design IK research projects that are participatory in nature, with the community becoming a partner in a cooperative process of knowledge creation rather than merely being the object of research.

Such partnerships require agreement on how to make sense of IK findings. IK is contextual knowledge; all systems of knowledge have their own cultural context and assumptions. The researcher comes into an IK project with his/her background knowledge and worldview—usually that of Western science or social science. Thus the model of participatory research has to address the dilemma of how to combine two different knowledge systems or to ponder the question of whether the two can be combined at all. We would argue that two different knowledge systems cannot be melded or synthesized. They can be combined only in the sense of dialogue, mutual learning, and sharing across systems of knowledge, as in the two cases we have summarized.

The intersection of post-normal and post-colonial science explored here leads us into new territory that is not clearly charted (Davidson-Hunt and O'Flaherty 2007). Perhaps ecosystem science and management can chart a course that considers the cultural processes that allow for sharing across knowledge systems. We offer these concluding thoughts on the ontological and epistemological shifts that we have identified as problematic. Two ontological foundations that we have inherited as "Western" scientists impede our ability to share across knowledge systems. First, meaningful dialogues with IK holders will continue to be difficult if we cling to dualistic assumptions such as the separation of "nature" and "culture" (Bateson 1972; Ingold 2000). Second, the assumption that science is "value free" is problematic. All knowledge is embedded in, and emerges from, interwoven forms (i.e., institutions, organizations) and processes (life experiences, scientific research) that are cultural, political, social, and ecological in nature. These ontological shifts would allow for a position of respect to be built for "other" ways of knowing the environment.

A more helpful epistemology for sharing across knowledge systems might start with the idea that knowledge is a process of knowing that can emerge from pluralistic and cooperative efforts to understand ecosystems and regions (Reid et al. 2006). Operationalizing this idea calls for place-specific research that brings together people with different kinds of knowledge into a process of dialogue in which the legitimacy and authority of different ways of knowing are mutually respected. The IK holder and the scientist can be brought into a process of coproduction of knowledge in which the research questions and methods may be developed and the results evaluated cooperatively. This is the essence of civic science or sustainability science in the terminology of Kates et al. (2001). The epistemology of ecosystem science and management should be ecological, but with a focus on research processes that allow relationships, networks, and trust to be built out of long-term, place-specific, and cooperative research processes. Out of this process of learning from each other about the ecology of a place and the relationship between that place and the larger systems in which it is embedded, appropriate institutions for ecosystem management might emerge.

References

Anderson, E. N. 1996. *Ecologies of the Heart: Emotion, Belief, and the Environment.* New York: Oxford University Press.

Balick, M. J. and P. A. Cox. 1996. *Plants, People, and Culture: The Science of Ethnobotany.* Cambridge, UK: Cambridge University Press.

Bateson, G. 1972. *Steps to an Ecology of Mind.* New York: Ballantine.

Berkes, F. 2008. *Sacred Ecology.* 2nd ed. New York: Routledge.

Berkes F. and C. Folke (eds.). 1998. *Linking Social and Ecological Systems.* Cambridge, UK: Cambridge University Press.

Berkes, F., K. Kislalioglu, C. Folke and M. Gadgil. 1998. Exploring the basic ecological unit: Ecosystem-like concepts in traditional societies. *Ecosystems* 1:409–415.

Berkes, F., J. Colding and C. Folke. 2000. Rediscovery of traditional ecological knowledge as adaptive management. *Ecological Applications* 10:1251–1262.

Blann, K., S. Light and J. A. Musumeci. 2003. Facing the adaptive challenge: Practitioners' insights from negotiating resource crises in Minnesota. In *Navigating Social-Ecological Systems*, eds. F. Berkes et al., 210–240. Cambridge, UK: Cambridge University Press.

Bradshaw, G. A. and M. Bekoff. 2001. Ecology and social responsibility: The re-embodiment of science. *Trends in Ecology and Evolution* 16:460–465.

Callicott, J. B. 1994. *Earth's Insights: A Survey of Ecological Ethics from the Mediterranean Basin to the Australian Outback.* Berkeley: University of California Press.

Capra, F. 1982. *The Turning Point.* New York: Simon and Schuster.

Clement, D. 1998. The historical foundations of ethnobiology. *Journal of Ethnobiology* 18:161–187.

Conklin, H. C. 1957. *Hanunoo Agriculture: A Report on an Integral System of Shifting Cultivation in the Philippines.* Rome: Food and Agriculture Organization of the United Nations.

Cotton, C. M. 1996. *Ethnobotany: Principles and Applications.* New York: Wiley.

Davidson-Hunt, I. and R. M. O'Flaherty. 2007. Researchers, indigenous peoples and place-based learning communities. *Society and Natural Resources* 20:291-305.

Davidson-Hunt, I. J. 2003. Journeys, plants and dreams: Adaptive learning and social-ecological resilience. Ph.D. thesis. Winnipeg, Manitoba, Canada: Natural Resources Institute.

Davidson-Hunt, I. J. and F. Berkes. 2003. Environment and society through the lens of resilience. In *Navigating Social-Ecological Systems*, eds. F. Berkes et al., 53–82. Cambridge, UK: Cambridge University Press.

Davidson-Hunt, I., L. C. Duchesne and J. C. Zasada (eds.). 2001. *Forest Communities in the Third Millennium: Linking Research, Business and Policy Toward a Sustainable Non-Timber Forest Product Sector.* St. Paul, Minn.: U.S. Department of Agriculture, Forest Service.

Feyerabend, P. 1987. *Farewell to Reason.* London, UK: Verso.

Flannery, K. V., J. Marcus and R. G. Reynolds. 1989. *The Flocks of the Wamani. A Study of Llama Herders on the Punas of Ayacucho, Peru.* San Diego, Calif.: Academic.

Ford, J. and D. Martinez (eds.). 2000. Invited feature: Traditional ecological knowledge, ecosystem science and environmental management. *Ecological Applications* 10(5):1249–1340.

Ford, N. 2000. Communicating climate change from the perspective of local people: A case study from Arctic Canada. *Journal of Development Communication* 1(11):93–108.

Glacken, C. 1967. *Traces on the Rhodian Shore: Nature and Culture in Western Thought from Ancient Times to the End of the Eighteenth Century.* Berkeley: University of California Press.

Gunderson, L. H. and C. S. Holling (eds.). 2002. *Panarchy. Understanding Transformations in Human and Natural Systems.* Washington, D. C.: Island Press.

Gunderson, L. H., C. S. Holling and S. S. Light (eds.). 1995. *Barriers and Bridges to the Renewal of Ecosystems and Institutions.* New York: Columbia University Press.

Hviding, E. 1996. *Guardians of Marovo Lagoon: Practice, Place and Politics in Maritime Melanesia.* Honolulu: University of Hawaii Press.

Ingold, T. 2000. *The Perception of the Environment: Essays on Livelihood, Dwelling and Skill.* New York: Routledge.

Kates, R. W., et al. 2001. Sustainability science. *Science* 292:641–642.

Lansing, J. S. 1993. *Priests and Programmers.* Princeton, N. J.: Princeton University Press.

Lansing, J. S., J. N. Kremer and B. B. Smuts. 1998. System-dependent selection, ecological feedback and the emergence of functional structure in ecosystems. *Journal of Theoretical Biology* 192:377–391.

Latour, B. 1986. *Laboratory Life: The Construction of Scientific Facts.* Princeton, N. J.: Princeton University Press.

Leopold, A. 1949. *A Sand County Almanac.* Oxford, UK: Oxford University Press.

Levin, S. 2000. *Fragile Dominion: Complexity and the Commons.* Reading, Mass.: Perseus.

Martin, G. J. 1995. *Ethnobotany: A Methods Manual.* London, UK: Chapman and Hill.

Merchant, C. 1980. *The Death of Nature: Women, Ecology and the Scientific Revolution.* San Francisco, Calif.: Harper & Row.

Olsson, P. and C. Folke. 2001. Local ecological knowledge and institutional dynamics for ecosystem management: A study of Lake Racken watershed, Sweden. *Ecosystems* 4:85–104.

Posey, D. A. and G. Dutfield. 1996. *Beyond Intellectual Property: Towards Traditional Resource Rights for Indigenous Peoples and Local Communities.* Ottawa, Canada: International Development Research Centre.

Rappaport, R. A. 1984. *Pigs for the Ancestors: Ritual in the Ecology of a New Guinea People* (2nd ed.). New Haven, Conn.: Yale University Press.

Reichel-Dolmatoff, G. 1976. Cosmology as ecological analysis: A view from the rain forest. *Man* 11:307–318.

Riedlinger, D. and F. Berkes. 2001. Contributions of traditional knowledge to understanding climate change in the Canadian Arctic. *Polar Record* 37:315–328.

Reid, W.V., F. Berkes, T. Wilbanks and D. Capistrano (eds.). 2006. *Bridging Scales and Knowledge Systems: Linking Global Science and Local Knowledge in Assessments.* Washington DC: Millennium Ecosystem Assessment and Island Press. [online] URL: http:// www.millenniumassessment.org/en/Bridging.aspx

Salmón, E. 2000. Kincentric ecology: Indigenous perceptions of the human-nature relationship. *Ecological Applications* 10(5):1327–1332.

Scott, C. 1996. Science for the West, Myth for the rest? In *Naked Science*, ed. L. Nader, 69–86. London, UK: Routledge.

Wilkins, D. 1993. Linguistic evidence in support of a holistic approach to traditional ecological knowledge. In *Traditional Ecological Knowledge: Wisdom for Sustainable Development*, eds. N. M. Williams and G. Baines, 71–93. Canberra: Centre for Resource and Environmental Studies, Australian National University.

Williams, N. M. and E. S. Hunn (ed.). 1982. *Resource Managers: North American and Australian Hunter-Gatherers.* Washington, D. C.: American Association for the Advancement of Science.

8

A Family of Origin for an Ecosystem Approach to Managing for Sustainability

Martin Bunch, Dan McCarthy, and David Waltner-Toews

A Family of Origin

The ecosystem approach that we articulate in Part III draws on several schools of thought and practice in both the natural and social sciences. The intent of this chapter is not to provide an in-depth examination of these various schools of scholarship; that task would require several libraries. Our intent here is to acknowledge the intellectual parentage of the ecosystem approach. The roots of this approach in ecology and environmental management have been well-reviewed by Bocking (2004). In this chapter, we review some of the origins of our approach in the social science and management literature. In psychosocial terms, we might refer to these as an extended "family of origin." We can learn a great deal about present and future behaviors of the children by studying the parents and relatives in an extended family. While we cannot present all the origins, we present brief biographies of some of the extended family below. These are not intended to be critical reviews but rather an acknowledgement of some of the key ideas that have influenced the methods that have emerged in our work. Our review includes soft systems methodology, social systems design, interactive planning, Beer's viable systems model and team syntegrity, participatory action research, and adaptive management.

Soft Systems Methodology

Soft systems methodology (SSM) is a methodology for making sense of real-world "messy" or "problematic" situations. It is a qualitative action-research approach that was developed in the early 1970s by Peter Checkland and his colleagues at Lancaster University in response to the failure of so-called hard systems approaches to deal effectively with organizational and institutional change (Flood and Carson 1993). Such "soft" problems are usefully discussed as ill-defined "problematic situations" in which the same problem may be perceived differently by various people (Flood and Carson 1993).

An important idea in SSM is that human beings form intentions according to how they interpret the situations they experience, and they act on these intentions. We are continually taking "purposeful action" related to experiences of situations and the knowledge generated by our experiences. Experience-based knowledge informs purposeful action, which creates new experience of the world, yielding further experienced-based knowledge. SSM is a methodology for formally operating this learning cycle (Checkland and Scholes 1990a).

SSM deals with human activity systems—sets of interrelated human activities acting together as a whole for a particular purpose. Woodburn (1991:30) indicates that a useful way to think of a human activity system is as "the expression of a level of order (or purpose) higher than that contained in its component parts." Thus a house as a dwelling place, viewed as a system, is not merely a physical structure with people or a family living and performing daily functions and having particular interactions. It has meaning attributed to it by actors or observers of the system—it is a home.

Early expressions of the approach (e.g., Checkland 1976, 1981) outlined a seven-stage methodology that guided practitioners through the following: (1) identification of a problematic situation, (2) expression of the situation in nonsystems terms, (3) initiation of systems thinking through the development of root definitions of purposeful human activity, each based on an explicitly declared world view, (4) development of conceptual models of relevant systems based on the root definitions developed in step 3, (5) comparison of conceptual models with a model of a formal system and with the expression of the real-world situation developed in step 2, (6) identification of change that would improve the situation, and (7) taking action in the real world.

Action taken in the real world will change a situation and result in new experience of the world, requiring new expression of the problem, and so on. Thus the process should be iterative and ongoing, such that it actively operates a learning cycle.

Tools associated with various stages in the methodology include "rich pictures" to express problematic situations, a CATWOE (customer, actor, transformation, *Weltanschauung*, owner, environmental constraints) analysis, to construct root definitions, sequencing of verbs and action statements to build conceptual models, and comparison by formal questioning. Descriptions of such tools and their use can found in Checkland (1979a,b, 1981), Checkland et al. (1990), and Checkland and Scholes (1990a,b). Some of them are applied in the study by Bunch described in chapter 10. Use of SSM in the early 1970s and 1980s often consisted of applying these tools in sequence. In fact, Naughton (1981) developed a set of constitutive rules that outlined the use of such tools within the seven-stage process in order to be said to be doing SSM. This has become known as "mode 1" of SSM (Checkland and Scholes 1990a).

More recent development and applications of SSM ("mode 2") have expanded this methodology to be less prescriptive and more flexible (Atkinson 1986; Kreher 1994). In mode 2, SSM applications, the seven stages, and attendant methods are contingent upon the *context*, the *use*, and the *users* of that methodology (Atkinson 1986). Mode 2 applications of SSM are characterized by three general phases (Woodburn 1991): (1) building a rich picture of a problematic situation, (2) developing models of human activity systems that are relevant to the situation, and (3) using the models to stimulate thinking about organizational change. Figure 8.1 presents a mode 2 conception of SSM and its main principles.

Rich pictures portray actors and elements in a situation and indicate relationships among them. From this, important themes may be identified and further studied. It has been found by SSM practitioners that diagrams are more effective than linear prose in presenting relationships and that pictorial representation of multiple interacting relationships promotes holistic thinking (Checkland 1999).

The second phase of SSM involves selection of perspectives from which to construct models of relevant human activity systems. Typically, this begins with development of a set of "root definitions." A root definition is a "core description of purposeful activity taken from a specific point of view" (Flood and Carson

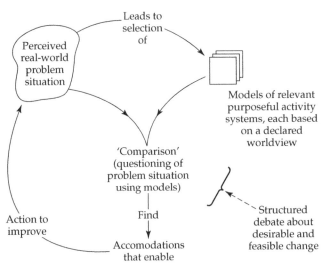

Principles
- Real world: a complexity of relationships
- Relationships are explored via models of purposeful activity based on explicit worldviews
- Inquiry is structured by questioning a perceived situation using the models as a source of questions
- 'Action to improve' is based on finding accomodations (versions of the situation which conflicting interests can live with)
- Inquiry is in principle never-ending; it is best conducted with a wide range of interested parties; give the process away to people in the situation

Figure 8.1   The inquiring-learning cycle, "mode 2," of soft systems methodology (after Checkland 1999).

1993:111). Root definitions may be constructed to reflect the components of the mnemonic CATWOE—customers (who are victims or beneficiaries of the system?), actors (who performs the activities?), transformation (what input is transformed into what output?), Weltanschauung (what view of the world makes this system meaningful?), owner (who could abolish this system?), and environmental constraints (what in its environment does this system take as given?). Root definitions act as core expressions upon which to build conceptual models of purposeful human activity that are relevant to debate about change in the real world (Checkland 1999). A conceptual model is a description (not prescription) of "what the system *must do* in order *to be* the system named in the root definition" (Flood and Carson 1993:114).

Comparison of conceptual models to the "real-world situation" is intended to generate debate about desirable and feasible change. Such comparison is at the heart of SSM and is central to structured systems thinking (Ledington and Ledington 1999). Informal discussion, formal questioning, and scenario writing based on operating models and attempting to model the real world into the structure provided by conceptual models are all ways of generating debate. The most common practice, formal questioning, may proceed, for example, by asking the following of each activity in the conceptual model: Does it exist in the real situation? How is it done? How is it judged? (Checkland and Scholes 1990a).

Regardless of how comparison of models with the real world is undertaken, the aim is not to improve the models but to "find an accommodation between different interests in the situation, an accommodation which can be argued to constitute an improvement of the initial problem situation" (Checkland and Scholes 1990a:44).

A primary strength of SSM is its recognition of the existence and validity of multiple perspectives on any situation. Development of conceptual models based on various perspectives results in different idealized models of the same situation. The methodology is best undertaken with as wide a range of pertinent stakeholders as possible. Comparison of the models with the existing situation, and with each other, generates debate about systemically desirable and culturally feasible change (Woodburn 1991).

SSM is also important in the way that it uses systems concepts. "Hard systems" approaches typically generate visions of the future as blueprints. They set system objectives and develop comprehensive plans to achieve them. SSM, in contrast, is about process. To the soft systems thinker, there are not systems "out there" in the real world to be engineered. Instead, there is a complex set of relationships of which one might make some sense by applying systems concepts. The process of attempting to do so is participatory, and the process itself is given away to the participants (Checkland 1999). Ownership of the process invests stakeholders' interest in the accommodations for change in the real world that it produces.

Ecosystem approach practitioners have recognized the potential usefulness of SSM to inform the ecosystem approach in general (Wilson and Morren 1990; Allen and Hoekstra 1992; Allen et al. 1994; Brown and MacLeod 1996) and a few have attempted to apply the methodology in environmental and resource management situations (e.g., McLeod and Van Beek 1993; Ison 1993; Whittaker 1993; Bunch

2001, 2003; Bunch and Dudycha 2004). A more pervasive influence of SSM is evidenced in the way systems concepts have been used to make sense of complicated and complex situations, in the recognition of multiple interests, the legitimacy of multiple perspectives, and the operation of ecosystem approaches as ongoing programs that manage ongoing and ever-changing situations.

Although SSM has had a large influence over how sustainable development is practiced, it also comes with some serious faults, which is at least one reason why environmental managers have not simply adopted it without reservation or adaptation. The most serious shortcoming is that the notion of what is "real" is left implied, with no guidance as to how one might assess that reality and what to do if it contradicts the agreed upon consensus of the actors. The social construction of reality by a set of peers is all very well, but what if they are running out of water or if people are dying of a particular disease? This fault may have arisen in part because Checkland and his associates were responding to the inability of hard, engineering systems approaches to effectively incorporate social realities and the ways in which social organizations interact with biophysical realities. In looking for ways to adapt SSM to questions of sustainable development and the ecosystem approach, many of the authors in this book have sought to find ways to better link SSM's organizational insights with methods to ground the rich pictures in some observable external reality.

Social Systems Design

C. West Churchman (1979), one of the founders of operations research (OR) originally developed in the World War II era, in the 1960s and 1970s became disillusioned with the objective and expert-oriented "hard" systems approaches of OR. He set out to develop a more "interdisciplinary, holistic, and experimental" approach to address complex problems in social systems. His social systems design approach was premised on the following four aphorisms:

1. The systems approach begins when first you see the world through the eyes of another.
2. The systems approach goes on to discovering that every worldview is terribly restricted.
3. There are no experts in the systems approach.
4. The systems approach is not a bad idea.

On the basis of the work of such philosophers as Hegel and Kant, through the social systems design approach, Churchman asserted that every view of the world is based on a set of assumptions (after Kant) and that there are as many of these worldviews as individuals (after Hegel). As such, there can be no objective, or omniscient "super-observer." Thus, subjectivity is an inherent and integral component of the systems view of the world. In order to gain a holistic view of a social system, Churchman argues that one must look at the system from as many viewpoints as possible. By bringing together multiple subjectivities one approaches a more "objective" holistic view of a system. Following Hegel, Churchman advocates a dialectical approach to integrating such subjectivities and indicates that it is the role of the systems practitioner to ensure stakeholders involved in such a process become aware of the nature of their worldview and its associated limitations. In order to do this, Churchman (1970) developed a process of dialectical debate based on the themes of thesis, antithesis, and synthesis:

1. Thesis: understand decision makers' proposals and understand the worldview that underlies these proposals.
2. Antithesis: develop a polarized alternative (a "deadly enemy") and then develop proposals based on it.
3. Synthesis: evaluate data and information based on both worldviews and arrive at a more rich appreciation of the system.

While Churchman acknowledges that a truly holistic view of a system is not feasible, he argues that it is the role of systems practitioners to maintain a "heroic" holistic vigilance in their work.

If SSM's contribution to systemic interventions comes from management and offers demonstrably useful managerial techniques, then Churchman's work emphasizes the philosophical assumptions and roots of ecosystemic scholarly investigations, and the degree to which many scientists rarely question their own assumptions. This raises the tentative nature of the knowledge generated through this engagement. Churchman's dialectical approach does not appear to answer some of the critical issues related to how one assesses quality of information, what the rules of engagement are, and if some kinds of information are privileged over others.

Interactive Planning

Russel L. Ackoff shared with Churchman the view that subjectivity was indeed a characteristic of both science and social science, despite scientists' attempts to be "objective." That is, purposeful behavior cannot be value-free. Ackoff (1999:312) noted that,

> objectivity cannot be approximated by an individual investigator or decision-maker; it can be approached only by groups of individuals with diverse values. It is a property that cannot be approximated by individual scientists but can be by science taken as a system.

Ackoff was another forefather of operations research in the United States. He, too, realized that the "hard" science approach, with its emphasis on quantification and elaborate mathematical models, being adopted by the OR researchers of the day, was, in fact, modeling the complexity of social issues right out of the research. The emphasis on the collective expertise of science and the critique of privileging mathematical modeling by some scholars have been important in the development of a practice of post-normal science for environmental management and public health.

Ackoff's methodological outlet for this view of social systems science was the interactive planning approach. On the basis of the three principles of participation, continuity, and holism, Ackoff's interactive planning approach represents another precursor to the self-organizing, holarchic, open (SOHO) systems framework. Ackoff's participative principle states that the *process* of planning should be seen to be more important than the resulting plan itself. Central to the approach is the integration or synthesis of the various perspectives of those affected (and so, those involved). Ackoff's second principle, continuity, can be likened to the notions that underlie Holling's adaptive management (below). Essentially, continuity highlights the need for a plan to continually change and evolve as the context changes and knowledge about it improves. Finally, Ackoff's holistic principle parallels Churchman's notion of holism—interactive planning requires understanding and integration of as many components and scales in a system as possible.

Elements of Ackoff's five-phase interactive planning approach bear some similarity to the SOHO and adaptive methodology for ecosystem sustainability and health (AMESH) approaches. Paralleling the flow of AMESH or the SOHO diamond diagram, Ackoff (1999) describes five stages in interactive planning: (1) formulating the mess, (2) ends planning, (3) means planning, (4) resource planning, and (5) design of implementation and control.

Formulating the mess is essentially a system identification. Ends planning can be related to a visioning or goal setting exercise, and the final three stages deal with issues of management, monitoring, and governance.

Beer's VSM and Team Syntegrity

Stafford Beer, another seminal systems thinker and a pioneer in the field of organizational cybernetics also developed a model and approach that can be seen to be a precursor to the SOHO or AMESH protocol. On the basis of his viable systems model (VSM) (Beer 1972, 1979, 1985) Beer's "team syntegrity" (Beer 1995) is a process designed to promote nonhierarchical, participative, and effective decision making in organizational, nonorganizational, or multiorganizational contexts. The term itself is the result of a fusion of the terms Tensegrity (Buckminster Fuller's notion of tensile integrity) and Synergy. VSM is an organizational model that aids the definition of characteristics of (not surprisingly) a viable system. For Beer a "viable" system is one that is capable of adapting to changes in its environment even if these changes are unforeseen. Attempting to account for this kind of uncertainty in an organizational planning process, his team syntegrity approach is an attempt to synthesize the various perspectives of those affected in decision making through a protocol that is intended to be open, self-organizing, and nonhierarchical.

The process employs a metaphor based on Buckminster Fuller's geodesic dome shape and the tensions that hold it together. For Beer, ensuring that no individual dominates this type of process is key, and the metaphor of the dome is useful in this regard. The protocol has three stages, the "Problem Jostle," the "Topic Auction," and the "Outcome Resolve." The Problem Jostle allows those involved to identify the problem by identifying a "statement of interest." These are synthesized into 12 "Composite Statements of Importance" that then provide foci or topics for further critical discourse in the Topic Auction. Finally, through the Outcome Resolve several "final statements of importance" are developed. Beer's team syntegrity, like participatory action research, social systems design, and interactive planning, is in many ways a precursor to the mutliperspective, holistic, integrative, systems-oriented thinking that was key to the evolution of the SOHO and AMESH models discussed later in this book.

Participatory Action Research

Participatory action research (PAR) reflects a set of theoretical and practical developments which, in many ways, parallel those of SSM. In the late 1940s, Kurt Lewin, an experimental psychologist and one of the fathers of social psychology, advocated a "theory in action" approach that emphasized a more holistic, humanistic, and community-based perspective. Recognizing some of the limitations of analytical reductionist approaches when applied to complex social and psychological phenomena, Lewin highlighted the need to view humans in their environmental and community contexts. Lewin's (1951) field theory is premised on the notion that individuals are always functioning at the nexus of a set of interacting fields in which the individual and environment are interdependent parts of the whole. In Lewin's opinion, removing the subject of study from its context artificially fractures this interdependent whole.

Lewin adopted an action-based research approach. Rapport (1970:499) defines this Action Research approach as relating "both to the practical concerns in an immediate problematic situation and to the goals of social science by joint collaboration within a mutually agreed ethical framework." Lewin's work led to several related streams of action research, PAR (Whyte 1991), action science (Argyris and Schon 1974, 1978; Schon 1983), and Co-operative Inquiry (Reason 1988, 1994; Reason and Heron 1995; Heron 1996), among others. We focus here on PAR as it has been widely adopted in many fields of study, because of its importance in sustainable development programs in countries of the Global South and because of its explicit use to inform the ecosystem approach in some of our current work (e.g., Bunch et al. 2005).

According to Whyte (1991:20) "in participatory action research, some of the people in the organization or community under study participate actively with the professional researcher throughout the research process from the initial design to the final presentation of results and discussion of their action implications." This type of research contrasts with more conventional pure research approaches in which individuals or communities are seen as passive subjects. Instead, PAR "involves a sharing of knowledge, power, and skills in the research process that changes and builds as participants define research questions, shape the process, and interpret findings in collaboration with researchers" (Rosenberg 2003:61).

Beyond this version of PAR, that is rooted in industrial-type applications to organizational change, PAR is also associated with social activist work

in the Global South (Fals-Borda 1992:13) that is "biased in favour of dominated, exploited, poor or otherwise ignored women and men and groups"
(Hall 1992:16).

PAR's evolution out of frustration with the notion of narrowly defined (single
criteria) university-trained expert as authority, which often resulted in serious development failures, has shaped the process of its application. One especially useful
characteristic of the approach and one that aids in our attempts to escape narrow
disciplinary approaches is PAR's problem orientation (Whyte 1991:40):

> Instead of beginning in the conventional fashion with a review of literature,
> the specification of hypotheses, and the finding of a target organization to
> test out our design, we start by discovering the problems existing in the
> organization. Only as we work with members of the organization, diagnos-
> ing those problems, do we draw upon the research literature as well as our
> own past experience.

PAR has much in common with the ecosystem approach that we describe
in this book, especially in that the role of the "expert" in PAR is less a disciplinary scientific role but rather a role as a contributor to the collective expertise and
knowledge base and facilitator in the mobilization of local knowledge and of building capacity in communities and organizations to address their problems. On the
other hand, while PAR emphasizes stakeholder engagement, it does not necessarily
require ecosystemic—or even systemic—thinking. In our experience, PAR has to be
explicitly coupled with an ecosystemic perspective; without this, PAR is sometimes
reduced simply to public mobilization around issues of concern.

Adaptive Management

Adaptive environmental assessment and management (AEAM, also known as
adaptive management) is an approach to resource and environmental management that deals explicitly with uncertainty (Holling 1978; Walters 1986; Lee 1993;
Gunderson et al. 1995a). AEAM mobilizes best available knowledge about a situation by bringing together scientific experts, planners, and policy makers in a series of workshops to design management interventions so as to generate knowledge and facilitate learning. The approach is anticipatory (solutions are developed
based on predictable future events) and flexible (accommodating changes in goals,
revised predictions, and new evidence). It is a continuous process of learning.

Practitioners of AEAM know that they intervene in systems that are both complex and continually evolving and that they will never have complete knowledge
of the system. Because of this, AEAM approaches the management of these systems as an ongoing learning process. By designing interventions as experiments,
knowledge about systems is maximized and learning can occur from unexpected
events. This makes the approach adaptive. New information is generated from
the experience of intervening in and monitoring the response of the system. This
knowledge informs further management of the system, which generates new

information. Ongoing adaptive management of the system progressively reduces uncertainty about it.

The adaptive management approach can be traced to the Gulf Island Recreation Land Simulation study in 1968, which attempted to "bridge gaps among scientific disciplines, technical experts, and policy designers" (Gunderson et al. 1995b:490). Later, in the mid-1970s, an interdisciplinary team of biologists and systems analysts led by Holling, a Canadian ecologist, developed the basic concepts of AEAM (Lee 1993). Since that time many resource and environmental management programs have adopted the approach. Notable among these are adaptive management of water quality and living resources habitat in the Chesapeake Bay and its catchment basin (Hennessey 1994; Constanza and Greer 1995), the Great Lakes Program (Imperial et al. 1993; Francis and Reiger 1995), and adaptive management of salmon and power generation in the Columbia River Basin (Lee 1989; Lee and Lawrence 1986; Volkman and McConnaha 1993).

AEAM is an interdisciplinary approach that involves, for example, policy people, managers, and scientists from various backgrounds. If PAR is necessarily participatory, but not necessarily systemic, AEAM may be systemic without being participatory. However, although theoretically AEAM can be done by self-appointed experts, its success as a management program often depends on collaboration and communication among decision makers, scientists, managers, and the public. In fact, communication is so important that Holling (1978) indicates that it deserves the dedication of at least as much effort as analysis.

Analytic techniques employed in AEAM depend on the nature of the problem being addressed. However, commonly associated with adaptive management is the use of a series of workshops to bring together key experts and stakeholders, as well as the collaborative development and use of dynamic system (simulation) models.

The process of adaptive management involves two groups of people: (1) core analysts and support staff and (2) key cooperators in the management project (Holling 1978). The core group coordinates the project, integrates information using systems techniques such as simulation modelling, and brings together the second group in a series of workshops that are central to the approach. Workshops address such tasks as problem analysis, determination of goals and objectives of the management program, development of a framework for a dynamic system model, and allocation of tasks for subgroups (Holling 1978:51). Workshops are also an important means to facilitate communication among key actors and in the creation of an atmosphere conducive to the generation of creative management alternatives (Environmental and Social Systems Analysts 1982:2, 28). Following the initial workshop and between secondary workshops, the core group consolidates information by constructing and testing simulation models, evaluating management policies, and collecting data (Holling 1978:56). Workshops that involve the public in scenario analysis of management scenarios may be organized to facilitate public participation and communication (Environmental and Social Systems Analysts 1982:28).

Conclusions and Synthesis

Most of the approaches we have reviewed in this chapter are associated with the general field of management sciences. These, and approaches like them, are intended to manage human activity rather than to manage environments or ecosystems. Managing human activity is something that traditional approaches to managing for sustainability, which stem out of the biological and physical scientific disciplines, do not deal with well. It should not be surprising to the reader that we study and adapt these methods to inform our application of the ecosystem approach, complementing but not displacing the biophysical studies of energy, information, and nutrient flows. After all, as our late friend and mentor James Kay would say, it is not the biophysical environment per se, but our interactions with it that we need to manage.

It is quite apparent that the approaches described in this chapter do, indeed, comprise a family, with many characteristics in common. What should also be apparent is that the approaches we describe in the following chapters are themselves not "new" in any absolute sense but reflect a rethinking, or reorganization, of their origins and attempt to ground them in our best understanding of how social-ecological systems function. In so doing, beyond their systemic and holistic foundations, we draw upon lessons and techniques derived from approaches such as soft systems methodology, social systems design, interactive planning, the viable systems model and team syntegrity, participatory action research, and adaptive management that have to do with participatory and collaborative processes, adaptation, and the intentional operation of learning cycles. What we have attempted to do is to bring together the best notions of stakeholder engagement, which are emphasized by participatory research methods, with the latest theories of how best to frame our messy collective reality in terms of complex systems.

References

Ackoff, R. L. 1999. Ackoff's Best: His Classic Writings on Management. New York: Wiley.

Allen, T. F. H. and T. W. Hoekstra.1992. *Toward a Unified Ecology*. New York: Columbia University Press.

Allen, T. F. H., B. L. Bandurski and A. W. King. 1994. The ecosystem approach: Theory and ecosystem integrity. Initial Report. United States and Canada: Ecological Committee to the International Joint Commission's Great Lakes Science Advisory Board.

Argyris, C. and D. A. Schon. 1974. *Theory in Practice: Increasing Professional Effectiveness*. San Francisco, Calif.: Jossey-Bass.

Argyris, C. and D. A. Schon. 1978. *Organizational Learning: A Theory of Action Perspective*. Reading, Mass.: Addison-Wesley.

Atkinson, C. J. 1986. Towards a plurality of soft systems analysis. *Journal of Applied Systems Analysis* 13:19–31.

Beer, S. 1972. *Brain of the Firm*. London, UK: Allen Lane.

Beer, S. 1979. *The Heart of the Enterprise*. New York: Wiley.

Beer, S. 1985. *Diagnosing the System for Organizations*. New York: Wiley.

Beer, S. 1995. *Beyond Dispute: The Invention of Team Syntegrity*. New York: Wiley.

Bocking, S. 2004. *Nature's Experts: Science, Politics, and the Environment.* New Brunswick, N. J.: Rutgers University Press.

Brown, J. R. and N. D. MacLeod. 1996. Integrating ecology into natural resource management policy. *Environmental Management* 20(3):289–296.

Bunch, M. J. 2001. An adaptive ecosystem approach to rehabilitation and management of the Cooum River Environmental System in Chennai, India. Department of Geography Publication Series, No 54. Waterloo, Ontario, Canada: University of Waterloo.

Bunch, M. J. 2003. Soft systems methodology and the ecosystem approach: A system study of the Cooum River and environs in Chennai, India. *Environmental Management* 31(2):182–197.

Bunch, M. J. and D. J. Dudycha. 2004. Linking conceptual and simulation models of the Cooum River: Collaborative development of a GIS-based DSS for environmental management. *Computers, Environment and Urban Systems* 28(3):247–264.

Bunch, M. J., B. Franklin, D. Morley, T. Vasantha Kumaran and V. Madha Suresh. 2005. Research in turbulent environments: Slums in Chennai, India and the impact of the December 2004 tsunami on an EcoHealth project. *EcoHealth* 2(2):150–154.

Checkland, P. B. 1976. Towards a systems-based methodology for real-world problem solving. In *Systems Behaviour* (2nd ed.), eds. J. Beishon and G. Peters. London, pp. 51–77. UK: Open University.

Checkland, P. B. 1979a. Techniques in 'soft' systems practice Part 1: Systems diagrams —Some guidelines. *Journal of Applied Systems Analysis* 6:33–40.

Checkland, P. B. 1979b. Techniques in 'soft' systems practice Part 2: Building conceptual models. *Journal of Applied Systems Analysis* 6:33–40.

Checkland, P. B. 1981. *Systems Thinking, Systems Practice.* New York: Wiley.

Checkland, P. B. 1999. Soft systems methodology: A 30-year retrospective. In *Soft Systems Methodology in Action*, eds. Peter B. Checkland and Jim Scholes, pp. A1–A66. New York: Wiley.

Checkland, P. B. and J. Scholes. 1990a. *Soft Systems Methodology in Action.* New York: Wiley.

Checkland, P. B. and J. Scholes. 1990b. Techniques in soft systems practice Part 4: Conceptual model building. *Journal of Applied Systems Analysis* 17:39–43.

Checkland, P. B., P. Forbes and S. Martin. 1990. Techniques in 'soft' systems practice Part 3: Monitoring and control in models and in evaluation studies. *Journal of Applied Systems Analysis* 17:29–37.

Churchman, C. W. 1970. Operations research as a profession. *Management Science* 17:B37.

Churchman, C. W. 1979. *The Systems Approach.* New York: Dell.

Costanza, R. and J. Greer. 1995. The Chesapeake Bay and its watershed: A model for sustainable ecosystem management? In *Barriers and Bridges to the Renewal of Ecosystems and Institutions*, eds. L. H. Gunderson et al., pp. 169–213. New York: Columbia University Press.

Environmental and Social Systems Analysts. 1982. *Review and Evaluation of Adaptive Environmental Assessment and Management.* Vancouver: Environment Canada.

Fals-Borda, O. 1992. Evolution and convergence in participatory action-research. In *A World of Communities: Participatory Research Perspectives*, ed. James Frideres, pp. 14–19. Toronto, Canada: Captus University Publications.

Flood, R. L. and E. R. Carson. 1993. *Dealing with Complexity: An Introduction to the Theory and Application of Systems Science* (2nd ed.). New York: Plenum.

Francis, G. R. and H. A. Regier. 1995. Barriers and bridges to the restoration of the Great Lakes Basin Ecosystem. In *Barriers and Bridges to the Renewal of Ecosystems and Institutions*, eds. L. H. Gunderson et al., pp. 239–291. New York: Columbia University Press.

Gunderson, L. H., C. S. Holling and S. S. Light (eds). 1995a. *Barriers and Bridges to the Renewal of Ecosystems and Institutions.* New York: Columbia University Press.

Gunderson, L. H., C. S. Holling and S. S. Light (eds.). 1995b. Barriers broken and bridges built: A synthesis. In *Barriers and Bridges to the Renewal of Ecosystems and Institutions*, pp. 489–532. New York: Columbia University Press.

Hall, B. 1992. From Margis to Centre? The development and purpose of participatory research. *American Sociologist* Winter:15–28.

Hennessey, T. M. 1994. Governance and adaptive management for estuarine ecosystems: The case of Chesapeake Bay. *Coastal Management* 22:119–145.

Heron, J. 1996. Co-operative Inquiry: Research into the Human Condition. Newbury Park, Calif.: Sage.

Holling, C. S. (ed.). 1978. *Adaptive Environmental Assessment and Management.* Toronto, Canada: International Institute for Applied Systems Analysis.

Imperial, M. T., T. Hennessey and D. Robadue Jr. 1993. The evolution of adaptive management for estuarine ecosystems: The National Estuary Program and its precursors. *Ocean & Coastal Management* 20(2):147–180.

Ison, R. L. 1993. Soft systems: A non-computer view of decision support. In *Decision Support Systems for the Management of Grazing Lands, Man and the Biosphere Series*, vol. 11, eds. J. W. Stuth and B. G. Lyons, pp. 83–121. Carnforth, UK: Parthenon.

Kreher, H. 1994. Some recurring themes in using soft systems methodology. *Journal of the Operational Research Society* 45(11):1293–1303.

Ledington, P. and J. Ledington. 1999. Extending the process of comparison in soft systems methodology. *Journal of the Operational Research Society* 50(11):1149–1157.

Lee, Kai N. 1989. The Columbia River Basin: Experimenting with sustainability. *Environment* 31(6):6–11, 30–31.

Lee, Kai N. 1993. Compass and Gyroscope: Integrating Science and Politics for the Environment. Washington, D. C.: Island Press.

Lee, Kai N. and J. Lawrence. 1986. Restoration under the Northwest Power Act—Adaptive management: Learning from the Columbia River Basin Fish and Wildlife Program. *Environmental Law* 16(3):424–460.

Lewin, K. 1951. *Field Theory in Social Science; Selected Theoretical Papers*, ed. D. Cartwright. New York: Harper & Row.

MacLeod, N. D. and P. G. H. Van Beek. 1993. Rural research and development moving toward systemic approaches: The 'GLASS' Sustainable Grazing Project—A case study. In *Ethical Management of Science as a System*, pp. 536–545. Sydney, Australia: Society of the System Sciences, University of Western Sydney.

Naughton, J. 1981. Theory and practice in systems research. *Journal of Applied Systems Analysis* 8:61–70.

Rapport, A. 1970. Three dilemmas in action research. *Human Relations* 23:499.

Reason, P. 1988. Human Inquiry in Action: Developments in New Paradigm Research. Newbury Park, Calif.: Sage.

Reason, P. 1994. Three approaches to participative inquiry. In *Handbook of Qualitative Research*, eds. N. K. Denzin and Y. S. Lincoln, pp. 324–339. Newbury Park, Calif.: Sage.

Reason, P. and J. Heron. 1995. Co-operative inquiry. In *Rethinking Methods in Psychology*, eds. R. Harre et al., pp. 122–142. Newbury Park, Calif.: Sage.

Rosenberg, D. G. 2003. Education for a healthy future: Training trainers for primary prevention: Participatory action research and evaluation (PARE) project. In *Head, Heart and Hand: Partnerships for Women's Health in Canadian Environments, Volume 2*, ed. P. van Esterik, pp. 60–72. Toronto, Canada: National Network on Environments and Women's Health.

Schon, Donald A. 1983. *The Reflective Practitioner: How Professionals Think in Action.* New York: Basic Books.

Volkman, J. M. and W. E. McConnaha. 1993. Through a glass, darkly: Columbia River salmon, the Endangered Species Act, and adaptive management. *Environmental Law* 23(4):1249–1272.

Walters, C. 1986. *Adaptive Management of Renewable Resources*. New York: Macmillan.

Whittaker, D. A. 1993. Decision support systems and expert systems for range science. In *Decision Support Systems for the Management of Grazing Lands, Man and the Biosphere Series*, vol. 2, edited by J. W. Stuth and B. G. Lyons, pp. 69–81. Carnforth, UK: Parthenon.

Whyte, W. F. 1991. *Participatory Action Research*. Newbury Park, Calif.: Sage.

Wilson, K. K. and G. E. B. Morren Jr. 1990. *Systems Approaches for Improvement in Agriculture and Resource Management*. New York: McMillan.

Woodburn, I. 1991. The teaching of soft systems thinking. *Journal of Applied Systems Analysis* 18:29–37.

PART II

Case Studies
Learning by Doing

In this section, we introduce case studies from Africa, India, Canada, and Latin America that wrestled explicitly with the theoretical issues described in Part I and served to inform the further development of both the theory and practice of the ecosystem approach. In the case write-ups, some authors have emphasized social process, some more conventional scientific modeling issues, and some the interplay between the two. In part, this reflects the areas where the greatest challenges were faced and hence the best possibilities for learning—if not something new, then at least important. The case studies were not intended to be illustrations of a completed theoretical structure, nor, strictly speaking, were they tests of a hypothesis. Nevertheless, they did serve to test various aspects of the theory and to highlight weaknesses, strengths, and possible options for theoretical and practical progress.

9

Linking Hard and Soft Systems in Local Development

Reg Noble, Ricardo Ramirez, and Clive Lightfoot

Introduction

Environmental crises due to habitat degradation caused by human activity seem to be ever increasing as people seek to expand their control and exploitation of the Earth's natural resources. Although new concepts of social-ecological complexity and post-normal science are changing how we see the world, in many cases the default option still seems to be the pursuit of natural-science-based, narrowly focused technical solutions to problems seen in isolation from context rather than dealing with underlying social and cultural issues (Woodhill and Röling 1998).

In part, this reflects the disciplinary structure and outputs of scholarly inquiry over the past century (Adams 1992). Even where there is good information on ecosystems, social systems, and/or people-in-ecosystems, there has been a great reluctance for environmental scientists and their social counterparts to cross disciplinary boundaries, to share knowledge in order to address this human equation in environmental issues, and, ultimately, to acknowledge that social systems and biological systems are complementary "takes" on the same complex reality in which humans (as bio-psycho-social beings) are embedded (Chambers 1983; Adams 1992).

At the same time, it is becoming increasingly clear that the landscapes we live in are the visible expression of ". . . the struggles, compromises, and temporarily-settled relations of competing and co-operating social actors" (Cline-Cole 2000:109). Resolution of problems in those landscapes will require a deeper understanding of social structures in relation to the environment and the integration of scientific knowledge with cultural engagement and understanding (Reynolds and Busby 1996; Pretty 1998; Borrini-Feyerabend et al. 2000; Woodhill and Röling 1998; Röling and Jiggins 1998).

As Borrini-Feyerabend et al. (2000) explain ". . . in traditional societies, the units of natural resource management and units of social life tended to coincide." (p. 7) This is not the case in modern globalized society, facilitated by a mix of technology and socially constructed "markets," in which species, nutrients, and energy are moved at very rapid rates across ecological boundaries around the world. The disjunction between new, global networks and both evolutionarily and politically defined boundaries has hampered the development of place-based science.

While any effective ecosystemic approach requires us to integrate social and ecological perspectives in order to get a sense of the complex reality as a whole, premature closure into a kind of "fuzzy" holism may prevent effective learning. It is thus also useful to tease the system apart (systems analysis) and then identify points of intersection, trade-offs, and effective action (systems synthesis). There is no single correct way to do this analysis. However, strategies to promote sustainability have often been framed in terms of dual social-ecological (Berkes and Folke 1998) or "hard" and "soft" systems (Checkland 1981; Checkland and Scholes 1990). These approaches recognize that as there is such a diversity of people-environment systems, there can be no "blanket" technological solution to coping with environmental issues. In this chapter, we will explore the framing of these issues into hard and soft systems and how their integration can provide useful insights for creating innovative systems of environmental management.

For communities to design appropriate strategies for managing their local conditions, innovative changes in civil society will be required. This innovation will only be possible if people are prepared to adopt a learning approach whereby they test out new roles and ways to organize themselves to achieve their environmental goals. Such learning will be dependent on improved access to appropriate, understandable, and timely knowledge and information concerning both technological (hard system) options and organizational and institutional (soft system) possibilities. Such collectively self-aware learning assumes the necessity that many "heads" need to come together to share experience and reach common agreement on environmental action plans. Such a process requires creation of "learning spaces" for social actors to work together, build coalitions, and develop new skills to meet these challenges.

We have designed and tested a process in a variety of international settings, which has provided both tangible results and insights into social and ecological linkages. Other authors in this book have discussed the systemic aspects of understanding and learning. Our focus in this chapter is primarily on the upper right-hand side and "diamond" in fig. 14.1 and on the first several steps of the adaptive methodology for ecosystem sustainability and health (AMESH) process, as described in Part III.

Learning Approaches for Environmental Design and Management

Lightfoot et al (2001a,b), Lightfoot and Okalebo (2001), and Lightfoot and Noble (2001) provide an operational framework for designing such a collaborative learning process. On the basis of the creation of local multistakeholder learning groups, participants debate and reflect by sharing their different perceptions, knowledge, experience, interest, and values concerning the future of their local ecosystem.

This shared learning experience provides insights into what needs to change in the socioeconomic system in order to rehabilitate the biophysical system to meet humanly defined socioeconomic expectations. In some of the literature, socioeconomic and biophysical systems have been called soft and hard systems, respectively. As discussed in chapters 8 and 10, Checkland and his colleagues developed an explicit methodology and set of techniques, soft systems methodology, for operating a learning cycle within organizations and institutions. In this chapter, we are using soft systems in the more generic sense of social or human activity, systems. Checkland and Scholes (1990) note that these soft systems exist as a result of negotiation where people set agreed goals and boundaries for action with regard to each other and the physical system they are utilizing.

Daniels and Walker (1996) have documented such negotiation processes in protected areas and refer to them as "collaborative learning." By working as a collaborative learning group, people begin to appreciate how the human system they create and the activities and behavior it endorses will determine the nature of the biophysical environment in which they live. As learning proceeds, participants gain confidence in analyzing their human system, identifying areas for change, and negotiating new partnerships to achieve them. Lightfoot et al. (2001a) note that this learning process needs three dimensions that are important for its success:

1. Learning space should be created for multiple stakeholders to meet, argue, and reflect.
2. Soft system thinking and common future visions should form the basis for designing the framework for analyzing the hard system (see more discussions of this relationship in the context of the ecosystem approach diamond diagram and AMESH in chapter 14).
3. Local ownership of the process and the resources to act are essential.

The first dimension addresses the orientation of the process in terms of attitudes and behavior. Often people have great difficulty in organizing themselves and reconciling their conflicting agendas in order to create an effective learning space for innovative action. Hence there is the need for more resources to be allocated to establishing and maintaining collaborative learning groups.

The second dimension addresses the need for soft system thinking and future visions to drive the learning process. Human society is the dominant influence determining the nature of most ecosystems. Therefore improving skills in soft systems thinking will enable people to decide what kind of hard system analysis will be needed to measure impact of changes in human activity on their environment (Lightfoot et al. 2001a; Jiggins and Röling 1997).

The third dimension addresses the fact that ownership of the learning process and the power and resources needed to act must reside with representative coalitions of local individuals and interest groups. These are the social actors who are directly or indirectly having an impact on or have an agenda concerning the ecosystem resources in the local area.

This condition of local ownership is essential for effective learning. In addition, there must be institutional support and buy-ins from key stakeholders at other administrative and organizational levels in society to give assistance to this local multistakeholder learning coalition. The overall goals are to (1) strengthen the planning capacity of civil society (i.e., community-based organizations, interest groups, etc.) to generate future visions of sound ecosystem management; (2) promote more responsive service provision (in terms of social, economic, scientific information, and advice); and (3) influence policies and environmental legislation such that it is more consultative and thus supportive of the futures that local civil society envisions for its environment.

This learning process allows accommodation for different agendas and encourages participants to build their understanding of environmental issues from a systems perspective. This is achieved by learning coalitions undertaking an iterative planning process as summarized in fig. 9.1.

This iterative learning process can be started as shown in the next sections.

Visioning for the Future of One's Local Ecosystem

Visioning is the "small mental switch with large behavioral consequences. From looking at hillsides and seeing only soil erosion to looking at hillsides and seeing what you would like it to look like in ten years time" (Lightfoot and Okalebo 2001:1).

Visioning for the whole ecosystem rather than problem diagnoses of its components provides a holistic approach and a positive impetus for initiating a learning process. Envisioning a better future brings much needed energy, enthusiasm, and commitment in local people for their own development.

For example, Lightfoot et al. (2001b) demonstrate how Ugandan farmers used "mapping" to share their knowledge and experience on the current biophysical status of their local agroecosystem. Their maps became a focal point for discussion of how human activity has been instrumental in determining the state of the natural resources in the area. Farmers used their initial map to design a new one,

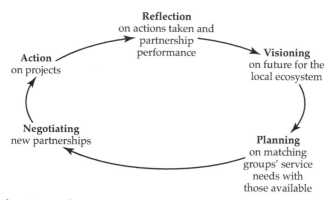

Figure 9.1   The learning cycle.

presenting a future vision for an improved agroecosystem. What is known and unknown about their natural environment could be identified along with additional knowledge and learning needed to achieve this future vision. In one exercise for a parish in Soroti District, eastern Uganda, the farmers demonstrated how they wanted to improve soil quality, water management, and conservation (more ponds, rice paddies, etc.) and rehabilitate riverbanks and hillsides (tree planting on hills and along water courses, zero grazing regimes, etc.). However, it soon became clear for the Soroti farmers that these desirable changes required considerable social change as well. Land tenure arrangements, exploitative agricultural practices, and abuse of natural resources had to be resolved to everyone's satisfaction first. Mutual agreements on soft system adjustments had to precede those of the hard system.

Mind mapping, idea diagramming, or "rich pictures" provides a way of further exploring and learning about a central theme or future vision of local communities (Checkland and Scholes 1990; Caley and Sawada 1994). This is an open-ended method whereby participants explore the extent of their knowledge and understanding about a particular topic. The objective is to understand the entire social and technical context that everyone thinks relates to the central theme. As with agroecosystem mapping, the group identifies where it should focus its attention and gain more understanding and information in order to realize their future vision.

Figure 9.2 illustrates a partial analysis of nutrient management issues in a Canadian agroecosystem. A multistakeholder learning group consisting of farmers, environmental interest groups, local conservation scientists, etc. from the Maitland Valley, southwest Ontario began developing this diagram. This was their first step in trying to understand the major issues affecting nutrients within the Maitland watershed so that common agreement could be reached on ensuring water quality for the future. They identified, from each of their perspectives, the constellation of issues relating to nutrient flow within the watershed. This information was organized as a diagram of ideas to show the system of possible interrelationships.

Figure 9.2 only shows a small part of this analysis. Much more detail, particularly of social issues, was added as the discussion proceeded. From the diagram, different areas for possible action could be identified (e.g., livestock farming and manure management, industrial and domestic waste management, water quality of river and lake ecosystems, etc.).

As with the Ugandans, the Ontario learning group realized that what appeared as a very technical issue was dependent on different socioeconomic approaches to farming and domestic water use. For example, the whole ethos of the farming systems needed examination (i.e., the spectrum from organic family farms to intensive livestock farms managed by corporations). Farming communities are disappearing as large agro-industrial conglomerates buy out small farms, and this is profoundly affecting the sense of local ownership and concern for the local environment. There is also the question of how to bring homeowners and representatives of the weekend cottage community into the learning group. If septic tanks are contributing to

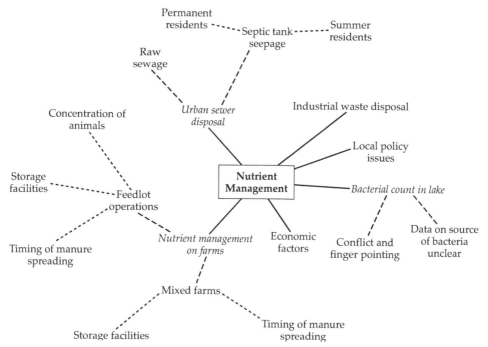

Figure 9.2   Diagram of some of the ideas and issues concerning nutrient management in the Maitland Valley Watershed.

bacterial loading in local rivers and lakes, then people's attitude to domestic water management needs to be addressed. Questions of local government policy also came to the fore with regard to bylaws on waste management. These issues arising out of the rich pictures demonstrated that the soft system presents as many or more learning challenges as the hard system. Moreover, the technological knowledge needed to be made available, and this did not just require communication products but necessitated people's involvement and agreement about what constituted relevant and trustworthy information.

Identifying Partnerships and Establishing Learning Networks for Innovation

In both the Ugandan and Canadian examples, the people influencing the structure of the ecosystem were using technologies and practices that arose out of the sociopolitical networks and knowledge systems of which they were members. It is important for any learning group to understand how these networks have influenced people's approaches to environmental planning and use of natural resources. Such analyses will help to highlight where changes are needed in current social networks so they better support collaborative learning and innovation.

Engel and Salomon (1996, 1997), Engel et al. (2000), and Ramirez (1997) used process-oriented methodologies of stakeholder analysis to identify the possible knowledge and information linkages that communities need to design and implement action plans. Rapid rural appraisal of agricultural knowledge systems (RAAKS) formed the basis of this approach. RAAKS elucidates how technical innovations arise from the way different social actors interact and poses questions such as: "How do we organize ourselves to create innovation?" "Is the way we organize ourselves now going to create the learning space for the innovation we will need?" Such self-analysis enables the learning group to place their current environmental knowledge and practices within a social context. People can then redesign old or create new social networks that are better able to support their action plans for environmental rehabilitation.

Lightfoot et al. (2001a,b) have taken partnership formation a step farther by having learning groups create a "marketplace" for comparing farmers' demands with what services are available from government, nongovernment, and private sector organizations. The learning group identifies initial actions for realizing future visions of sound ecosystem management and against them they list the services and information they will need. Likewise, stakeholders from the service sector list what they have on offer. Matches are then made between demands and services. Sometimes, there will be no matches. This arises because either a demand cannot be met or stakeholders are offering services no one needs or realizes were available. This matching process is a learning instrument for exploring and discussing the potential partnerships that will be necessary to fulfill the community's future vision for its environment. People can identify where services and information pathways need to be created and bring to their attention services they never realized are available and might find useful. Institutions, community interest groups, local government, etc. will also need to explore the implications of forming partnerships for how they organize themselves. Many might have to rethink their roles, change attitudes, and retrain in order to form effective partnerships. So an essential part of the learning process must be clarifying what the implications are for the social actors in terms of new activities. Figure 9.3 summarizes the process and is one that has been adopted in Uganda to develop environmental action plans at various administrative levels in starting at the Parish level (Lightfoot and Okalebo 2001).

Clarifying the Characteristics of Successful Partnerships

The underlying reasons for successes or failures of previous partnerships become attributes around which to negotiate criteria and build partnerships that work. For example, Tanzanian farmers listed trust, transparency, awareness building, cooperation, clear roles, and good communication as essential criteria for successful partnerships (Shao et al. 2000).

Establishing Indicators to Monitor the Soft and Hard Systems

This is probably the most difficult part of the learning process and initially will be "hit or miss." People need to hone their skills on doing the testing to choose appropriate indicators that provide useful information for refection on "next steps." Where does one start in this process? Monitoring for sustainability requires indicators that measure self-organization. In our experience, and in most of the case studies described in this book, short-term indicators are developed empirically. Some of these can be seen to reflect systemic variables, even if they were initially chosen for aesthetic or practical reasons.

For example, Lightfoot and Noble (1993, 1999, 2001) and Lightfoot et al. (2001b) describe indicators for agroecosystem performance that were jointly designed with African and Asian farmers. These indicators provide a rough guide as to the economic and ecological impact of new farming strategies on the environment. The ecological indicators are biological diversity (number of plant and animal species on farmland), productivity (biomass of crops, animals, etc. produced per annum), and recycling (number of internal bioresource flows on the farm). The economic

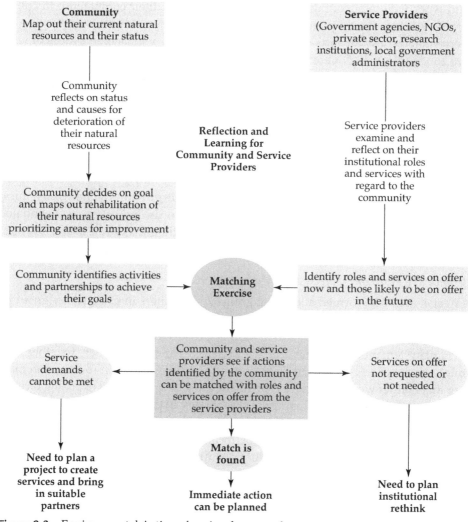

Figure 9.3   Environmental Action planning framework.

indicator is a simple profit-cost ratio for the farming year. Farmers compare the values of these indicators before and after changing their farming strategies. Increases in value over time demonstrate a general improvement in the economic and ecological health of their agroecosystem.

Soft system indicators to couple with those of the hard system arise out of the discussions of the characteristics that contribute to successful relationships (e.g., transparency in decision making, timeliness of delivering information, etc.). A qualitative scale can be designed to score performance for these indicators (Ramirez 1997). Each partner in the network scores the other and combines this information with that of the hard system indicators. When farming communities compare these soft and hard system indicators across a range of their farms, they can begin to draw out general lessons for creating a sustainable agroecosystem.

Simple indicators such as those described above are useful at the small scale and work well to encourage people to start building their competence in collaborative environmental planning. However, soft system indicators of partnership performance raise all sorts of issues such as women's versus men's ideas of what are good indicators. Different social actors within the learning coalition (including, as Gitau et al. (chapter 12) point out, researchers and local stakeholders) may want different types of indicators that suit their interests and so on. Likewise, there will be initial problems in deciding on suitable hard system indicators that accurately reflect the influence of social changes on the biophysical environment. Negotiation about suitable performance measures takes time but is very valuable as a learning process. It helps to forge understanding between disparate groups. What also arises from this negotiation is the realization that learning groups need to experiment to find the right mixes of indicators that will provide the necessary information to design effective action plans. In other words, indicators are negotiated rather than perceived as predetermined measures; their major role is as a means to help communicate complex changes to a wider audience (Guijt 1998).

Combining Hard and Soft System Methodologies

Engel et al. (2000) formed a learning group with farmers in Chile to design an integrated approach to analyzing agroecosystem performance. Their argument was that the qualitative nature of soft system approaches and the quantitative nature of hard system methodologies both have strengths and weaknesses. Together, they might have synergistic effects that bring out useful information for decision making that neither would do on their own. They also proposed that exposing farmers to this integration would help improve the quality of their strategizing on environmental issues. Likewise, the researchers and service providers on the learning team would gain understanding of how the local sociocultural and economic context influences hard system performance. Four different methods were combined to monitor their joint learning and research with the farmers. These were the Granjas simulator (assesses farmers' decisions on resource allocation), multiple-objective linear programming (uses farm-level information on resources and production), rapid appraisal of agricultural knowledge systems (RAAKS; uses information on the social and institutional components of the farming system), and participatory rural appraisal (PRA; used for working with farmers to analyze their agroecosystem in terms of natural resources and livelihoods).

The Chilean learning group found that this integration of methodologies provides a richer understanding of how soft and hard systems components interact. This enabled farmers to make better-informed decisions about managing their natural resources. The hard system tools allowed for ex ante analysis of the possible effects of new collective strategies on individual farms. The soft system tools provided input on the likely social and institutional influences that would affect those hard system outcomes.

Gargicevich and Maroni (2000) also integrated methodologies for their learning exercises with farmers in Argentina and had similar results. They used two quantitative tools, Granjas and representative sampling, and two qualitative tools, RAAKS and communicational diagnosis for their group learning. Such integration encouraged a move from generating data to make a diagnosis to searching for information that would help in promoting effective action to achieve a future goal. Gargicevich and Maroni (2000) comment that "The ex-ante analysis was participatory, negotiated, and agreed among all participants."(p. 105).

In both the Chilean and Argentinean examples, farmers began to understand the importance of social context in ensuring improvement in the status of natural resources. They also became more aware of the complexity of the relationships between the soft and hard components of their local agroecosystem and the knowledge networks that affected its design and management. Gargicevich and Maroni (2000) noted that researchers within the learning coalition realized they must continually adjust their hypotheses to reflect the field realities within each iteration of the learning cycle. They also needed to change their roles from external problem solvers to catalysts and consultants to multistakeholder learning groups.

Conservation authority scientists in the Canadian example mentioned earlier are members of a local multistakeholder group for the watershed within their jurisdiction. These scientists acted as catalysts for encouraging a local learning group on watershed planning. They won a grant from the federal government to build a local coalition of interest groups in their watershed as an experiment to see if community-based planning would be more effective than the old "top-down" approach. The scientists have adopted the role of consultants that provide technical input for the local learning coalition. For example, part of the group's future vision is "Working together to protect and improve water quality and quantity." The scientists provide information on possible ways to measure these parameters and the technical issues involved with water pollution and have put a great deal of effort into becoming better communicators. More important, they do not tell the group what they should do. The learning coalition decides on target goals for water quality and quantity and agrees on indicators to monitor progress and on the soft and hard system changes that will be needed to achieve them. Nonscientists in the group are learning new skills in handling and interpreting technical data to use for decision making. Scientists are learning that soft system thinking enables them to interpret technical data from different perspectives. In both cases, scientist and nonscientists are learning how to organize themselves into an effective environmental planning group. Although emphasis has been on the hard system indicators, the major accomplishments in terms of group formation today lie in the soft system in that they have strengthened their social capital (Pretty and Ward 2001).

Concluding Remarks

Knowledge of ecological processes and how they affect environmental sustainability will always be imperfect. The equilibrium of ecosystems is continually in flux, and it is difficult enough to pin down the biophysical, let alone the sociocultural contribution to these system dynamics. As Lee (1999) emphasizes, we need to admit to ourselves "humans don't know enough to manage ecosystems." We are very good at exploiting but much less adept at fostering our resources. Woodhill and Röling (1998) also add that with increasing rates of change and risks of environmental decay coupled with the escalating complexities of globalization, demands on societies' capabilities to evolve to meet these challenges become immense. Given this scenario, people will need to learn and adapt quickly to changing circumstances if they are to conserve and sustain their natural resources. A collaborative and iterative learning approach probably holds the most promise for generating the continual innovation that will be needed to respond to such rapid environmental changes.

In this chapter, we have provided a framework and examples of tools and methodologies that we found useful to facilitate social learning. The basis of this approach is equal partnership between social actors (community members, private and public sector service providers) in order to share experience and design collaborative action plans for their local environment. This process presents many challenges for the local social actors involved and for the people and institutions at other levels in society that need to support such learning initiatives.

A major challenge is who should be in a learning coalition. As Ramirez (2001) explains, the nature of the environmental issue will determine who must be included. If the nature or boundaries of the issue are not clear, then identifying the social actors will be difficult. For example, owners of the environmental issues of a watershed are much more difficult to determine than those of a small forest. It may require several iterations of identifying stakeholders to create a representative coalition for the larger ecosystem. Also, only social actors who have the power (to implement and/or convene), legitimacy (within the local social context), and urgency (have needs to be addressed quickly) are likely to get involved and play an effective role in any learning coalition. As many of the cases and discussions in this book make clear, who is in and who is left out are questions that can have profound effects on problem definition and design of solutions.

Learning approaches to environmental management will also necessitate a change in how social actors formulate action plans and set their goals. Allen and Bosch (1996) and Allen (1997) demonstrated in their studies of community-based research for environmental planning, that proposals for action must be continually adjusted as people's experience increases. Lightfoot et al. (2001a,b) also emphasize the need for people to move away from action plans for dealing with specific problems to holistic planning characterized by experiment, reflection, and redesign. Piecemeal planning to deal with specific environmental problems will not be effective in handling the multidimensional nature of soft and hard system interactions.

Learning needs to be supported by information networks. Reynolds and Busby (1996) noted that biodiversity projects were most successful when communities became involved as active participants in creating information networks to handle the issues most relevant to them. Basic design of these networks has to begin with the local learning group because they need to tailor it to support their needs for planning. These networks also have to be responsive to the evolving requirements of a learning group as they gain more experience of managing environmental issues.

As James Kay has repeatedly pointed out (see Part I), researchers find themselves taking on very different roles in this new context. Few have the communication and relationship building skills to work effectively with such participatory approaches to environmental planning. Engel et al. (2000) note, that it will take a certain kind of professional to be able to integrate effectively soft and hard system methodologies in order to assist local learning groups. Gargicevich and Maroni (2000) found that the scientists in their learning groups hotly debated how to balance the personal security offered by traditional scientific methods against the risks and uncertainties of using integrated approaches (see Berkes and Davidson-Hunt, chapter 7). In addition, environmental professionals will be challenged to change their mindset from driving the research agenda to taking the role of service provider to the client learning group. All of this will necessitate changes in approach not only by individual professionals but also by their institutions. This is no small task. One problem is that there are high transaction costs in establishing learning processes. It takes time, effort, and resources for the concerned organizations and institutions to meet and design learning instruments for working with local groups and creating effective interinstitutional cooperation. The same will be true for finding agreement on an appropriate operational process for learning. The importance of this learning dimension as a required component of effective ecosystem management is well documented in the literature where promising technological interventions failed because they did not engage the farming community in a social learning process (Pretty and Hine 2001).

A further challenge is the complexity of relationships between local and national policy makers and their clients, the local community. Government managers and policy makers may distrust local capacities. There is also the question of power. Who drives the environmental planning process? Who will allocate funds for learning processes and control how they are used? If learning approaches are to succeed, then local decisions have to be respected. The notion of interactive policy making will need more attention, and experience in the Netherlands suggests that this only happens when the power differences between rural groups and policy makers are reduced (Glasbergen 1995).

All of these challenges may seem daunting but not insurmountable. Dixon et al. (2001) in their survey of agroecosystems state that ways have to be found to unlock the potential of communities to become involved in environmental planning. In particular, there needs to be a legislative, political, and fiscal enabling environment that (1) provides incentives for sustainable resource management; (2) supports community-led learning and technology development; (3) decentralizes rural public services, responsive to local needs; (4) creates local development alliances and partnerships; and (5) delivers relevant education and training in rural areas.

The Canadian example on watershed planning from Ontario is a case in point where this is beginning to happen. Likewise in East Africa, several countries (Kenya, Tanzania, and Uganda) are initiating decentralization policies on environmental and agroecosystem planning. Learning approaches of the kind outlined in this book and elsewhere are being used to bring operational reality to these policies and build local capacity to take on responsibility for their own environmental planning and action. These changes reflect a growing realization that

- rehabilitation and long-term sustainable improvements in ecosystems will be reliant on social innovation to drive the technological development needed to support these changes;
- innovation is more likely to occur if soft systems thinking is integrated with that of hard systems;
- a learning process is probably best able to deal with the complexity and dynamism entailed in environmental planning; and
- the power to form policy and act should lie with local learning coalitions that have a vested interest in sustainability of their local environment.

If we are to succeed in restoring and maintaining our ecosystems, then people have do be brought back into the planning equation. The changes in attitude listed above provide a way forward for achieving this end.

References

Adams, W. M. 1992. *Green Development: Environment and Sustainability in the Third World.* London, UK: Routledge.

Allen, W. J. 1997. Towards improving the role of evaluation within natural resource management programmes: The case for "learning by doing." *Canadian Journal of Development Studies* 17:625–638.

Allen, W. J. and O. J. H. Bosch. 1996. Shared experiences: The basis for a co-operative approach to identifying and implementing more sustainable land management practices. In *Proceedings of a Symposium "Resource Management: Issues, Vision, Practice,* pp. 1–10. Christchurch, NZ: Lincoln University Press.

Berkes, F., and C. Folke. 1998. *Linking social and ecological systems: management practices and social mechanisms for building resilience.* Cambridge University Press, Cambridge, UK.

Borrini-Feyerabend, G., M. T. Favar, J. C. Nguinguiri and V.A Ndangang. 2000. *Co-Management of Natural Resources: Organising, Negotiating and Learning-by-Doing.* Heidelberg, Germany: Kasparek Verlag.

Caley, M. T. and D. Sawada. 1994. *Mindscapes: the Epistemology of Magoroh Maruyama.* Langhorne, Pa.: Gordon and Breach Science.

Chambers, R. 1983. *Rural Development: Putting the Last First*. London, UK: Longman.

Checkland, P. B. 1981. *Systems Thinking, Systems Practice*. New York: Wiley.

Checkland, P. B. and J. Scholes. 1990. *Soft Systems Methodology in Action*. New York: Wiley.

Cline-Cole, R. 2000. Knowledge claims, landscape and the fuelwood degradation nexus in Dryland Nigeria. In *Producing Poverty and Nature in Africa*, eds. V. Broch-Due and R. A. Schroeder, pp. 109–147. Stockholm, Sweden: Elanders Gotab.

Daniels, D. E. and G. B. Walker. 1996. Collaborative learning: Improving public deliberation in ecosystem-based management. *Environmental Impact Assessment Review* 16:71–102

Dixon, J., A. Gulliver and D. Gibbon. 2001. Improving farmers' livelihoods in a changing world. In *Farming Systems and Poverty Series*, ed. M. Hal. Rome, Italy: Food and Agricultural Organization.

Engel, P and M. Salomon. 1996. Dare to share: Networking on ecologically sound agriculture. Workshop Report, 60 pp. Navrongo, Ghana: LEISA Group of Northern Ghana.

Engel, P. H. and M. Salomon. 1997. *Facilitating Innovation for Development*. A RAAKS Resource Box. Amsterdam, Netherlands: Koninklijk Instituut Voor de Tropen.

Engel, P. H., I. Visser, I. Guijt, A. Alvarez and O. Melo. 2000. Integrating "hard" and "soft" system methods for assessing farmer strategies in Nûble, Chile. In *Deepening the Basis of Rural Resource Management*, eds. I. Guijt et al., pp. 39–55. Proceedings of a workshop. The Hague: International Service for National Agricultural Research. (Available at: http://www.isnar.cgiar.org/environment/nrm-basis.htm)

Gargicevich, A. and J. Maroni. 2000. Methodological complimentarity and learning processes: a survey into extensive agriculture in Argentina. In *Deepening the Basis of Rural Resource Management*, eds. I. Guijt et al., pp. 100–108. Proceedings of a workshop. The Hague: International Service for National Agricultural Research. (Available at: http://www.isnar.cgiar.org/environment/nrm-basis.htm)

Glasbergen, P. 1995. Environmental dispute resolution as a management issue. In *Managing Environmental Disputes: Networking as an Alternative*, ed. P. Glasbergen, pp. 1–17. New York: Kluwer Academic.

Guijt, I. 1998. Participatory monitoring and impact assessment of sustainable agriculture initiatives: An introduction to key elements. SARL Discussion Paper 1. London: International Institute for Environment and Development.

Jiggins, J. and N. Röling. 1997. Action research in natural resource management: Marginal in the first paradigm, core in the second. In *Pour une Méthodologie de la Recherche Action*, eds. C. Albadalejo and F. Casabianca, pp. 151–169. Versailles: French National Institute for Agricultural Research—Science for Action and Development.

Lee, K. N. 1999. Appraising adaptive management. *Conservation Ecology* 3(2):3, http://www.consecol.org/vol3/iss2/art3/.

Lightfoot, C. and R. Noble. 1993. A participatory experiment in sustainable agriculture. *Journal for Farming Systems Research-Extension* 4(1):11.

Lightfoot, C. and R. Noble. 1999. A farming systems approach to ecological agriculture research. In *Ecoagriculture: Initiatives in Eastern and Southern Africa, Environmental Capacity Enhancement Project*, Environmental Roundtable Series, eds. J. Devlin and T. Zettel, pp. 205–226. University of Guelph. Harare, Zimbabwe: Weaver.

Lightfoot, C. and R. Noble. 2001. Tracking the ecological soundness of farming systems: instruments and indicators. *Journal of Sustainable Agriculture* 19(1):9–29.

Lightfoot, C. and S. Okalebo. 2001. Vision-based action planning: report on a learning process for developing guidelines on vision-based sub-county environmental action planning. Report for Conserve Biodiversity Support Project, Uganda, 13 pp. Montpellier, France: International Support Group.

Lightfoot, C., R. Ramirez, A. Groot, R. Noble, C. Alders, F. Shao, D. Kisauzi and I. Bekalo. 2001a. *Learning Our Way Ahead: Navigating Institutional Change and Agricultural*

Decentralisation, Gatekeeper Series 98. London, UK: International Institute for Environment and Development.

Lightfoot, C., M. Fernandez, R. Noble, R. Ramirez, A Groot, E. Frenandez-Baca, F. Shao, G. Muro, S. Okelabo, A Mugenyi, I Bekalo, A Rianga and L Obare. 2001b. A learning approach to community agroecosystem management. In *Interactions Between Agroecosystems and Rural Communities*, ed. Cornelia Flora, pp. 131–155. Boca Raton, Fla.: CRC.

Pretty, J. 1998. Supportive policies and practice for scaling up sustainable agriculture. In *Facilitating Sustainable Agriculture*, eds. N. Röling and M. A. E. Wagenmakers, pp. 23–45. Cambridge, UK: Cambridge University Press.

Pretty, J. and R. Hine. 2001. Reducing food poverty with sustainable agriculture: A summary of new evidence. SAFE-World, The Potential of Sustainable Agriculture to Feed the World Research Project. Final Report. Colchester, UK: Centre for Environment and Society, University of Essex. (Available at: http://www2.essex.ac.uk/ces/ResearchProgrammes/CESOccasionalPapers/SAFErepSUBHEADS.htm)

Pretty, J. and H. Ward. 2001. Social capital and the environment. *World Development* 29(2):209–227.

Ramirez, R. 1997. *Understanding Farmers' Communication Networks: Combining PRA with Agricultural Knowledge Systems Analysis*, Gatekeeper Series 64. London, UK: International Institute for Environment and Development.

Ramirez, R. 2001. Understanding the approaches for accommodating multiple stakeholders' interests. *International Journal of Agricultural Resources, Governance and Ecology* 1(3/4):264–285.

Reynolds, J. and J. Busby. 1996. *Guide to Information Management in the Context of the Convention on Biological Diversity*, 84 pp. Nairobi, Africa: United Nations Environmental Programme.

Röling, N. G. and J. Jiggins. 1998. The ecological knowledge system. In *Facilitating Sustainable Agriculture*, eds. N. Röling and M. A. E. Wagenmakers, pp. 283–311. Cambridge, UK: Cambridge University Press.

Shao, F., E. Mlay and G. Muro (eds.). 2000. *Proceedings of a District Multi-stakeholder Workshop on Linked Local Learning. June 12–16 2000*. Mikumi, Kilosa. Dar es Salaam, Tanzania: Farm and Natural Resources Management Consultants.

Shanley, P. and G. R. Garcia. 2000. Break barriers, increase impact: Equitable generation and dissemination of natural resource information in the Brazilian Amazon. In *Deepening the Basis of Rural Resource Management*, pp. 181–197. The Hague, Netherlands: European Commission.

Woodhill, J. and N. Röling. 1998. The second wing of the eagle: The human dimension in learning our way to more sustainable futures. In *Facilitating Sustainable Agriculture*, eds. N. Röling and M. A. E. Wagenmakers, pp. 46–71. Cambridge, UK: Cambridge University Press.

10

Human Activity and the Ecosystem Approach

The Contribution of Soft Systems Methodology to Management and Rehabilitation of the Cooum River in Chennai, India

Martin Bunch

Soft systems methodology (SSM) was one of the major influences on our development of the ecosystem approach at the close of the twentieth century (Checkland 1979, 1981; Checkland and Scholes 1990). We introduced SSM in chapter 8. In this chapter, I describe the adaptation of SSM to the understanding and management of a problematic social-ecological situation in Chennai, India, and explore the methodology in more detail. The term "soft systems" is used here to denote use of systems thinking in an interpretive mode (using systems thinking to make sense of messy situations) as opposed to the functionalist "hard systems" approach that conceives of systems as real things that can, for example, be designed. This use should be differentiated from the more general use of soft systems to refer to social or other human systems (see, for instance, chapter 9).

The Study Area: The Cooum River, Chennai, India

The Cooum River is a highly polluted languid stream that flows through the center of Chennai. Chennai, with a population of 6.4 million in its urban agglomeration (Government of India 2001) is India's fourth largest metropolis. The population of Chennai is the primary source of pollution in the Cooum River. Although parts of the city are serviced by sewage collection and primary or secondary sewerage treatment, much raw sewage is diverted into the waterways and ocean (Government of Tamil Nadu 1981; Srinivasan 1991). In fact, the production of sewage and wastewater in the city accounts for most of the flow in the Cooum during the dry (nonmonsoon) season. A study by Wardrop Engineering in 1995 identified 116 wastewater outfalls into the river within the city and more in its tributary drains and canals (Government of Tamil Nadu 1997), and the most recent survey of slums (in 1986) indicated at least 37 unsewered hutment areas located along its banks (Bunch 2001). Debris dumping, open-air defecation, animal husbandry, clothes washing, and other activities are obvious along the urban course of the Cooum.

The river passes through a flat landscape on a coastal plain. On the coast, changes in littoral currents associated with the Madras Port have resulted in migration of sand bars that block the river mouth. This leads to stagnation in the dry season and impedes tidal mixing. Quality of water in many parts of the Cooum River system (e.g., as indicated by the 5-day biochemical oxygen demand) is comparable to raw sewage (Government of Tamil Nadu 1997; Gunaselvan 1999). Silt and organic (fecal) sludge, known to contain pathogenic parasites and enteric pathogens, have also accumulated along the bottom and banks of the river to depths up to a meter (Mott MacDonald 1994). The river is a foul-smelling open sewer and a multidimensional hazard to the health and well-being of Chennai citizens.

The Cooum problem is well known and long standing. Many government agencies have attempted to improve the situation, but the problem has worsened. This is partly because previous attempts to address the problem have been piecemeal and used engineering-oriented approaches based in a rational-positivist paradigm. Various reaches of rivers and canals within the city have been dredged of sludge, their banks lined, and once a (short-lived) sand pump and regulator were installed at the mouth of the Cooum River to keep it clear of sand. Such interventions resulted in short-term improvement of the situation but have repeatedly failed to solve the problem (Appasamy 1989; Sahadevan 1996).

Management efforts have also been constrained by jurisdictional boundaries and a mechanistic management environment. In Indian institutions an orientation toward programmed approaches, rigid hierarchical authority and structural organization, vertical communication channels, and centralized control are typical. Rondinelli (1993) has argued that such an organization is incapable of effectively dealing with complex and uncertain situations.

Soft Systems Methodology

Despite the physical manifestation of the Cooum River problem, human activity is obviously at its core: people produce and dispose of waste, coexist with the waterway and environs, and are affected by them. SSM, as a means to deal with problematic situations characterized by human activity, informed the approach to the Cooum River situation as a mode 2 application of the methodology, as described in chapter 8.

In the Cooum River situation, we applied SSM techniques where they were deemed useful, while other tools were adapted to the process where appropriate.

The Cooum River Environmental Management Research Program

This research operated a participatory process in which, during workshops in 1998 and 1999, stakeholders identified and expressed the Cooum River problem situation; undertook conceptual modeling of relevant systems, generated and debated goals, objectives, and interventions for management of the Cooum River and its environs; developed a framework for a decision support system (DSS) that was based on a coupled geographic information systems and environmental model; and used the DSS to develop exploratory management scenarios. Workshops were also intended to promote less jurisdictional, less disciplinary understandings and to guide exploration of the cultural climate, including the expression of values and norms that characterize the current situation and that influence stakeholders' preferences for desirable futures.

Both workshops combined participant paper presentations (that provided background about selected aspects of the issue) and working sessions (in which participants identified, scoped, conceptualized and debated the problem situation and potential action to improve it). Approximately 50 people participated in each of the workshops in some way (e.g., attending inaugural or valedictory sessions and various paper presentations). In each year, there were about 25 stakeholders from government agencies, nongovernmental organizations (NGOs), and academe who formed the core group of participants throughout the entire workshop.

Problem Definition and System Identification

The ecosystem approach we used drew on two streams of activity for identification and description of a "socioecological" system. These guide development of a conceptual understanding of the situation as a "system" (ecosystem understanding) and comprehension of the social, institutional, cultural, and political context with which it is associated (issues framework). This informs an understanding of possible and desirable future states of the system. These two streams of activity correspond also to the two streams of activity described in second-generation models of SSM. Checkland (1999:A14) describes these as a "logic-based stream of analysis" (that is, systems analysis), and a "cultural and political stream which [enables] judgements to be made about the accommodations between conflicting interests." This research pursued both of these streams simultaneously. This represents a more detailed and nuanced consideration of the upper right quadrant of the "diamond diagram" (fig.14.1).

Much of the development of a socioecological system description in this work focused on the current and historical situation of the Cooum River and its environs, beginning with the development of a shared understanding of "the problem." Box 10.1 outlines this initial exercise. The extreme variation and number of stated "problems" that resulted from this exercise demonstrate that the Cooum River situation is extremely complicated, multidimensional, and ill-structured. A particularly interesting outcome of the initial problem identification exercise was a shift in perception about the problem stimulated late in the exercise by the question "Really now, what is the problem?" Upon reflection, participants tended to indicate that their initial concerns (e.g., water quality, hydrology, odor) were actually symptoms of problems rooted in the political, social, and management realm.

Primary elements, actors, and relationships in the situation identified in initial problem identification exercises were organized as a rich picture (a diagrammatic technique borrowed from SSM). This diagram (fig. 10. 1) represented participants' shared understanding of the situation. It acted as a touchstone throughout both

BOX 10.1   Expression of the Problem Situation

Expression of the problem situation began with a problem identification exercise [from United Nations Centre for Human Settlements (Habitat) 1991], consisting of written responses to the questions:

1. What is the problem?
2. Why is it a problem? What would the problem look like if it were solved?
3. Whose problem is it? Who owns it?
4. Where is it a problem? Is it localized and isolated, or is it widespread and pervasive?
5. When is it a problem?
6. How long has it been a problem?
7. Really now, what is the problem?
8. Finally, what would happen if nobody did anything to solve the problem?

Participant responses to these questions demonstrated a wide range of perceptions on the Cooum problem, demonstrating that the situation is ill-defined and extremely complicated. Responses to problem identification questions (1, 2, and 7 above) identified 71 distinct problems in the situation! These were grouped into eight thematic categories:

1. sensory aspects (3), e.g., visual eyesore, foul smell
2. health hazards (11), e.g., mosquito breeding, habitat for rodents
3. objectionable land & land use (6), e.g., illegal encroachments, location of slum developments
4. hydrology (8), e.g., slow flow, stagnation (dry season); sandbar blockage at mouth
5. pollution and related factors (20), e.g., heavy pollution load; illegal sewage outfalls
6. population (4), e.g., population growth, densification
7. tourism and recreation (4), e.g., no walkways, lawn, gardens, parks; denial of a tourism asset
8. political, social and management aspects (15), e.g., lack of political will, inability to solve environmental problems.

Almost all issues identified by participants in this exercise were physical, observable manifestations of the problem situation. Participant responses in the eighth category, which consisted primarily of responses to question 7 above, highlighted issues such as lack of political and public will, poor coordination and communication of agencies, inappropriate models for environmental problem solving, and basic uncertainty about the situation. Many reports cite untreated sewage being routed to the river (e.g., Srithar 1982; Appasamy 1989; Sahadevan 1995; Government of Tamil Nadu 1997) and physical and hydraulic conditions (e.g., Mott MacDonald Ltd. 1994; Sahadavan 1996; Government of Tamil Nadu 1997; Inland Waterways Authority of India 1998) as the root of the Cooum problem. Until this point, however, none identified issues such as coordination and communication among government agencies, absence of data sharing, and inadequate approaches to dealing with environmental problems as critically important to the Cooum problem.

workshops, aided the development of themes to be modeled as relevant systems, and evolved to represent participants' shared concept of the overall "Cooum system." An important revelation to participants, stimulated by this exercise, was that most previous work regarding the Cooum targeted only parts of what participants conceived as the Cooum system. This is particularly true for consultancy studies, which have been the primary source of information. Although participants already understood that the problem was multidimensional, there was a great deal of enthusiasm for this approach, which could make connections between the most important elements and actors in a coherent way.

Following this initial expression of the Cooum River situation, participants engaged in exercises to identify important themes and model them conceptually as systems relevant to the problem situation. This involved a shift from nonsystems to a systems mode of thinking. Initially, this was done using facilitated discussion based on CATWOE analysis from SSM. Later exercises refined the root definitions, built them into conceptual models of human activity systems, and used them to stimulate debate about change.

One of the insights resulting from this system analysis was the development of what, from all indications, was a new understanding of the situation. This can be summarized in the description of emergent properties of the system that arose from attempting to describe important interrelationships, elements, and actors in the system and in attempting to identify relevant spatial and temporal scales for them. These efforts produced a shared understanding of a system that was variously characterized as a "river system *cum* sewer system," an "urban system," and a "waste disposal system." This system was identified as operating in the built-up areas of the city and was described as distinct from the upper Cooum for which a different set of actors and processes were seen to exist. That is, the lower and upper Cooum systems were identified as subsystems within a wider system. This wider Cooum system was set within a still wider system encompassing the interconnected waterways, tanks, and canals in the Chennai region. Both the identification of the system as one which is primarily urban (characterized by sewage production, its disposal, and transport) and the location by participants of the lower Cooum system within a hierarchy of systems are examples of insight into the problem situation stimulated by systemic analysis.

These results are typical of systems-based studies but were novel in the Chennai context. Most significantly, the system itself was seen as having an urban character. Rather than being merely a "natural" biological and physical system, it was seen also as a social system. It was *characterized* by human activity rather than merely affected by human activity. Instead of seeing sewage merely as an input into the system, the population of Chennai and its role in producing sewage are understood to be *part of* the system. Similarly, rather than merely attempting to manage the biophysical system from the "outside," the various government agencies were understood to be *inside* the system.

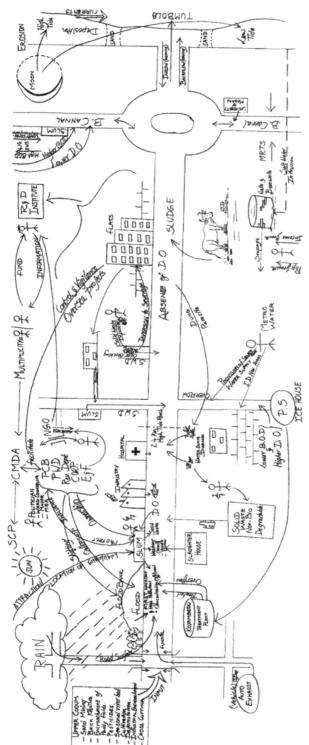

Figure 10.1   A rich picture of the problem situation developed by participants at the workshops of the Cooum River Environmental Management Research Program. Development of this diagram was facilitated by an interactive working session and discussion that was intended to provide an overview of the main components, actors, and interrelationships in the situation. This diagram was not constructed as an explicit system model, but as its construction progressed, and especially later in the workshop, it began to be used as a representation of a "system" —a bounded whole with a particular spatial and temporal scope and having an environment that influences and is impacted by the system. The representation of the upper Cooum as a separate system that provides inputs into the lower Cooum system exemplifies this. Representations of the moon having influences on tidal activity, that, in turn, has some flushing action in the river system: the Sun, with relevance to rates of evaporation and periodic drought; rain clouds representing the impact of monsoons; and a population figure indicating population growth and changing levels of income in the population at large, are examples of important influences from the environment of the system as perceived by workshop participants (Bunch 2001).

This contribution to the understanding of the situation represented a shift in the way participants thought of the problem. It has implications for how they perceived that such a situation might be alleviated. In this research, for example, a holistic understanding of the nature of the problem led to discussion of potential interventions that were, in essence, aimed at changing the waste production and disposal nature of the system. That is, in addition to the traditional engineering interventions to deal post hoc with the presence of pollution in the Cooum River (such as the dredging of sludge and flushing the Cooum), participants more and more began to propose systemic interventions targeted at altering characteristics of the system that underlie its current organizational state. These included educational awareness campaigns to change attitudes toward the environment and modify (polluting) behavior of citizens, public participation in management programs, rainwater harvesting by house owners, and promotion of tourism and recreation.

Further systems-based analysis in the first workshop (primarily influenced by SSM) provided a framework on which to base a simulation model. Facilitated discussion and working sessions, based on CATWOE (customer, actor, transformation, *Weltanschauung*, owner, environmental constraints) analyses from SSM (see chapter 8), led to the identification of subsystems in the Cooum system described in terms of primary actors and elements, transformations occurring in each subsystem, inputs and outputs, system environments, and control. For example, in an attempt to understand the current state and dynamics of the Cooum system, participants discussed and conceptualized subsystems focusing on slums, the physical hydrology of the river, the population at large, the sewerage system, the storm water drainage system, the provision of sewerage and water supply, tidal action, animal husbandry, politics, and government agency intervention and control. From such analysis, and through discussion where these conceptual models were contrasted and compared to the real-world problem situation, a deeper understanding arose.

Out of the understanding of the Cooum system represented by the rich picture and further investigation of subsystems, a general consensus arose as to the core structure of the overall system of interest. Primary elements and processes such as the population of Chennai, their activities in transforming water, food, and other goods into waste, the routing of sewage *via* the sewerage system, the monsoon and the routing of storm water *via* the storm water drainage system, the treatment of sewage at the Koyembedu Sewage Treatment Plant, and the disposal of waste and storm water into (and its transport by) the Cooum were brought into the foreground to provide a structure around which to build the Cooum decision support system (fig. 10.2).

A multitude of interrelated elements and processes were identified as impacting on, and being interrelated with, this basic structure. Importantly, the system structure demonstrated that the overall condition of the system was seen by participants to be indicated by the quality of water in the Cooum River.

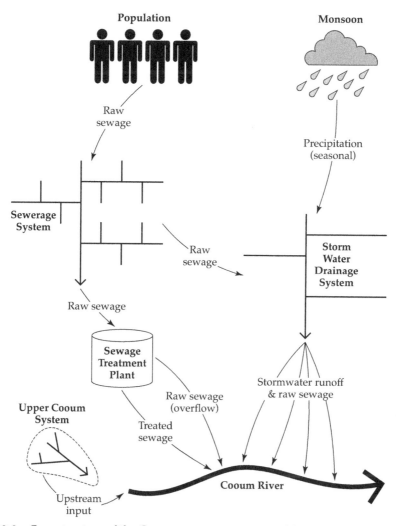

Figure 10.2   Core structure of the Cooum system as it emerged from the first workshop of the Cooum River Environmental Management Research Program. This provided a framework for the construction of a computer simulation model.

The System "As It Is" Versus the System "As It Could Be"

A common criticism of forecasting techniques, which are grounded in an understanding of a system "as it is," is that perceptions of future states of the system that are possible and reasonable are constrained by the known current state and dynamics of the system (Dreborg 1996). This can lead to identification of the most probable evolutions of the system, when what we really need to identify are desirable and feasible future states (Robinson 1990). Such an approach may preclude planning for significant change in the structure and processes of the system itself and has led to the use of techniques such as backcasting (Robinson 1990; Dreborg 1996) and future search visioning (FSV) (Weisbord 1992; Weisbord and Janoff

2000) to develop creative solutions not constrained by the preconditions of causal models. Such methods make desired future states of the system a primary focus.

From a complex systems perspective, there is an explicit recognition that the evolution of a system is often discontinuous (Gunderson and Holling 2002; Kay et al. 1999). Accordingly, this work encouraged the development of visions of desirable future states of the system that were not necessarily causally linked to the current system via projections of current trends and system configurations. To do this, from the very first working session of the first workshop, elements of future visioning were incorporated into the exercises. Statements such as those presented in box 10.2 represent images of the future that describe a system operating in a

---

BOX 10.2   What Would the Problem Look Like If It Were Solved?

"The river will be pleasant, attracting tourism."

"The city will look beautiful."

"Serious flooding problems can be avoided and spreading of epidemics will come under control."

"The river or river bed could be free from flow of sewage and the entire area serve as a[n] open space to provide free-flowing fresh air and add to the acceptability of the city as a city of people who are prepared to solve environmental problems."

"Slum dwellers will feel healthier."

"a cleaner, stench-free waterway, perhaps even permitting navigation"

The initial working session of the first workshop included a question that initiated visioning of desirable futures for the Cooum system. This question elicited statements, of which those above are a sample, about the aesthetic beauty of the river and the city, the river as a resource for tourism and recreation, a freely flowing and navigable river, the absence of slums, a healthy environment, absence of sewage and solid waste, flood protection, and an environmentally friendly city. The responses, however, did not come equally from the government, nongovernment organization (NGO), academic, and "other" groups. Thirty-four percent of these partial visions of a problem-free future came from NGO representatives, who accounted for only 18 percent of those participating in the exercise. "Other" citizens made 19 percent of the statements while accounting for 18 percent of participants. However, government delegates (38 percent of participants) and academics (27 percent of participants) were underrepresented in their responses, accounting for only 33 and 14 percent, respectively, of these statements.It may be that NGO delegates can more easily remove themselves from the context of the problem to envision a desirable future state than can government representatives who must deal with the day-to-day management of the problem or academics concerned with understanding and developing "solutions" to specific aspects of the current problem situation. Also, certain NGOs are in the business of expressing and promoting visions of desirable futures. For example, one of the participant NGOs in this research was "Exnora", whose name stands for "EXcellent NOvel and RAdical ideas," and which has the stated goal of "the generation of innovative ideas and implementing them, [to] help transform the society" (Exnora International 2000). This is also an indication of the importance of including such groups in management efforts.

different organizational domain than at present. They certainly have no direct causal root in the current (stagnant, odorous, repugnant) state of the Cooum. Further sessions in the first workshop explored objectives representing aspects of visions of desirable future states (e.g., beautification and ecological enhancement, maintenance of a navigable waterway), loosely tying these to indicators (e.g., number of trees planted, flow and depth of water in the waterway) and to management interventions that might encourage the system to evolve in a desired direction (such as greening and landscaping, regulation of flow, etc.).

This process was also taken up in the second workshop that employed simulation tools to explore aspects of the system dynamics of such desired future states. Thus images of desirable future states and future dynamics were expressed initially as "problem solved" statements. They were later manifest in the expression of objectives for rehabilitation and management of the system (box 10.3), in

---

BOX 10.3   Objectives for Management of Subsystems

Objectives for management of subsystems (themes in the rich picture) were generated by way of a brainstorming exercise. Once generated, questions were asked of each objective with respect to the problem situation: Is it specific? Is it measurable? What is the variable measured? Is it results-oriented? Is it realistic and attainable (Within what time frame?) Who will do it? Who will benefit? When will it happen? How will you know if it has been successful? In this way conceptual models were used as a source of questions to ask of the real-world situation. Objectives generated in this way included the following:

1. Improved land use and enforcement of zoning regulations
2. Maintenance of the river course as an operational waterway
3. Increased recreational use of the Cooum and environs
4. Groundwater recharge
5. Prevention of encroachment
6. Beautification and ecological enhancement
7. Improvement of surface water quality in the lower Cooum system (within city limits)
8. Improved flood defenses
9. Decreased waterborne disease
10. Increased public awareness and political will regarding rehabilitation of the Cooum system
11. Increased physical quality of life
12. Control of mosquito vectors
13. Environmental improvement of slum areas

Participants also ranked objectives, related these to indicators in the system, and discussed interventions to achieve them. The purpose of these exercises was not merely to produce and record answers to formal questions but to generate debate about change in the situation. Generation of objectives, for example, illuminated aspects of desirable future states of the system. Discussion about objectives led participants to express visions of an aware and involved citizenry, the river as a recreational resource, the river as a navigable waterway, and the Cooum as a clean river.

interventions in the system that might encourage the realization of such desirable system states, in informal narratives that arose out of group discussion and debate throughout the workshops, and in aspects of potential future states represented as simulation scenarios in the Cooum DSS.

Because participants' visions of desirable futures were not always causally linked to the current state of the Cooum system, discussion of "solutions" to the Cooum problem were often fundamentally different from interventions tried in the past. Contrast, for example, the characterization of the Cooum system as a "waste disposal system" to a system characterized by the role of the river and environs in promoting and supporting tourism, recreation, health, and happiness of the population and in presenting a positive image of the city. In discussions about this vision, participants indicated that complementary but qualitatively different alternatives (from standard physical interventions) were thought to be necessary. Such alternatives as educational campaigns aimed at modifying attitudes and behavior toward the environment and the participation of the population in management programs would be targeted at changing the nature of the system rather than at remedial pollution control.

Similar to the investigation of subsystems of the Cooum system, images of desirable future states arose as "types" (Allen, chapter 3; Allen and Hoekstra 1992). That is, in workshop debate and discussion, desirable future states of the system tended to be defined by a single theme or perspective. These included visions of the future physical hydrology of the system that described free flow and navigable depth of water within the city limits; radically different attitudes of the population toward the environment and environment-friendly behavior; a tourism and recreation economy based on the river and riverside parks as a sustainable tourism resource; serviced and hygienic communities for low income groups; and sewerage collection and treatment for all citizens. Description of such potential systems arose in discussion from exercises and paper presentations that stimulated future visioning. Such discussions produced informal narratives of the state and dynamics of such futures. Because of the focus on investigating the present situation explicitly as a system, such narratives tended to model systems as well.

An example of such a narrative is one that might be labeled a "tourism system." Discussion by participants, primarily in the second workshop, produced a narrative that described an attractor state or domain of organization for the Cooum system in which tourism is a primary activity. In this context, participants discussed the possibility of intervening to make the system more amenable to the tourism industry (by way of interventions such as increase of treatment plant capacity, slum improvement, landscaping, etc.). If this could be done, they postulated that more tourists would visit Chennai, and the tourism industry would begin to flourish. This would lead to more improvement in the system, which would stimulate increased tourist activity, which would promote maintenance and improvement of a system amenable to tourism in a type of positive feedback referred to as a morphogenic causal loop (Kay et al. 1999). In this example, increased revenue to the government from the tourism industry would be reinvested into the system to continue to promote tourism. As well, it was thought that employment and income to entrepreneurs and individuals from the tourism trade would stimulate behaviors that would maintain and improve a tourism-friendly system.

Soft Systems Thinking and Complex Adaptive Systems

The development of plans to achieve future visions as explicit targets was not an (overt) objective of this work. Instead, the research focused on learning about the system rather than its design or optimization. The implication of comprehensively planning an envisioned future system state and charting a path to its realization (Checkland 1981:256),

> is that there are systems to be engineered and the way to do this is by defining system objectives. But the context...is explicitly one of soft ill-structured problem situations in which the planning process is more important than any plan and in which 'problems do not stay solved'.

In the context of "participatory" or "complex planning" processes, generating visions of the future to be used as blueprints, Checkland (1981) has argued that the needs of such a situation cannot be met by designing "an idealized future for the system being planned for." This view provides further insight into the failure of past attempts to engineer solutions in the Cooum River situation. A plan implemented in the late 1960s and early 1970s (Appasamy 1989) involving an engineered system of lined banks, flow regulation, clearance of the mouth of the river, dredging of sludge, and construction of recreational amenities, for example, failed badly. The problem was more complex and had too much to do with human activity (grounded in norms, values, and intentions) to allow such a solution to be anything more than short-lived. The system could not be adequately defined functionally (i.e., the situation was ill-structured) and the solution depended on an attempt to understand and control causal linkages, while not allowing for the role of human activity in determining the nature and operation of the system.

Thus in turbulent situations of this nature there are not "systems to be engineered." Rather, there are only real-world problematic situations in which to intervene. We use systems thinking to help analyze and understand the situation, build models (conceptualize systems), and compare them with each other and to the real-world situation for the purpose of stimulating insight and provoking debate about desirable and feasible change. This informs action in a problematic situation, creating new experience of the world, producing new experience-based knowledge, and further informing action. The methodology is about the continual process of learning and adaptation. Mitchell (1997) drew a similar lesson for management of turbulent environments when he noted that such situations are not amenable to the use of a master plan or "blueprint" of a future state and that the *process* is as important as the product in managing them. Box 10.4 demonstrates operation of such a learning cycle in this research.

An approach that is potentially more useful and, in the long term, more effective than attempting to engineer envisioned future states is one that employs

BOX 10.4   A Process for Learning

One of the outcomes of the first workshop in this program of research was the expression of a framework for a system model of the Cooum system. Such an expression is one way to map themes drawn and systems conceptualized from the rich picture, back onto the real-world situation. The process was intended to generate debate about critical activities and processes in the situation. Formal (algebraic) expression of such activities and processes highlighted areas of uncertainty and stimulated further debate over assumptions made about the Cooum system. For example, this process led workshop participants to revise their understanding of relationships among income, consumption of water, and sewage generation. Their modification allowed for spatial variation in water quality characteristics of sewage.

   In the interval between workshops, the Cooum system framework was used to develop a Geographic Information System database and decision support system, loosely coupled with an environmental simulation model (the ACooum DSS@). Using water quality as an indicator of overall system health, the Cooum DSS was employed in the second workshop to develop exploratory management scenarios. Participants developed scenarios to explore interventions such as provision of sewerage to nonserviced slum areas, population increase, improved sewerage system technology, increased treatment capacity, artificial flushing, and runoff of the first monsoon storm flush. It was found the process of scenario building based on the Cooum system model generated debate, for example, about data quality, institutional issues related to data sharing, access to information, agency cooperation, and public participation in management programs. Participants considered such issues so important that the final workings session was usurped in favor of the inaugural meeting of a multistakeholder working group. The group's mandate is to undertake data collection and guide research on the Cooum system and to continue model development and system exploration using the Cooum DSS.

images of future states to describe alternative domains of organization of the system. In this context, propensities of the system that maintain its organization in the current domain may be identified and discouraged, while propensities that would encourage its evolution toward an attractor that characterizes a desired future might be promoted. This approach depends on an understanding of the system as a self-organizing entity (Kay et al. 1999). A central component of such an approach is the development of descriptions ("narratives") of the system that focus on a "qualitative/quantitative understanding" of the system (Kay et al. 1999; Allen et al. 2001; Waltner-Toews 2004). Kay et al. (1999) indicate that these should describe:

- the human context for the narrative,
- the hierarchical nature of the system,
- the attractors which may be accessible to the system,
- how the system behaves in the neighborhood of each attractor, potentially in terms of a quantitative simulation model,
- the positive and negative feedbacks and autocatalytic loops and associated gradients which organize the system about an attractor,

- what might enable and disable these loops and hence promote or discourage the system from being in the neighborhood of an attractor, and
- what might be likely to precipitate flips between attractors.

Although systems language was not employed in the Cooum workshops, the similarities of these guidelines with the activities undertaken in this work are obvious. An example of some of the results of this work, seen from the point of view of self-organization around particular domains of behavior or attractor states, is evident in workshop participants' analysis of the current character of the system. For example, some of the current socioecological system characteristics that could be considered to be such propensities may be identified as

- governance and management characterized by disjointed jurisdictional environments,
- mechanistic management cultures in agencies and institutions,
- overriding predominance of approaches to problem solving that assume simple linear causality and ignore social and ecological context,
- widespread ignorance and disregard of environmental consequences of personal actions, and
- corruption.

Similarly, participants discussed the organization of the system in ways that often highlighted causal loops. An example is provided by a summary of participant discussions on the cumulative effect of individual behavior and the polluted state of the system. Participants in the workshops noted that in their experience, residents typically believe that individual behavior is insignificant compared to the scale of the problem and that therefore there is no point in going to extra effort or cost to avoid contributing to this pollution. Thus polluting behavior is accepted, resulting in widespread polluting activity. The continuation of the problem at such a scale and the acceptance of such polluting behavior reinforces the belief that individual efforts will not make a difference.

This kind of thinking about feedback loops by workshop participants led to the identification of interventions such as educational campaigns and public participation in programs for rehabilitation and management. Such interventions would weaken some of the propensities in the system that lead it to organize in its current domain and strengthen others so that this human activity system may be encouraged to organize around a different attractor state (such as that characterized by the tourism system, discussed above). The evolution or reorganization of such a behavioral subsystem will alter inputs to the physical subsystem such that it may also "flip" between attractor states. A desired "flip" would see reorganization from the current domain, characterized by high levels of organic pollutants, absence of dissolved oxygen, the presence of anaerobic bacteria, and the emission of noxious gases, to one in which inputs to the river are much less polluted, dissolved oxygen is present and so decomposition of organic matter is done *via* aerobic processes, and the system can even support the presence of fish. It is a small step to make the connection of a positive feedback loop between the physical system and an envisioned system state such as that described by the tourism system.

This discussion promotes a conclusion that it is the *propensities* for systems to self-organize around particular attractor states that should be targeted by interventions in the system. This is likely to lead to qualitatively different kinds of intervention in the system. This understanding, although not explicitly expressed by participants in systems jargon, is reflected in the recommendations of the second workshop. While typical engineering interventions such as "regulation of flow," "maintenance of depth," and "construction and maintenance of intercepting sewers" had been considered and were seen as potentially useful, workshop participants did not propose them. Instead, the two workshop recommendations recognized that (1) management of the system as a whole needs to be addressed by "an overarching agency" to coordinate, monitor, and control efforts of agencies to intervene in the evolution of the Cooum system and (2) such a process is ongoing and should involve stakeholders, and this can be done in part through the involvement of a working group consisting of representatives from pertinent government agencies, academia, NGOs and interested parties, which will undertake to research and monitor the system in support of efforts for its rehabilitation and management.

Conclusions

In their report to the International Joint Commission for the Great Lakes, Allen et al. (1994) recommend SSM as a particularly appropriate methodology "for making operational the ecosystem approach." This research supports their recommendation. Not only has SSM contributed a useful set of tools, such as CATWOE and rich picture techniques, to support problem identification and system conceptualization but, at a more fundamental level, SSM also informs a soft systems approach in which to use all techniques in the ecosystem approach "toolbox." In this research the ecosystem approach, influenced by SSM, was operated as a system of learning about the Cooum problem situation. The research was oriented to the generation of debate about desirable and feasible change and accommodation of interests rather than on forecasting and comprehensive engineering of future systems.

This approach was found also to be amenable to incorporation of complex systems ideas and is complementary to emerging techniques in ecosystem science, such as the use of narratives (Kay et al. 1999; Allen et al. 2001; Waltner-Toews 2004), to deal with the problems that complexity poses for anticipatory science. As this research program progressed, for example, participants began to propose interventions that would have the effect of weakening propensities of the current organizational system state and that would encourage evolution toward alternative domains of behavior. Contrast this to past engineering-oriented solutions that merely targeted symptoms or manifestations of complex systems behavior.

Such proposed interventions (e.g., educational campaigns, improved communication, and coordination and stakeholder participation) are representative of a fundamental shift in understanding of stakeholders about the problem situation. Initial conceptions of the problem were rooted in a biophysical conception of the Cooum system. The holistic approach to the problem embodied in the ecosystem approach and the capability provided by soft systems methodology to deal with human activity in the system fostered the development of an understanding of the situation in which human activity is central. As the research program progressed, participants began to characterize the system more as an urban system and a waste disposal system than as a natural river system. This, in turn, influenced objectives for rehabilitation and management of the Cooum system and proposed interventions to achieve them. Assuming that such a participatory, adaptive ecosystem approach can survive in the kind of command and control, programmed management environment that exists in India, the potential for management of severe environmental problem situations characterized by human activity, such as that of the Cooum River in Chennai, can be greatly improved.

Acknowledgments

Logistical and financial support for parts of this research was provided by the Madras-Waterloo University Linkage Program (CIDA) and a CUC-AIT travel grant. The author wishes to express his appreciation to Dr. S. Subbiah and his colleagues at the Department of Geography at the University of Madras for hosting field activities related to this work, to participants in the research program for their individual and collective contributions, and to Dr. D. Dudycha, Dr. B. Hyma, and Dr. J. Kay at the University of Waterloo for their guidance and support for this work. This chapter summarizes work presented in greater depth in the book by Bunch (2001).

References

Allen, T. F. H. and T. W. Hoekstra. 1992. *Toward a Unified Ecology*. New York: Columbia University Press.

Allen, T. F. H., B. L. Bandurski and A. W. King. 1994. The ecosystem approach: Theory and ecosystem integrity. Initial report. United States and Canada: Ecological Committee to the International Joint Commission's Great Lakes Science Advisory Board.

Allen, T. F. H., J. A. Tainter, J. C. Pires and T. W. Hoekstra. 2001. Dragnet ecology—"Just the facts, Ma'am": The privilege of science in a postmodern world. *BioScience* 51(6):475–485.

Appasamy, P. 1989. *Managing Pollution in the Waterways of Madras City: An Initial Assessment*. Working Paper 88. Chennai, Tamil Nadu, India: Madras Institute of Development Studies.

Bunch, M. J. 2001. *An Adaptive Ecosystem Approach to Rehabilitation and Management of the Cooum River Environmental System in Chennai, India*. Geography Publication Series 54. Waterloo, Ontario, Canada: Department of Geography, University of Waterloo.

Checkland, P. B. 1979. Techniques in 'soft' systems practice part 2: Building conceptual models. *Journal of Applied Systems Analysis* 6:33–40.

Checkland, P. B. 1981. *Systems Thinking, Systems Practice*. New York: Wiley.

Checkland, P. B. 1999. Soft systems methodology: A 30-year retrospective. In *Soft Systems Methodology in Action*, eds. P. B. Checkland and J. Scholes, A1-A66. New York: Wiley.

Checkland, P. B. and J. Scholes. 1990. *Soft Systems Methodology in Action*. New York: Wiley.

Dreborg, K. H. 1996. Essence of backcasting. *Futures* 28(9):813–828.

Exnora International. 2000. "Exnora International - An Acquaintance" (Mar.15, 2000), at (http://exnorainternational.org/).

Government of India. 2001. Urban Agglomerations/Cities having a population of more than one million in 2001. Census of India (Provisional). New Delhi: Office of the Registrar General, India. (Available at: www.censusindia.net)

Government of Tamil Nadu. 1997. Terms of Reference for Consultancy Services for Preparation of Master Plan, Immediate Works Programme and Bid Documents for the Chennai Waterways Rehabilitation and Reclamation Project. Chennai, Tamil Nadu, India: Tamil Nadu Public Works Department.

Government of Tamil Nadu. 1981. Census of India, 1981: Series 20, Tamil Nadu, Part XIII A and B; District Census Handbook, Town Directory and Divisionwise Primary Census Abstract, Madras. Chennai, Tamil Nadu, India: Government of Tamil Nadu.

Gunderson, L. H. and C. S. Holling (2002) *Panarchy: Understanding Transformations in Human and Natural Systems*. Washington, D. C.: Island Press.

Gunaselvam, M. 1999. *Preserving the Identity of Waterfronts in Chennai City*. Chennai, Tamil Nadu, India: Department of Geography, University of Madras (unpublished).

Inland Waterways Authority of India. 1998. Report on Reconnaissance Hydrographic Survey and Pre-Feasibility Study on Cooum River and South Buckingham Canal. New Delhi: Government of India.

Kay, J. J., M. Boyle, H. A. Regier and G. Francis. 1999. An ecosystem approach for sustainability: Addressing the challenge of complexity. *Futures* 31(7):721–742.

Mitchell, B. 1997. *Resource and Environmental Management*. Edinburgh Gate, Essex, UK: Addison Wesley Longman.

Mott MacDonald. 1994. Sludge Disposal Consultancy, Madras: Final Report. (August 1994), A consultancy under consignment from the Overseas Development Administration for the Department of Environment and Forests, Government of Tamil Nadu. Chennai, Tamil Nadu, India: Government of Tamil Nadu, Department of Environment and Forests.

Robinson, J. B. 1990. Futures under glass: A recipe for people who hate to predict. *Futures* 22(8):820–842.

Rondinelli, D. A. 1993. *Development Projects as Policy Experiments: An Adaptive Approach to Development Administration* (2nd ed.). New York: Routledge.

Sahadevan, P. V. 1995. A comprehensive land and water resources development plan for flood mitigation and for sustainable healthy environs of Madras City. In *Water Energy 2001*, 709–723. New Delhi, India.

Sahadevan, P. V. 1996. Artificial floods to Flush Cooum: A multi-purpose reservoir at Korattur-Anicut for implementation. Discussion document. Chennai, Tamil Nadu, India: Tamil Nadu Public Works Department.

Srinivasan, S. 1991. Water supply and sanitation for Madras: The 2011 context. In *Madras 2011: Policy Imperatives - An Agenda for Action, Volume III: Research Papers*, ed. The Times Research Foundation, III.1–III.28, Chennai, Tamil Nadu, India: Madras Metropolitan Development Authority.

Sridhar, M.K.S.. 1982. A Field Study of Estuarine Pollution in Madras, India. *Marine Pollution Bulletin* 13(7):233 236.

United Nations Centre for Human Settlements (Habitat). 1991. *Guide for Managing Change for Urban Managers and Trainers*. Nairobi, Africa.

Waltner-Toews, D. 2004. *Ecosystem Sustainability and Health: A Practical Approach.* Cambridge, UK: Cambridge University Press.

Weisbord, M. B. 1992. *Discovering Common Ground: How Future Search Conferences Bring People Together to Achieve Breakthrough Innovation, Empowerment, Shared Vision, and Collaborative Action.* San Francisco, Calif.: Berrett-Koehler.

Weisbord, M. B. and S. Janoff. 2000. *Future Search: An Action Guide to Finding Common Ground in Organizations and Communities.* San Francisco, Calif.: Berrett-Koehler.

11

Landscape Perspectives on Agroecosystem Health in the Great Lakes Basin

Dominique Charron and David Waltner-Toews

Sustainability and Health in Great Lakes Basin Agroecosystems

Over about two centuries, the Great Lakes Basin (GLB) has changed from old-growth forest systems and areas of prairie supporting hunter-gatherer societies and a diversity of other life to a densely populated, urbanized, industrial, and richly agricultural system supporting very affluent, technological societies. There remain only vestigial remnants of the old forest and eastern prairie ecosystems.

This profound and rapid change of the landscape has meant that all levels of ecosystem are deeply affected by human activity and that human activity is, in turn, affected by those changes. The shift in Lake Erie from a turbid-water pelagic system to a clear-water benthic one, after the apparent success of pollution control programs in the GLB and the invasion of the Great Lake by zebra and quagga mussels, has resulted in reduced stocks of some species that favor a pelagic system. As a result, fishermen have needed to adapt by abandoning some favorite species and pursuing those more abundant. Such adaptation incurs retooling costs and learning of new skills (Regier and Kay 1996; J. J. Kay, personal communication). The introduction and spread of West Nile virus and Lyme disease have both been facilitated by social-ecological changes in the GLB (Charron et al. 2003).

Although agricultural practices in the GLB have improved in recent times, economic intensification and land pressures from urban sprawl are potential obstacles to sustaining this improvement. Issues surrounding the carrying capacity of the GLB's agroecosystems are emerging. These include threats to drinking and recreational water quality, food safety, biodiversity conservation, and others. The sustainability of livestock farming in the GLB is an important issue among these. Yet, few examples of indicators of livestock carrying capacity can be found in the literature. Studies on the topic have not utilized an ecosystemic approach with humans and agriculture intrinsic to the system. Rather, they have viewed agriculture as a stress from outside on nondomesticated biota and on physical components of ecosystems. As a result, these studies have investigated responses to this agricultural stress on populations or organisms but not systemically.

Agriculture is an integral aspect of southern GLB landscapes. In fact, many remnants of ancient forest ecosystems are only preserved today because they are a deliberate part of agricultural landscapes and thus protected from urbanization. As is the case in agroecosystems everywhere, livestock contribute to agroecosystem cycles and processes by recycling organic matter and nutrients. Livestock agriculture also represents a diversification of land use in the cultivation of forage and feed crops absent where the intensive cultivation of grain by monoculture is practiced. Certainly, many rural livelihoods in the GLB depend on livestock agriculture, and urban residents of the GLB rely on neighboring farms for food and benefit from the lower transportation costs associated with locally produced food. These interdependencies tend to support more exploitative, less sustainable dynamics within agroecosystems. A shift in focus may help identify some central issues of agricultural sustainability, water quality, and biodiversity conservation.

Are Great Lakes Basin Agroecosystems Sick?

The relationships between agricultural activity, underlying ecosystems, and society are complex. Despite trends toward economic consolidation, individual agricultural enterprises tend to be economically small compared to other industries. Cumulatively, however, agricultural enterprise has a profound and complex impact on large landscapes. Agriculture fills a basic human need—food—and thus is viewed by society as being different from other industries. Agricultural effects—both positive and negative—tend to be geographically and economically diffuse. Impacts of agriculture, including the familiar environmental issues as well as social and economic ones, also tend to be highly controversial given the complex role of agriculture in society. These circumstances may make the agricultural sector more sensitive to criticism and perhaps less willing or able to respond.

Livestock farming affects many aspects of aquatic ecosystems, with biophysical, human health, and economic consequences. In the GLB these effects are detectable in groundwater, in headwaters, and in the Great Lakes themselves. The most direct and proximate of such livestock impacts will be seen in rural streams because these gather surface runoff. The deterioration of water quality and of stream ecosystems is embedded in a system in which agricultural land is being lost to urban sprawl, and agriculture itself is becoming industrialized, with associated threats to rural communities and the family farms that support them. All of this is, in turn, embedded in larger, systemic environmental changes, both regional (such as air quality deterioration) and global (such as climate change).

It is apparent that livestock farming is an integral part of GLB agroecosystems and may contribute to long-term agricultural sustainability. It is also apparent that stream ecosystems and water quality, for example, are suffering from negative effects of livestock production and other agricultural practices. An assessment of these problems requires an integrative approach that will take into account multiple scales and effects while still providing solutions.

An Ecosystem Approach to a Quantitative Analysis of Indicators

An ecosystem approach, with an emphasis on multiple scales, perspectives, and feedback loops, would seem to provide an ideal way of framing these complex issues. In this case study, an "agroecosystem health" approach—applying community health concepts of characterization and evaluation to agroecosystems—was used to develop a theoretical model of the livestock farming-streams subsystem in the 1990s GLB. This project, part of the Agroecosystem Health Project at the University of Guelph (Smit et al. 1998), was primarily scholarly and expert-driven, without major local stakeholder involvement.

The Guelph project research structure focused on issues of scale, perspective, and evaluative criteria (health attributes), one version of which was applied to the GLB livestock project (fig. 11.1). Perspective referred to something akin to Allen's "type." Researchers tended to divide "perspective" into various categories; in the GLB livestock project, we used what we regarded as the minimum set: ecological (in this case used synonymously with biophysical) and socioeconomic. Health attributes can be thought of as vectors that express variability between one level usually associated with poor health and one or several other levels usually associated with good health. These vectors represent distinguishable elements that, taken together, reflect the status of the system under study. Integrity as a health attribute (see chapter 19 for a more detailed consideration of the notion of integrity) refers to good condition and diversity of natural, social, economic, and cultural elements of the agroecosystem. Efficiency reflects output per unit input.

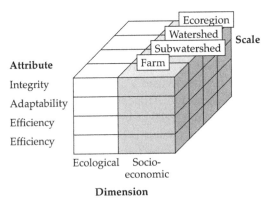

Figure 11.1   A three-dimensional schematic of the agroecosystem health framework. Attributes of health are measured in both biophysical and socioeconomic dimensions and over many scales. Four superattributes are illustrated. These attributes vary, within constraints, over time.

Adaptability includes response capacity, stability, and resilience. The balance be-tween stability in production and resilience to stress in agroecosystems is included in this attribute because anything outside the optimal range of balance may result in catastrophic events (which are generally undesirable in agroecosystems). The capacity of the system to respond and adapt to disturbance is a product of the rela-tionship between stability and resilience. Whether the response is desirable or not, however, is a goal-related interpretation that falls into the effectiveness attribute of health. Effectiveness refers to the performance of the agroecosystem in relation to goals attributed to the system by a given group of stakeholders (Smit et al. 1998). Health is a value judgment of system status based on an overall interpretation of many different indicators seen as a whole, usually in relation to some goals or targets.

Using Allen and Hoekstra's (1992) terminology, this project focused on eco-system criteria at the large subwatershed scale. The ecosystem approach was also used in the selection of indicators from multiple dimensions and scales, in inform-ing statistical models of those indicators, and in the framing of the interpretation of those models. The theoretical dynamics of the system are described later.

The Face of Agriculture in the Great Lakes Basin: System Description

Agriculture in the GLB is diversified, with intensively managed crop or livestock farms, mixed farms, extensive grazing operations, and, where climate and soils permit, specialized fruit, vegetable, and tobacco cultivation. Crop farming is both economically important and a major disturbance of the biophysical dimension of GLB agroecosystems. Demand for feed and availability of manure fertilizer influ-ence a farmer's choice of crops and fertilizer and closely link crop farming to live-stock agriculture. Livestock rearing practices in all sectors of the industry have un-dergone major changes since the Second World War. In their report on agriculture in the GLB, Jarvis et al. (1996) made several observations regarding the economic and mechanical intensification over this period, including high-production dairy cattle raised on large farms, large beef feedlots, large segregated herds of swine, and poultry managed intensively. These changes in the livestock industry have accompanied changes in markets and policies, and have affected land use patterns in the GLB accordingly. In the early 1990s, agriculture in the GLB was experiencing a period of rapid change, which persists more than a decade later. Intensification of livestock production, an emphasis on cash crops, and farmland conversion to urban land use are among the economic responses to this change, eventually be-coming forces of change themselves.

A Biophysical Perspective

The GLB measures 8.9 million hectares (fig. 11.2). The five lakes, Superior, Huron, Michigan, Erie, and Ontario (one third of this area), and their watersheds drain into the North Atlantic Ocean via the St. Lawrence River. Each Great Lake has its own watershed, with typical soils, native flora, and fauna. These lake water-

Figure 11.2   Map of the Great Lakes Basin—political boundaries and major cities. Map credit: National Oceanographic and Atmospheric Administration, Great Lakes Environmental Research Laboratory (www.glerl.noaa.gov/pubs/photogallery/albums/Misc/pages/1094.htm)

sheds are subdivided again into major river watersheds, which contain smaller river and stream watersheds. The GLB includes part of the Canadian province of Ontario and parts of eight U.S. states (Wisconsin, Minnesota, Michigan, Illinois, Indiana, Ohio, Pennsylvania, and New York). The soils of the northern portion of the GLB are thin or peat, overlying the Precambrian bedrock of the Canadian Shield. The southern soils are rich sedimentary deposits and glacial till and well suited to agriculture. Agriculture occupies over one third and forest just under two thirds of the GLB land mass. Most of the agricultural portion of the GLB enjoys a mild, humid, continental climate, with cool-cold snowy winters and warm summers.

A Socioeconomic Perspective

The population of the GLB in 1991/1992 was over 39 million, an increase of 5.7 percent over the preceding decade. Urban land use accounted for 1 percent of the total basin area, mostly in the south, where 80 percent of the Basin's population lived in only seventeen cities (Jarvis et al. 1996). The rest of the population lived in smaller cities and in rural areas.

Agriculture is an economically important industry in all GLB jurisdictions, with its eight American states (including portions outside the GLB) in 1991–1992 producing 30 percent of all U.S. sales in agriculture, for a return of $45 billion USD. Major products included corn, soybeans, and milk. There were 5 million hectares in agricultural production in Ontario (all of these in the GLB except the eastern tip of the province), and these produced 22 percent of all Canadian agricultural revenue. There were nearly 204,000 farms in the GLB in 1991–1992, of which 29 percent (59,000) were Canadian. Proximity to urban centers is typical of GLB farms, with 64 percent of farmland within 50 km of a major urban center (Jarvis et al. 1996).

A Model of Livestock Impacts on Streams in the Great Lakes Basin

Agroecosystems in the GLB share elements of geological history, climate, and an industrialized, profit-oriented agricultural sector. The system is intensively managed to achieve many different agricultural, economic, and industrial goals. The GLB is also organized into subsystems, defined by both spatial-temporal scale and type criteria, such as urban or rural land use, intensive or extensive farming practices, natural or exploited ecosystems, etc. There are innumerable interactions and feedback loops between the socioeconomic elements and the biophysical elements of the GLB. Although the focus of this case study was on the biophysical aspects of the livestock subsystem, a socioeconomic perspective was also considered.

A theoretical agroecosystem model was elaborated for the livestock production-streams subsystem in the GLB (fig. 11.3). The theoretical subsystem model was built around three linked nodes: livestock production, stream health, and farm economic health. In this model, the livestock node affects stream health through three major pathways: manure pollution, feed crop management, and grazing. Contained within the feed crop management pathway are the effects of tillage differences between crops, pesticide application, chemical fertilizer application, and tile drainage. Grazing impact includes riparian zone degradation, as well as compaction of the soil, a factor in run off. Some feedback loops are evident. A simple example is seen where farm economic health affects livestock management, which affects corn production (corn is produced to feed livestock), which feeds back on farm economic health (the sales of corn increase farm income and potentially, improve economic health). With this model as a backdrop, this project focused primarily on relationships between indicators of livestock and cropping intensity and biotic indicators of stream health.

As a measure of the biophysical health (particularly integrity) of the system, stream health provides an integration of the cumulative, mitigated, or buffered effects of livestock production. Streams respond to livestock production by collecting surface and subsurface runoff from crop fields, livestock yards, pasture, and tile drains. Streams also reflect grazing pressure, indirectly through erosion and increased runoff, or directly from overgrazing in the riparian zone.

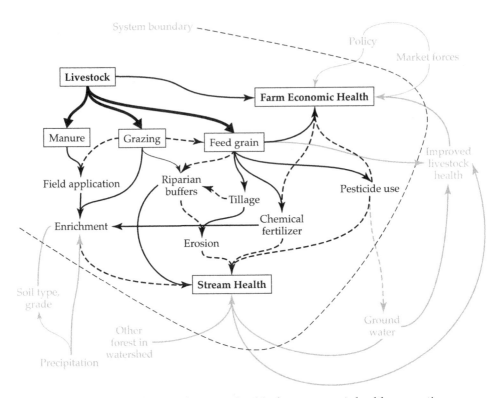

Figure 11.3   A model of livestock–stream health–farm economic health connections within the watershed and beyond. These three nodes of interest are shown in boxed, black boldface. Relationships between these nodes and through many other variables are shown as positive associations (solid arrows) and negative associations (dashed arrows). Lines without arrowheads indicate an association that could be either positive or negative. Gray font identifies variables outside the model that influence model parameters.

For this model, farm economic health is characterized in terms of financial and other measures that would modify the effects of livestock production on stream health. For example, financially marginal farms may not afford riparian buffers, intensive cash-crop operations may exploit the riparian zone, and other affluent farms may choose to invest in stream protection measures. Both very poor and very strong economic health conditions may be associated with poor stream health but for different reasons. In this model, livestock production indicators measure only the stress or pressure in the system, and stream indicators, the response to this stress. Economic indicators could lie on either side of the equation, i.e., as stressor or as response.

The theoretical livestock-streams subsystem model was analyzed at two scales: an in-depth case study of a single stream (Carroll Creek, in the Ontario portion of the Lake Erie watershed) and adjoining farms, and a GLB-wide large subwatershed-level multivariate analysis of aggregate indicators of livestock and cropping pressures, farm economic health, and stream health. Many complex relationships and feedback loops occur both within the individual watersheds and between them. Within any given watershed, livestock operations will occur heterogeneously and streams will exhibit a range of health states, potentially from very poor to excellent. The research was conducted in two parts, details of which are given below. A more detailed account of methods and results is given by Charron (2001).

What We Found: System Dynamics

Farm Scale: Carroll Creek

Carroll Creek provided an opportunity for a detailed study at the farm and stream scale of the associations between livestock agricultural and financial indicators and stream health indices (an adapted index of biotic integrity (IBI) (Karr 1981) and a habitat index). The study was conducted in two parts. In the first part, an analysis was made of fish community trophic structure, using a form of Karr's (1981) IBI adapted to Toronto area watersheds by Steedman (1988). The IBI score is the sum of values allocated to several different criteria, including proportion of carnivores, presence of sensitive species, proportion of tolerant fish, fish health, etc. The higher the IBI, the better the biotic integrity of the stream. The IBI has been used successfully to assess livestock impact on streams (Karr et al. 1985). Correlation was measured between IBI, stream habitat score (a high score indicates an intact riparian zone and good in-stream fish habitat), and livestock access to the streams at ten representative sites along the creek. In the second part of the study, a detailed questionnaire on livestock and crop management was used to gather information on livestock and crops. These data were then examined for possible links to the IBI data.

The results of the first part showed that stream ecosystem integrity (as measured by IBI), the number of trout and bass species, and stream habitat score worsened where cattle had access to the creek. In the second part of the study, an exponential decay function of livestock density along Carroll Creek provided a model of how livestock density indicators behave with stream health indicators downstream. Prevalence of black spot disease (a parasitic skin infestation of fish) tended to increase where the decay model indicated higher livestock density effects. A higher proportion of carnivores in the fish community tended to be linked to lower livestock density effects in the decay model. The IBI was less informative in the decay model of livestock density effects than it had been for livestock access.

Results of the economic survey were limited by the small sample size ($n = 9$). However, the finding of wealthier farms having better stream health was suggestive of a greater ability to protect streams from the negative impacts of livestock farming. Although livestock density (all species) was higher on these farms, less grazing occurred.

Some of the linkages suggested by the Carroll Creek study are illustrated in fig. 11.4. The impact of livestock on fish-based stream health indices (IBI and its components) was paralleled by impact on an index of stream habitat quality. Live-

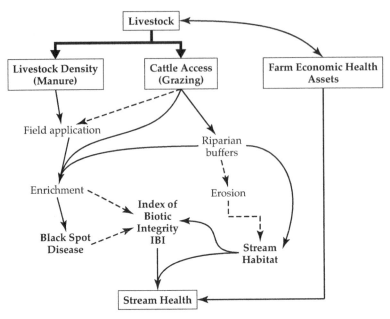

Figure 11.4   Results of the theoretical model of livestock production and farm economic indicators on stream health in Carroll Creek. This figure is adapted from fig. 11.3. Three linked nodes of interest (stream health, livestock production, and farm economic health) are boxed in thick black boldface. Their interrelationships through many other variables are shown as positive associations (solid arrows) and negative associations (dashed arrows). Variables significant in the analysis included: livestock density, cattle access, farm assets, black spot disease, IBI, and stream habitat.

stock grazing in riparian zones, riparian buffer width, quality of in-stream habitat, and the integrity of the fish community in the creek, seem to be involved in a key dynamic that may be an agroecosystem health determinant. The influence of farm economic health on elements of this dynamic was not clear from the Carroll Creek study but warrants further investigation.

Great Lakes Basin Watersheds

The Carroll Creek study confirmed some aspects of known negative impacts of livestock on stream ecosystems. It also suggested that stream ecosystems could be resilient or adapt to some livestock impacts. Was Carroll Creek representative of other GLB streams? What levels of livestock activity might a watershed withstand without jeopardizing stream ecosystems? How did crop farming and urbanization fit in? The approach used to answer these questions focused on the large watershed-scale, based on previous work suggesting that watershed dynamics provided an integrative view of agricultural impacts (Lotspeich 1980; Richards et al. 1996). An aggregate watershed-level analysis of indicators of livestock intensity and stream health indices was undertaken using existing census and stream monitoring data from multiple jurisdictions in the GLB.

Digital census data from the GLB for the years 1991–1992 were reaggregated from census division to watershed using the proportional area of each census division within a watershed. Two stream health indices were standardized, verified against existing indices, and then applied to existing GLB streams data. One index was based on aquatic macroinvertebrate tolerance to organic pollution [a taxonomic order-level Hilsenhoff biotic index (HBI)]. The HBI is calculated by summing the products of the proportion of each invertebrate order in the benthic macroinvertebrate sample and an organic pollution tolerance value. Higher HBI values signify high tolerance to organic pollution, thus poor integrity (Hilsenhoff 1987, 1988). The other index was a new modification (different from the IBI used for Carroll Creek) of the widely used index of biotic integrity (Karr 1981). This IBI was still based on scores for fish population trophic structure criteria, with higher scores indicating better biotic integrity. It was developed, using existing data from all parts of the GLB, to be robust to differences in species composition, stream thermal regime, and other differences between watersheds of the GLB (Charron 2001).

Aggregate, watershed-level analyses were performed on indicators of livestock pressure, farm financial indicators, and each of the two stream health indices (HBI and IBI). Mixed effects regression models were used to assess the associations between indicators of livestock production and each index. Several models were needed to describe the pathways of livestock impact (manure, crops, grazing) and the role of farm economic indicators. A more detailed description of the rationale and methodology is given by Charron (2001).

Different livestock production indicators were associated with HBI and IBI, consistent with previous findings that macroinvertebrate and fish communities do not reflect the same aspects of stream health (Charron 2001). The simplest models were of livestock effects on HBI—possibly because Hilsenhoff indices are designed to reflect organic pollution in streams (Hilsenhoff 1988). Given that livestock density, crops, and farm financial conditions are not independent, their associations with stream health were modeled separately. Swine density, corn area, and farm assets were each separately and inversely associated with HBI (fig. 11.5). One interpretation of the results of HBI models is that each model reflects one aspect of the same intensive livestock farming system (swine production, in this case). That swine density and corn area were stronger predictors of HBI than grazing livestock (cattle) or forage is a novel perspective on previously held ideas of livestock impacts on streams.

The associations between livestock production indicators and IBI were more complex. Again, intensive livestock system indicators (poultry density, soybean area, and capital value) were negatively associated with this fish community index. Further modeling and perhaps small area case studies are required to understand the apparently beneficial effect of higher dairy cattle density in these models. From fig. 11.5, the central importance of corn is evident though not solely as a degrading influence. Other cash crops (soybeans, wheat) were more important (statistically) than corn, however. These crops are grown as part of intensive (soy-

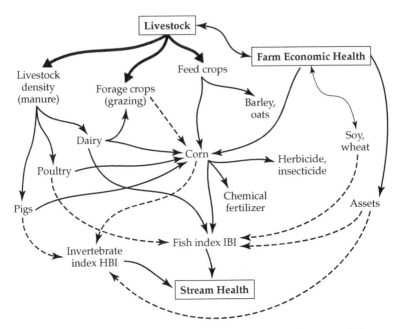

Figure 11.5 Results of the theoretical model of livestock production and farm economic indicators on stream health between watersheds in the Great Lakes Basin. This figure is adapted from fig. 11.3. Three linked nodes of interest (stream health, livestock production, and farm economic health) are boxed in thick black boldface. Their interrelationships through many other variables are shown as positive associations (solid black arrows) and negative associations (dashed arrows). All variables in the figure were assessed in the multivariate regression analysis.

bean) or less intensive (wheat) agricultural systems to generate income rather than feed livestock but are an integral part of many livestock rearing systems. Corn is also an important cash crop (for feed or for the production of ethanol). Another important statistical finding lay in that much of the variability in stream health indices occurred within a single stream watershed rather than between streams in a larger watershed. Further investigation of livestock impacts on streams might focus at a finer scale than the large watershed.

What Do These Models and Indicators Tell Us?

The Carroll Creek findings suggested that grazing livestock (in particular, cattle access to the creek) was detrimental to the fish community (measured by IBI) and in-stream habitat and that these two outcomes (fish and habitat) were highly correlated. The watershed-scale regression analysis clearly identified the important role of corn in many of these models. This suggests that practices surrounding corn cropping may be considered determinants of stream health, in particular, stream biophysical integrity, and that corn area may be an indicator of these effects. Corn is grown mostly for animal feed in the GLB (Jarvis et al. 1996). Dairy cow, pig, and poultry are corn-fed species and were important animal density variables in both IBI and HBI models. Other crop variables (wheat, soybeans), significant in the regression models, are often grown in rotation with corn. These interrelationships may, in part, explain the complex statistical behavior of the regression models. At the stream level, effects of cattle access to streams were significant, but at the larger watershed level, grazing effects were not as important as overall swine or poultry density. Dairy cattle may be pastured, yet dairy density was found to be less degrading than poultry density in an IBI model. These conflicting findings do not easily fall within a common vision. Rather, they highlight the nature of trade-offs that are likely within agroecosystems—be they across levels of the holarchy or between dimensions (economic versus ecological) of the agroecosystem.

Some of the findings suggest that, relative to more intensive livestock systems such as swine production, grazing cattle systems may be less harmful to stream health. Others have identified intensive livestock farming as a threat to drinking water (Ignazi 1987). The Environment Commissioner for Ontario in 2000 tabled a report calling for a provincial strategy to address groundwater quality and intensive farming (Miller 2000). High cattle density has been linked to increased risk of human disease in many of these same GLB agroecosystems (Michel et al. 1999; Valcour et al. 2002). This paradox provides further interesting research questions. Investigations that are more detailed could be made into the stream health effects of each farming system (extensive grazing, moderately intensive dairy, and intensive swine poultry crop), including microbiological water quality and human health indicators.

Did This New Approach Work?

This agroecosystem health framework provided a set of parameters for a complex research problem and was the organizing principle of this study. The choice of stream health as outcome, the selection of biotic indicators of stream health over chemical or physical measures, the watershed scale of analysis, and the inclusion in the analysis of farm economic indicators would not necessarily have occurred using a more conventional approach to the study of livestock impacts on the environment. Given constraints of money, personnel, time, and the analytical tools available, some other advantages of the framework had to be left aside; these include stakeholder participation and measuring effectiveness and adaptability of agroecosystem components. The framework could (theoretically) accommodate multiple temporal and spatial scales, as well as the feedback loops that either stabilize or destabilize these agroecosystems. However, the data to identify these loops, if measured at all, are not measured in a reliable or consistent manner. Epidemiological, ecological, or other models have yet to be developed to comprehensively analyze and synthesize such data. For instance, over the long term, certain types of subwatersheds may attract particular kinds of farmers who then have impacts on the streams, which then influence the structure of the farming systems. Debt loads of farmers may influence economic and fiscal policies that would, in turn, influence both the financial and ecological integrity of farming areas. At this time, the framework as presented here contributes most on conceptual and heuristic levels. As our understanding of agroecosystems improves and modeling approaches evolve, more applied aspects of the framework can be developed.

This case study of the impacts of livestock production on stream health indicators in the GLB identified several research needs. Additional information on the socioeconomic dimension of livestock production agroecosystems would be useful in understanding livestock management decisions that affect stream health. We need validation and scoring trials for the two stream health indices used, a better understanding of the degree to which such indices are scale and context dependent, and an understanding of what they actually indicate with regard to stream integrity and health. Given an apparent trend toward increased specificity of these types of indices, what role is there for indices that may be applied more generally?

The linear multivariate regression modeling approach of GLB watersheds, while analytically and statistically powerful, could not capture the interrelatedness and feedback mechanisms that are reflected in this dynamic, nonlinear system. The regression models showed that some indicators of intensive livestock production are more strongly associated with stream health indices (IBI and HBI). These findings relate mostly to the biophysical dimension of agroecosystems, although the economic indicator, farm assets, behaved consistently with other indicators of intensive agricultural production. IBI and HBI are indicators of biophysical integrity, and, to a lesser extent, efficiency. Though they might be very useful as measures of agroecosystem adaptability, the data did not permit the verification of this by assessing changes in the indices over time nor would the statistical methods we used have been appropriate for assessing temporal effects. We did not, in this GLB project, take any measures of social capital, although social capital was addressed within the larger AESH Project (Smit et al. 1998).

Without specific goals and stakeholder participation, agroecosystem effectiveness could not be quantified. The identification of agroecosystem stakeholders is difficult in a geographic area the size of the GLB, in particular, because the systems are under multiple levels of governance and land tenure. Agriculture, urbanization, and stream monitoring generally fall under different government agencies. In the GLB study, relevant levels of government included two nations, nine states and provinces, and many smaller regional and municipal governments. In Ontario, watersheds are also managed to some degree by another level of governance, the Conservation Authorities (CA). The scope of watershed management varies widely between CAs. Because of the complexity of the governance structure, responsibility for the problem (livestock impacts on stream ecosystems) is fragmented, and stakeholders are dispersed and disorganized. Clearly, the logistics of mounting thorough stakeholder consultations are prohibitive at this scale. Without strong stakeholder engagement, the project tended to emphasize the quantitative, scientific side of the diamond diagram (fig. 14.1); this has been a weakness of many of the early ecosystem health studies, particularly at large spatial scales. Identifying appropriate scales and points of engagement and intervention is one of the major challenges for these approaches if they are to be relevant to public policy. In

Ontario, the Conservation Authorities, being local watershed-based organizations and having responsibilities for land use and stream ecosystems, might provide an excellent starting place for future stakeholder involvement in this issue.

Degraded streams are clearly undesirable to most, if not all, stakeholder groups. These models have shown that there is a trade-off to be made between intensive agricultural productivity and biophysical integrity. This trade-off is detectable in analytical models at the watershed level, and, in 1991–1992, seemed to indicate a preference for economic health (higher assets) rather than stream health in the intensively agricultural sections of the GLB. A separate analysis of extensive agricultural systems (beef, and to a lesser extent, dairy) would provide additional information on the relationships between livestock production indicators and stream health.

This case study highlighted both the strengths and weaknesses of quantitative indices and their use in multivariable statistical models to assess progress toward sustainability or health. It is clear that considerable work remains to set such work into the more complex ecosocial context from which such indicators have emerged and within which they must be interpreted.

Conclusions

Agroecosystem health provided a useful framework for organizing our understanding of the well-being and dynamics of agroecosystems. Allen and Hoekstra (1992:9) have argued that, "for adequate understanding, it is necessary to consider at least three levels at once: 1) the level in question; 2) the level below that gives mechanisms; and 3) the level above that gives context, role, or significance." The Carroll Creek study provided some insight into local stream-agriculture dynamics, but multiple sites studied in greater depth would have provided a stronger basis for understanding watershed dynamics. Furthermore, participatory work at local levels would have enabled us to better identify important social issues, which drive many of the biophysical dynamics of the GLB. Indeed, the social visioning side of the diamond diagram (fig. 14.1) was largely missing or merely inferred from socioeconomic trends. This was in part at least a logistical issue: how does one do that kind of participatory work at a GLB scale? A larger overview of the GLB would have also provided a stronger context. Many of the issues were addressed in other projects that grew out of the Guelph Agroecosystem Health Project (e.g., in Kenya, Peru, and Nepal) and are discussed in this book. It is not accident, however, that the follow-up projects tended to be locally focused. The complexity of the issues being addressed, combined with the large watershed and basin scales at which we were working, quickly overwhelm the financial and intellectual resources of any single research team. As many of authors in this book have pointed out, political-economic critiques, including those that consider stakeholder engagement, "voice," empowerment, and related governance issues, need to be more thoroughly developed and linked to complex systems and conventional ecosystem studies in order to carry the ecosystem approach forward beyond local development.

References

Allen, T. F. H. and T. W. Hoekstra. 1992. *Toward a Unified Ecology*. New York: Columbia University Press.

Charron, D. F. 2001. Livestock production and stream health in the Great Lakes Basin: An agroecosystem health approach. Ph.D. dissertation. Guelph, Canada: University of Guelph.

Charron, D. F., D. Waltner-Toews, A. R. Maarouf and M. Stalker. 2003. A synopsis of known and potential diseases and parasites of humans and animals associated with climate change in Ontario. In *A Synopsis of Known and Potential Diseases Associated with Climate Change*, eds. S. Greifenhagen and T. L. Noland. Information Paper 154. Sault Ste. Marie, Canada: Ontario Ministry of Natural Resources, Ontario Forestry Research Institute.

Hilsenhoff, W. L. 1987. An improved biotic index of organic stream pollution. *The Great Lakes Entomologist* 20:31–39.

Hilsenhoff, W. L. 1988. Rapid field assessment of organic pollution with a family-level biotic index., *Journal of the North American Benthological Society* 7:65–68.

Ignazi, J-C. 1987. Intensive agricultural practices and the quality of drinking-water. *Fertilizers and Agriculture* 41:3–15.

Jarvis, I. E., K. B. MacDonald, D. B. Gleig and Y-T. Kang. 1996. Natural resources, production and environmental dimensions of Great Lakes Basin agriculture. In *Great Lakes Commission, An Agricultural Profile of the Great Lakes Basin: Characteristics and Trends in Production, Land-use and Environmental Impacts*. Ann Arbor, Mich.: Great Lakes Commission.

Karr, J. R. 1981. Assessment of biotic integrity using fish communities. *Fisheries* 6:21–27.

Karr, J. R., L. A. Toth and D. R. Dudley. 1985. Fish communities of midwestern rivers: A history of degradation. *BioScience* 35:90–95.

Lotspeich, F. B. 1980. Watersheds as the basic ecosystem: This conceptual framework provides a basis for a natural classification system. *Water Resources Bulletin* 16:581–586.

Michel, P., J. B. Wilson, S. W. Martin, R. C. Clarke, S. A. McEwen and C. A. Gyles. 1999. Temporal and geographical distributions of reported cases of *Escherichia coli* O157:H7 infection in Ontario. *Epidemiology of Infection* 122(2):193–200.

Miller, G. 2000. The protection of Ontario's groundwater and intensive farming. Special report. Legislative Assembly of Ontario, Toronto, Canada: Environmental Commissioner of Ontario. (Available at: http://www.eco.on.ca/english/publicat/sp03.pdf)

Regier, H. A. and J. J. Kay. 1996. An heuristic model of transformations of the aquatic ecosystems of the Great Lakes-St. Lawrence River Basin. *Journal of Aquatic Ecosystem Health* 5:3–21.

Richards, C., L. B. Johnson and G. E. Host. 1996. Landscape-scale influences on stream habitats and biota. *Canadian Journal of Fisheries and Aquatic Sciences* 53:295–311.

Smit, B., D. Waltner-Toews, D. J. Rapport, E. Wall, G. Wichert, E. Gwyn and J. Wandel. 1998. *Agroecosystem Health: Analysis and Assessment*. Guelph, Canada: University of Guelph.

Steedman, R. J. 1988. Modification and assessment of an index of biotic integrity to quantify stream quality in southern Ontario. *Canadian Journal of Fisheries and Aquatic Sciences* 45:492–501.

Valcour, J. E., P. Michel, S. A. McEwen and J. B. Wilson. 2002. Associations between indicators of livestock farming intensity and incidence of human Shiga toxin-producing *Escherichia coli* infection. *Emerging Infectious Diseases* 8(3):252–257.

12

An Agroecosystem Health Case Study in the Central Highlands of Kenya

Thomas Gitau,[1] David Waltner-Toews, and John McDermott

While the Guelph Agroecosystem Health project described in chapter 11 was nearing its completion, several new initiatives, involving some of the same researchers, were started in different parts of the world. The project described in this chapter was one of those initiatives.

Introduction

The central highlands of Kenya have a moderate climate and highly productive agricultural lands. Kiambu District is one of the four districts in the central highlands. Since Kenyan independence, the human population of Kiambu District has increased dramatically. Although there are considerable human and natural resources in the area, these are increasingly under pressure. Agricultural plots are sequentially subdivided into smaller and smaller units with increasing intensification and integration of crop and livestock enterprises in these smaller plots. However, this increasing intensification is not always feasible or successful so that in some areas productivity has declined because of land degradation and the loss of the traditional balance between people, natural resources, and socioeconomic systems. Thus there are serious concerns of the sustainability of the highland agroecological system, both from a biophysical and human activity perspective. On the other hand, management options for self-sustenance are high, providing a suitable venue for testing methods of implementing health and sustainability improvement.

Given this background, we initiated a project with the overall purpose of determining if an agroecosystem health assessment was feasible (Smit et al. 1998). At the end of our agroecosystem health research project in Canada (see Charron and Waltner-Toews, chapter 11), some of us concluded that the notion that we could come up with a single verifiable systemic model and that we could assess its sustainability using objectively verifiable indicators was simply not feasible. In Kenya therefore we decided to work both inside and outside the system, with outside scholars and local stakeholders as partners and using participatory methods as well as standard normal scientific research protocols. The aim was to help the local communities improve the well-being and sustainability of villages and farms within a high-density agroecosystem in the central highlands of Kenya. Specifically, we planned to enable farmers and communities to assess the health of their own agroecosystems, institute actions to improve the situation, and monitor their progress. We would work with them to develop indicators for assessing the sustainability of their smallholder farming systems. Finally, as researchers, we wanted to determine the social, economic, and biophysical factors that most influenced the continued sustainability of this system.

In this chapter, we describe the research process and outcomes and draw some conclusions for this kind of work in general. Several aspects of this project are reported by Gitau et al. (2000b) and Waltner-Toews et al. (2000) and in a series of reports to the International Development Research Centre, in the project library of the Network for Ecosystem Sustainability and Health, available at http://www.nesh.ca.

General Process

The general process is depicted in fig. 12.1. We engaged three groups of actors in this project: (1) communities in six study sites distributed across the district, (2) resource persons comprising extension and technical staff from divisional administrative offices, and (3) researchers. Our research team included agronomists, economists, engineers, medical personnel, sociologists, and veterinarians. Additional personnel, including district staff and experts from governmental and nongovernmental organizations were included when need arose. All people living within the study sites were invited to participate in most of the village workshops. However, communities decided to elect a contact group (committee) to serve as the focal point for communication between the community and other actors in the project. There was a resource persons' team in each division of the district. Each team served as the main link between the research team and the communities. A group of six to eight people was selected from a divisional team to be facilitators in participatory workshops organized in study sites within their jurisdiction.

Conceptual Framework

The conceptual framework used to guide this project was a variant of the AMESH process that was emerging simultaneously in several case studies in different parts of the world [see especially Murray et al. (chapter 13) and Neudoerffer et al. (chapter 15)]. The initial step was to collate background information (secondary data). The purpose of this was to frame the system (see chapter 2). We created a conceptual hierarchical structure of the Kiambu agroecosystem and identified the scales (in these hierarchies) at which the project would best be carried out (step 2). Once

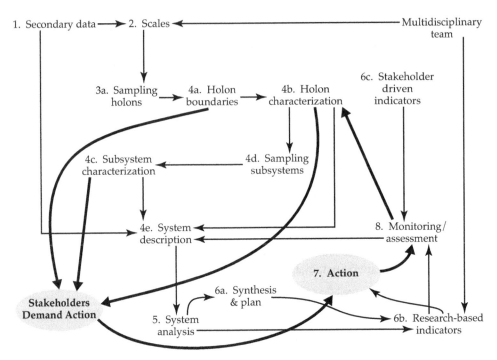

Figure 12.1   Flow chart of the research process used to assess and implement health and sustainability of a smallholder-dominated, tropical-highlands agroecosystem.

the target hierarchical scales were identified, we devised a sampling strategy for each scale (step 3). The sampling units depended on the levels selected. In this study, the first sampling units were study sites corresponding to villages on the human activity scale and catchments on the biophysical scales. The second sampling units were land use units (biophysical scale) corresponding to farms, households, and homesteads in the human activity scale.

After selecting units we wanted to study, we developed a systemic description of the agroecosystems (step 4). The description covered domains such as available biophysical resources, institutions, historical background, social structures (age, gender, and poverty) and trends in health, agricultural production, and human welfare. The objective here was to gain insight—from a systemic perspective—of the constraints, goals, value systems, and structure of the human activity systems in the target agroecosystems. This step (the right-hand arm of the diamond diagram, fig. 14.1) required the use of a variety of participatory, action research, and conventional data gathering tools.

In theory, we expected systems analysis and synthesis to follow the description and data-gathering step. This was difficult to follow in practice because the process of description was an intervention in itself and resulted in a demand for remedial action. Initial descriptions were therefore followed by community action planning and some limited remedial actions. The objective of system analysis is to elucidate how the various components interacted with one another to (1) generate and maintain constraints to health, sustainability and productivity or to (2) facilitate coping strategies, (3) increase management options or potential for increased productivity. Once this was done, remedial actions could be planned (step 6c) and executed (step 8), indicators of health and sustainability could be developed (step 6a and 6b) and used for monitoring the progress of remediation, evaluating the remedial processes and assessing the health and sustainability of the overall system (step 7). Information from the monitoring, evaluation, and assessment was used to refine the system analyses and, subsequently, the indicators and action plans.

Chronology of Activities

Table 12.1 shows a chronology of the main activities carried out in the project from inception to the end of the allocated project time span. Initial activities included (1) collation of background information, (2) training of researchers and their assistants in participatory methods, and (3) initial village workshops. The initial village workshops had three objectives: (1) a systemic description of the agroecosystem, (2) participatory problem analysis, and (3) community action planning. The workshop participants elected a committee consisting of members of the community to facilitate the implementation of the action plans and to serve as the link between the community and other actors. In nearly all villages, implementation of action plans proceeded immediately after the initial workshops.

On the basis of the initial secondary information and workshops, the research team created influence (loop or spaghetti) diagrams (Checkland and Scholes 1990; Flood and Carson 1993; Caley and Sawada 1994). In these diagrams, problems and issues of concern, as described by communities, were linked based on what we thought were the most likely causal relationships. The research team suggested that each community should make similar diagrams to help clarify the relationships more. Communities were able to produce detailed influence diagrams and use them to reevaluate their action plans. As a result of this reevaluation, all six communities altered the priority order of actions in their action plans.

The objective of the land use unit census was to better understand the typology of the scale of organization penultimate to the village. The transect walks and farm visits we did during the initial village workshops showed a "layer" comprised of several types of conceptual units ranging from typical nuclear households in the human activity scale to a multihousehold setup under a form of unified management. Data from the census were analyzed using a variety of statistical methods including multiple correspondence analysis (MCA). Details of this method are provided in conference papers (Gitau et al. 2000a, 2000b). This provided a basis for selecting land use unit level indicators. The census data also served as the sampling frame in subsequent data-gathering activities.

On the basis of influence diagrams and analytical work, the research team generated a list of potential indicators, each of which we evaluated in terms of

TABLE 12.1	Summary of Problems and Concerns as Prioritized by the Main Stakeholders in Six Villages of Kiambu District, Kenya, 1997					
Rank	Githima	Gitangu	Kiawamagira	Mahindi	Gikabu-na-buti and Itungi	Thirika
1	Water not easily accesible	Water not easily accesible	Water not easily accesible	Poor roads	Water not easily accesible	Security inadequate
2	Poor human health and health care	Poor roads	Water shortage	Unemploy-ment	Security inadequate	Poor human health and health care
3	Illiteracy	Poor human health	Low farm productivity	Water not easily accesible	Poor health and health care	"Grab-bing" of public land
4	Poor roads	Unemploy-ment and crime	Fuel shortages	Poor human health and health care	Unemploy-ment	Poor quality seeds
5	Fuel short-ages	Secondary school and polytechnic needed	Security inadequate	Nursery school needed	Crop diseases	Lack of unity and solidarity
6	Lack of A.I. services	Crop diseases, pests and poor seed quality	Inadequate A.I. services	Lack of knowledge	Outlet for tea produce needed	Lack of extension services
7	Security inadequate	Animal diseases and poor quality feeds	Poor human health and health care	Livestock disease	Lack of exten-sion services	Poor leadership
8	"Ignorance"	Soil erosion and infertility	Poor commu-nication		Alcoholism and drug abuse	Improper use of ag-rochemical
9		Lack of market and shopping center			Lack of school fees	Soil erosion
10		Inadequate extension services			Food shortages	Crop diseases

validity, ease of measurement and interpretation, and the usefulness of the information it supplied relative to other, similar, indicators. We also talked about methods for introducing the indicators concept to the communities and facilitating them to develop a suite of indicators for their own use.

These concepts were introduced in an intervillage workshop organized to train community members on community leadership—a workshop requested by the community agroecosystem health committees. When they realized that, under this new way of doing research, they were responsible for devising their own strategies and evaluating them, they requested a course on leadership training skills. Participatory workshops were then held in each of the villages to facilitate the development of community-driven indicators. Each village developed its own suite of indicators and described the method to be used to measure each of the indicators. They selected people in the village and assigned them the tasks of measuring groups of related indicators and to report back to the village committee. The measurement of community-based indicators was carried out without the researcher's input, but researchers were invited as observers in the meeting called to discuss the results.

Concurrent with the development and measurement of community-driven indicators was the development of tools for measuring research-based indicators, which were more oriented to laboratory and long-term (slow-moving) (Gunderson and Holling 2002) variables. The research team developed questionnaires, semi structured interviews, and participatory and action research tools for use at both the land use unit level and the village level. Land use unit-level measurements were carried out first followed by the village-level measurements. The former were carried in about twenty lands use units per village selected at random from the census data. The latter were carried out in six villages involved in the development of indicators as well as in six other "nonintensive" villages. We analyzed the data gathered in this process using MCA among other methods to help further distill the suite. The distilled suite was used in a subsequent assessment.

We are still analyzing the field data to determine how the distilled suite relates to the larger (unrefined) suite and how these can be best used to monitor health and sustainability of the Kiambu Agroecosystem.[2] The field phase of the project was concluded with a wrap-up workshop, in which community leaders, resource persons, and some members of the multidisciplinary teams discussed the problems, advantages, and disadvantages of the AESH approach. At the final workshop, the researchers asked the villagers to present their view of the work and to challenge us with their views as to how it might have been done differently (better).

Some Methodological Issues

Set out this way, the project appears straightforward and the integration of systems thinking with participatory research uneventful. Neither of these is true. We were, in fact, challenged each step of the way in our thinking and practice; in effect, the villagers pushed us to be innovative in our research even as we were encouraging them to be so in their practice. It really was a co-learning experience.

Sampling of Study Sites

We sampled each of six administrative divisions of Kiambu District. Within each of these, we randomly selected one sublocation to be "intensively" studied using a full participatory approach and one sublocation to be "extensively" studied using indicators. Within these randomly selected sublocations, villages were chosen based on a scoring matrix with villages having a preponderance of smallholder farmers and those not having other governmental or nongovernment organization projects being favored. It turned out that there were no formal governance structures or boundaries at the village level. Initial workshops were held in community halls or churches that, by "common knowledge," were in certain villages. The participants at those meetings were able to define the boundaries and governance quite precisely. Within communities, farms were sampled on the basis of stratification criteria developed from initial village and researcher workshops.

Initial Village Workshops and Problem Analyses

A central focus of project activities was to facilitate the implementation of action plans by the communities and to institute measures for better management of their agroecosystem. None of the white outsider researchers were allowed into the study area at this time, as it was felt that this would bias the stakeholders toward a conventional, money-driven "development mode" of thinking. The underlying assumption was that health and sustainability depended on community participation and the community ownership of the problems, the remedial processes, and the outcomes. Because of their central role in the AESH approach, the first step of developing a systemic description of the Kiambu ecosystem was the convening of participatory community workshops. Four to six days were taken to develop a comprehensive description of each village.

The first activity in these workshops was participatory mapping. Both social and resource maps were drawn by the community members, indicating the village (itura) boundary, households, resources (rivers, forests, etc.), and infrastructure. During this process, discussions on various issues relating to particular aspects of the maps were encouraged. Such issues included population size, household sizes, household heads, soil types, soil erosion, and availability of water for domestic use. This proved to be a crucial step as it placed the initial focus on the community's point of view. It allowed communities to begin an objective self-assessment of its own health, before moving into a more subjective and value-laden problem analysis.

The second activity consisted of the compilation of a narrative history of the village and surrounding area. At the time, this was included because it is considered important in some of the participatory research and social learning literature (Freire 1972; Whyte 1991). However, this kind of storytelling has subsequently become an essential part of the AMESH process because of its importance in structuring the temporal relationships between variables and, indeed, in enabling marginalized, therapeutic sustainability stories to be expressed (White 2003; Waldegrave et al. 2003). Often, the old men and women narrated the history while the younger people asked questions on events that they had heard about. Subsequently, specific natural resource, socioeconomic, and health information were obtained using participatory methods from daily activity charts, seasonal calendars, access and control matrices, health analysis, and trend analysis. These were usually done by three different focus groups, one each for women, men, and youth. Each group would present their results to the other groups. These sessions were very useful in stimulating discussion and generating new insights between the different community groupings.

Community-based problem analysis was an important component of agroecosystem health assessment. Problem analysis was carried out using various participatory tools. Lists of problems, needs, and constraints were developed in focus group discussions and from triangulation of data generated in the mapping, historical, trend analysis, and health analysis processes of the workshop. These lists were discussed, and opportunities for adding to them were availed to the participants. Pairwise ranking was then used to score the problems based on their severity (effect on health and well-being of individuals in the community). The perceived causes and effects of the identified problems were then discussed, as were the coping strategies and opportunities for resolution.

Research-Based Indicators

The research-based indicators were based on both the qualitative and quantitative models developed during system analyses. Indicators were defined as variables that reflect (1) changes in key system attributes or (2) changes in the degree of risk or potential of the system. Indicators were selected based on the ease of measurement and interpretation, validity, cost effectiveness, and usefulness of the information gathered to researchers and policy makers. Initial systems descriptions were developed using a wide-variety of modeling strategies including soft systems analysis (Checkland and Scholes 1990), qualitative modeling such as loop analysis and time averaging (Puccia and Levins 1985), and time and space dynamic models (Hannon and Ruth 1994).

Community Actions, Monitoring, and Evaluation

Initial community action plans were developed based on the descriptive and problem analysis phases. In support of community interventions, the project provided analytical, management, and participatory skills to the communities to enhance their capacity for collective action, problem identification and analyses, consensus building, conflict resolution, action planning, monitoring, evaluation, and assessment. Training programs were organized in each of the six intensive study sites and at the district level.

Community leaders and select groups of ordinary members from the intensive villages were trained on participatory approaches, management methods, community mobilization, gender issues, community-based leadership, action planning, monitoring, and evaluation. To further support the implementation of the community action plans, communities were assisted, as required, with information and skills (e.g., proposal writing) for seeking technical and financial help from government and nongovernmental organizations and other development agencies.

Community leaders were expected to initiate participatory processes and to develop activity schedules, delegate duties, monitor progress, and evaluate the progress of individual projects. The research team attended some of these meetings as observers. In this aspect, the role of the research team was to identify experts, resource persons, or institutions that the communities might need for successful implementation of a project.

The communities in each of the six intensive villages developed a process for identifying and evaluating indicators during follow-up village workshops. Community members were asked to develop indicators they thought would help them measure success in achieving their community action plans and improving the health and sustainability of their village. For each potential indicator, likely outcomes were predicted, measurement tools and methods considered, and their interpretation debated. Finally, the relative usefulness of the information provided by each indicator was assessed. Indicators that were seen not to provide any additional information (as compared to the others) were dropped.

Process Summary

The most unique feature of the AESH process was that community, researchers, and development agents all played complementary roles. Using Biggs's (1989) framework, which describes relationship between researcher partners in terms of the extent to which local opinion and practice is given recognition, this process would fall into the category referred to as collegial. The community role was crucial in understanding the system and posing the key questions of interest. Through participatory problem analysis and action planning, the community's informal research and development system was actively encouraged. Researchers and resource persons (divisional officers, governmental and nongovernmental agencies) played an important role of facilitating the implementation of the community's action plan (e.g., technical advice, research activities to answer key community questions, facilitating contacts with outside agencies and proposals to investors, leadership training). Research questions of broader interest, such as social analysis of communities associated with AESH research and development and determinants of sustainability were also investigated with community input and collaboration.

The key outputs of the community-driven processes included (1) a systemic description of the agroecosystem, (2) a demand for action, (3) community collective action, (4) suites of community-driven agroecosystem health and sustainability indicators, and (5) a community monitoring, evaluation, assessment, and action program. Outputs resulting from research-based activities included (1) qualitative and quantitative system models of the agroecosystem, (2) a suite of research-based indicators, and (3) refinement of methods and processes of agroecosystem health and sustainability assessment. The main output from the interaction between the two (communities and researchers) was a synergy that augmented both the communities' and the researchers' ability to first detect and then investigate and act on agroecosystem health concerns.

Communities

Communities in all selected study sites agreed to participate. Community participation was high, with 75 to 100 percent of the households and homesteads being represented in all the participatory workshops held in the study sites. In all the communities, the concept of participation and action research was new. All communities, however, expressed the sentiment that this was a much better approach compared to the other forms of research they had come in contact with or heard about. The use of tools that removed the need for literacy in order to participate in the process was also highly applauded by participants.

Communities were able to supply most of the information requested. Certain cultural structures, however, influenced the quality and detail of data on some topics. The most affected of these were causes and degree of mortality and wealth ranking. In all the communities, the participants conceded that they were unable to discuss in detail issues related to mortality. Participants were reluctant to talk about wealth (common and individual), as this was tantamount to "telling God that you have had too much to eat."

The concept of agroecosystem health was well understood by most community members as evidenced by their use of health language, images, and concepts throughout the participatory workshops. This understanding provided a common thread that underpinned both their problem analysis and development of a community action plan.

Systemic Description of the Agroecosystem

The key outputs from the initial village workshops were a narrative report (Gitau 2001) of the communities' description of biophysical resources available in their village, institutions, their historical backgrounds, social structure (age, gender, and poverty), and trends in health, agriculture, and human welfare. Communities used this report as a guide to dialogues with other agencies and potential investors. Communities were interested in the description of other villages and made comparisons between themselves and the others.

The participatory descriptions of the village led naturally into discussions of the community's well-being and how this could be improved. These led to the other key outputs of the initial workshops: (1) an analysis and ranking of problems in the village (table 12.1) and (2) a plan of action for the communities. Concerns common to all villages included availability of water, poor roads, and poor health and health facilities. Only one village (Kiawamagira) had access to piped water, and this was only available for a half day per week. Roads were all loose surfaced and often became impassable during heavy rains. Flooding and gully formation blocking roads during the rainy season were common. The agriculture-related problems highlighted included low crop yields, poor soil fertility, lack of markets, lack of extension support, crop and animal diseases, and poor artificial insemination service for cows associated with poor roads.

On the basis of their initial agroecosystem health diagnosis, communities developed action plans and the organizational structures to carry these out. The action plans developed by the six intensive villages are summarized in table 12.2. In all but one intensive study site, consensus on needs and goals was achieved. In most villages, the majority of the actions planned were implemented. Only those that required high capital outlay failed to be implemented mainly for lack of funds and institutional support. Application of participatory tools, stakeholder analysis, and soft system methodologies resulted in high community participation in implementing the action plans as evidenced by their commitment of funds and labor.

The influence diagrams were an innovative methodological output from the communities in this project. They provided not only a summary of the systemic description of the villages but also clarified the causal relations between various elements in both the biophysical and human activity systems. Communities

TABLE 12.2   Actions Planned by Communities in the Six Intensive Villages to Ameliorate Problems Identified During the Participatory Village Workshops

Githima	Gitangu	Kiawamagira	Mahindi	Gikabu-na-buti	Thirika
Start a self-help medical clinic	Rehabilitate Gitangu water project	Carry out road repairs and regular maintenance	Carry out road repairs and regular maintenance	Construct a village dam	Organize security meetings
Carry out road repairs	Carry out regular maintenance	Start a water supply project	Start income generating activities	Organize security meetings	Organize health training, rehabilitate health center
Rehabilitate water system	Start a mobile medical clinic	Request for extension services	Start a water supply committee	Improve existing dispensary	Request for extension services
Add classrooms to secondary schools	Organize community security groups	Promote energy savers and agroforestry	Start a village medical clinic	Start self-help projects	Seek title deeds for public utility lands
	Start a village polytechnic	Start village security groups	Start a village nursery school	Seek extension services	Start a water supply scheme
	Start village extension programs	Start a community dispensary			From small marketing cooperatives

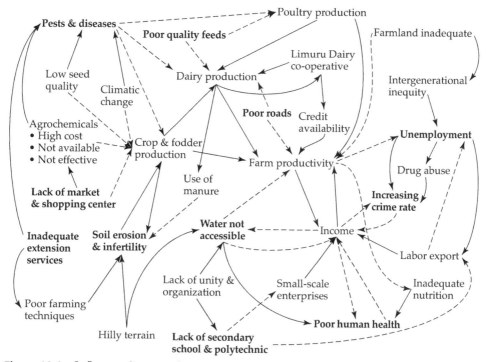

Figure 12.2   Influence diagram showing relationships among issues of concern as perceived by community members in Gitangu village. Dotted lines indicate an inverse relationship, while solid lines indicate a direct relationship. In the original drawings, lines were also colored red or black, depending on whether or not they were thought to have a negative or positive (regenerative) impact on the agroecosystem; elements in the system were classified as needs, inputs, institutions, and resource states.

revised their action plans immediately after analyzing their influence diagrams. Research analyses of these diagrams indicate that they may be useful in predicting the system response to certain interventions and in interpreting indicators to determine the overall system status in terms of health and sustainability. Figure 12.2 illustrates an influence diagram produced by the community in Gitangu village. It shows how dairy production is central to the well-being of this village.

Table 12.3 shows the list of revised action plans and the progress in their implementation. Githima village revised their action plans to begin with road rehabilitation, electrification, water supply, expansion of school, and then development of an extension program. Gitangu village opted to leave water supply as the first priority, but extension was moved to the second place. Soil conservation was given first priority in Kiawamagira followed by road rehabilitation and control of flooding. Mahindi retained only two items in their action plans: (1) rehabilitation of the access road and (2) development of an income-generating project. Gikabuna-buti village revised their action plans to electrification, water supply, income generation, and extension. Thiririka reordered their action plans to begin with water supply, followed by development of a market for farm produce, extension, and then road rehabilitation.

Demand for Action

At the end of the initial village workshops, all communities expressed a profound demand for action to ameliorate the problems identified. Formation of the village committees was evidence of their desire to implement the action plans. Five of the six villages proceeded with implementation of the action plans immediately after the workshops mostly without further contact or consultation with the research team. In nearly all the cases, this led to some degree of frustration on the part of the communities as they were ill-prepared in terms of organization and community leadership to carry out many of the tasks. However, there were some success and failure, and frustration did not deter most of the communities to keep trying.

Community Monitoring, Evaluation, and Assessment

Communities understood the concepts of health and indicators as applied to agro-ecosystem and accepted the notion of using indicators to evaluate the status of their agroecosystem. This was seen not as an innovation but as a revisiting and modernizing of the traditional methods of agroecosystem management. There were important contrasts between the research-based indicators and those selected by the stakeholders. In addition, stakeholder-driven indicators differed between villages. Table 12.4 provides a list of indicators selected by communities in three of the six intensive villages. These can be compared to the researcher-proposed indicators listed in table 12.5.

Communities in all six intensive villages have used and are using indicators to carry out evaluations and assessments of their agroecosystems. Evaluations are carried out by community leaders and selected groups of community members. In most cases, these evaluations are carried out jointly between villages. The communities initiated intervillage collaboration because they felt that peers provided a critical but constructive grounding to the assessment.

Researchers

Figure 12.3 illustrates a conceptual framework of the hierarchical structure of the Kiambu agroecosystem. On the biophysical scale, the agroecosystem was seen to consist of farms, catchments, agroecozones, and geoclimatic zones. The human activity holarchical structure was modeled to follow administrative boundaries up to the sublocation but included villages and households at the lowest level. Results from this study, however, showed that the lowest level in both the biophysical and human activity systems is the land use unit, which might correspond to a farm, a homestead, or a home for a nuclear family (with little or no farming activity).

For a health and sustainability assessment of the Kiambu agroecosystem the village (level $n + 1$) and the land use unit (level n) were the main holons focused on. Village-level assessments were crucial because ecological, economic, and social factors all combine at this level to form a system with unique properties that needs

TABLE 12.3   Progress in the Implementation of Revised Action Plans for Six Villages in Kiambu District, Kenya

Village	Githima	Gitangu
Planned activities in the revised action plans	5 (road, electricity, water, school, extension)	3 (water, extension, security, medical)
Number of projects past initiation stage by October 1998	5 (all)	2 (security, extension)
Number of activities completed by October 1998	2 (electricity, water)	2 (security, extension)
Number of projects past initiation stage by August 2000	1 (expansion of school)	1 (install water meters)
Activities current as of August 2000	School expansion	Water meters
Stage of project considered current as of August 2000	Near completion	Beginning
Number of planned activities completed by August 2000	All	All
Recurrent activities as of August 2000	Road maintenance	Extension meetings, M&E meetings
Frequency of self-initiated participatory meetings	Frequent[1]	Very frequent[2]
Attendance to meetings	High[4]	High
Linkage with other intensive villages	Very high[7]	Very high
Funds generated (by community) to support activities (in Kenya shillings)	1.2 million	120,000

[1] Roughly once every two months.
[2] Twice a month.
[3] Not regular. Only when need arises.
[4] More than half of the households represented.
[5] More than a quarter but less than a half of the households represented.
[6] No more than a quarter of the households represented.
[7] Have initiated visits to other villages and to their own village.
[8] Have initiated and hosted visits to their village.
[9] Have participated in all intervillage meetings.

Kiawamagira	Mahindi	Gikabu-na-buti and Itungi	Thirika
5 (conserve soil, road, control flood, extension, security)	2 (road, income generate)	4 (water, electricity, security, income generation, extension)	4 (water, market, extension, road)
2 (security, soil cons.)	1 (income generate)	0	2 (road, extension)
2 (security, soil conservation)	0	0	1(road)
2 (road and flood control)	0	0	0
Road and flood control	None	Electricity supply	Water supply
Beginning		Planning	Planning
2 (extension, soil conservation)	1 (income generation)	1 (extension workshops)	1 (road)
Road maintenance, soil cons. mtgs	Road maintenance	Nil	Nil
Rare[3]	Rare	None	Rare
Moderate[5]	Poor[6]	Poor	Moderate
Moderate[8]	Moderate	High[9]	None
10,000	6,000	Nil	20,000

TABLE 12.4	Indicators Selected by Stakeholders in Three of the Six Villages for Community-Based Agroecosystem Health Monitoring in Kiambu District, Kenya, July 1998		
Attribute	Mahindi	Kiawamagira	Gitangu
Lifestyle	Number of people with proper personal hygiene, types of diets dress habits	Farming techniques (new versus old) Types of houses	Personal hygiene Types of crops and livestock Time usage
Social organization	Number of completed community projects Number of people a attending meetings	Frequency of meetings Number of community plans executed Number of people gainfully employed	Number and severity of needs in the community Number of needs met over the past one year
Equity	Distribution of work by age and gender	Meeting attendance by age and gender Distribution of chores, household incomes Unfair cultural practices	Distribution of leadership positions by gender and age
Quality of environment	Distance to water Coloration of water Smell of water	Frequency of water-borne diseases Air quality (bad odors) Personal and homestead hygiene Garbage dumps in public places (road, river)	Types of chemicals used on farm Storage of chemicals in homestead Disposal of containers
Soil fertility	Color of soil Types of weeds	Quantity of harvest Soil color and texture Types of weeds	Soil erosion measures by farms Number of livestock per farm Quantity of harvest taken to market
Farm productivity	Number of homesteads with granaries Expected yields of crops	Types and quantity of foods bought from market	Quantity of produce sold versus pur-chases
Pests and diseases	Visits of hospitals Number of livestock deaths	Human mortality Human morbidity	Human morbidity
Markets	Location of nearest market Quantity of farm pro-duce going to market	Variety of goods available in the shopping center	Variety of goods in the market

(continued)

TABLE 12.4   Indicators Selected by Stakeholders in Three of the Six Villages for Community-Based Agroecosystem Health Monitoring in Kiambu District, Kenya, July 1998 *(continued)*

Attribute	Mahindi	Kiawamagira	Gitangu
Savings/ wealth	Types of houses Number of livestock per homestead	Number of cattle per homestead	Increasing or decreasing needs in homesteads
Knowledge	Types of skills	Farming techniques Behavior of youth and children	Knowledge of current affairs Frequency of extension visits
Infrastructure	Distance to primary schools Status of access road		Status of schools, medical facilities, and roads

TABLE 12.5   A Summary List of Researcher-Proposed Indicators with Reference to the Kiambu Agroecosystem

Biophysical	Social
Environmental degradation • Rainfall pattern • Aforestation/agroforestry • Chemical pollution	Aspirations • Level of satisfaction • General goals • Achievements
Farm efficiency • Technical • Allocative	Attitudes • Education • Employment • Health • Wealth • Work
Pests, diseases, and health • Demographics • Human health and nutrition • Human diseases • Animal pests and diseases • Crop pests and diseases	Equity • Roles • Ownership and control • Social values
Soil fertility • Physical fertility • Chemical fertility	Knowledge and information • Innovation and technology • Information sources • Informal education • Formal education

(continued)

TABLE 12.5   A Summary List of Researcher-Proposed Indicators with Reference to the Kiambu Agroecosystem (*continued*)

Biophysical	Social
Water • Water quality • Availability for domestic use	Linkages • Outmigration • External contacts • Familial ties
Economic	Organization • Leadership • Family structure • Reciprocity • Social control • Organizations and associations
Capital • Credit • Investments	Preferences • Leisure activity • Foods • Occupations
Farm profitability and economic • Farm inputs • Profitability • Farm outputs	Values • Behavioral • Wealth related • Well-being
Income • Sources of income • Average income • Employment • Savings	
Infrastructure • Accessibility • Status • Employment • Availability	

to be considered as a unit. In addition, trade-offs among land use units within a village are essential for the continued sustainability of the system. Likewise, balancing farm enterprises (the $n - 1$ level) within a farm has important implications for the sustainability of the farm. Broader-scale watershed level and neighboring village assessments were sometimes required. The characteristics of these agricultural systems in each village have been described in detail (Gitau 2001). Interestingly, three villages seem to have retained agricultural practices used over the long term and well adapted to local environmental conditions, while three villages have changed their systems and used inputs to do agricultural activities that are less naturally suited to environmental conditions in their villages.

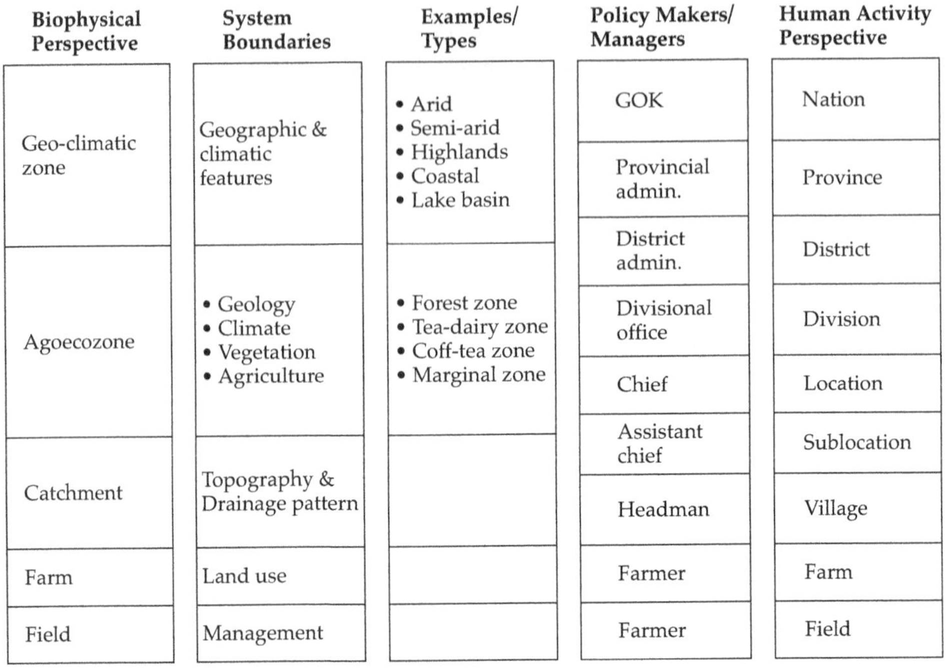

Biophysical Perspective	System Boundaries	Examples/Types	Policy Makers/Managers	Human Activity Perspective
Geo-climatic zone	Geographic & climatic features	• Arid • Semi-arid • Highlands • Coastal • Lake basin	GOK	Nation
			Provincial admin.	Province
			District admin.	District
Agoecozone	• Geology • Climate • Vegetation • Agriculture	• Forest zone • Tea-dairy zone • Coff-tea zone • Marginal zone	Divisional office	Division
			Chief	Location
			Assistant chief	Sublocation
Catchment	Topography & Drainage pattern		Headman	Village
Farm	Land use		Farmer	Farm
Field	Management		Farmer	Field

Figure 12.3   An illustration of the hierarchical structure of the Kiambu agroecosystem from both the biophysical and the human activity perspectives. Kiambu is within the central highlands geoclimatic zone and comprises four major agroecozones.

Systems Analyses

The most useful of the systems analyses carried out was that involving the influence diagrams drawn by the communities. In the Gitangu influence diagram (fig. 12.2), there were seventeen feedback loops of which four were negative feedback. Of the positive feedback, only two regenerative. Overall, nine loops were first and second degree, while the remainder were third degree and higher. Higher number of loops and preponderance of higher degree loops were considered indicative of a more complex causal structure but also of a functional monitoring subsystem. Preponderance of degenerative positive feedback loops was considered indicative on a degenerative system spiral. On the basis of this analysis, it would be expected that this village would have a stronger collective action but remediation of most of the problems would take longer. In contrast, the diagram drawn by residents of Kiawamagira village (not shown here) had eleven feedback loops, all of which were positive. Only two of these loops were regenerative. Eight of the loops were first degree, while the remainder was second degree. Preponderance of low degree loops was interpreted as indicative of simple causal structures and/or higher potential for remediation. The preponderance of positive feedback loops appeared to indicate overall system dysfunction and entrapment in a degeneration cycle. On the basis of these, resolution of the problems in the village was expected to be difficult or to remain generally unresolved. Comparison of the outcomes of this analysis with the results presented in table 12.3 reveals the potential of this analysis in predicting the success of community-collective action—a process that is key to implementation of agroecosystem health and sustainability.

Research-Based Indicators

Researchers developed a list of indicators for both short- and long-term assessment. These are summarized in table 12.5. Two assessments of these indicators have been conducted in the six intensive and six extensive villages. Methods of analysis of these indicator data are still being refined. Correspondence analysis (Gitau et al. 2000a, 2000b) appears to be a promising approach to summarize the key trends.

Interactions Between Communities and Researchers

The interactions between communities and researchers were an essential feature of the agroecosystem health process in this project. Communities have the closest perspective and greatest stake on the agroecosystem health of their village. Thus the village research and development process is crucial to well understand the community system, analyze problems, and develop meaningful action plans and realistic monitoring, evaluation, and adaptation methods, so that village initiatives and strategies can be sustained. Community input into the research process was crucial to better understanding of village-level agroecosystems.

Researchers effectively complemented community-based actions. Essentially all community-based actions listed in table 12.2 required technical expertise and links to technical and administrative organizations outside the village. Project researchers facilitated these contacts, provided support for letter and proposal writing and supported training and intervillage management, monitoring, and evaluation activities. In addition, researchers could offer a broader perspective on agriculture, health, environmental, and other issues to support community problem analysis, monitoring, and evaluation activities.

Lessons Learned

Main Benefits of this Approach

In general, the agroecosystem approach had many attractions from both the research and development perspectives. The health paradigm was easily understood and conceptually amenable to initial diagnosis and treatment and follow-up monitoring and evaluation. Because health assessments are value-laden, self-analysis and actions to establish a healthier community require that they are community led if they are to achieve meaningful and lasting results. In addition, analyses at different holarchical scales are helpful for communities because development requires cooperation across households, villages, and larger levels of organization such as governments and other agencies.

We noted a number of practical implications during the project. The first was that this research paradigm allows for the development of an effective forum for community research collaboration. The second was that integrating participatory and standard research approaches to address community concerns can address specific questions and achieve tangible results. The research input helped communities to better understand the choices to be made in developing and adapting actions. For researchers, there were real benefits in communities generating research questions based on the real needs of the community. Research results, in this context, are more likely to be adopted and sustained.

Main Difficulties

The main difficulties in the agroecosystem approach related to its time horizon, broad perspective, and location-specific nature. As the process is open-ended, only its initiation and early development fits into a standard project time frame. Longer-term issues, such as assessing sustainability, require longer-term mechanisms of support. The holistic view adopted in this process, while essential to establish the crucial context for decisions, means that sometimes there is a lack of decision-making focus. Lastly, from a research perspective, it is not yet clear how generalizable the lessons learned in one group of communities are to other locations. In our view, the process is reasonably transferable but we are not yet sure about results.

Lessons for Communities

The key lesson for communities is that the health approach to community description, problem analysis, and action planning only works if the community is committed to and leads the process. All communities had some success with this approach—mainly related to their organizational ability and commitment. The participatory techniques for analyzing, planning, and monitoring were effective and contributed to community mobilization and action. Communities also discovered that they could learn effectively from the experiences of other communities. Thus strategies to foster intervillage collaboration need to be an important feature of such efforts.

Lessons for Researchers

As researchers, we developed a profound appreciation for the ability of communities to formulate "research" questions and analyze constraints. We found that new research questions were opened up, and we were pushed to explore new analytical and synthetic methods.

Notes

[1] Thomas Gitau died on March 8, 2005, at the age of 38.
[2] As of this writing, the data remain only partially analyzed because of the death of Dr. Gitau.

References

Biggs, J. B. 1989. Approaches to the enhancement of tertiary teaching. *Higher Education Research and Development.* 8(1):7–25.

Caley, M. T. and D. Sawada. 1994. *Mindscapes: The Epistemology of Magoroh Maruyama.* Langhorne, Pa.: Gordon and Breach Science.

Checkland, P. B. and J. Scholes. 1990. *Soft Systems Methodology in Action.* New York: Wiley.

Flood, R. L. and E. R. Carson. 1993. *Dealing with Complexity* (2nd ed.). New York: Plenum.

Freire, P. 1972. *Pedagogy of the Oppressed.* Harmondsworth, Middlesex, London, UK: Penguin.

Gitau, T. 2001. Report of participatory action research workshops held in six villages of Hannon. In *Dynamic Modeling* (2nd ed.), eds. B. Hannon and M. Ruth. New York: Springer-Verlag.

Gitau, T., J. J. McDermott and B. McDermott. 2000a. Methods for epidemiological use of correspondence analysis in agroecosystem health assessments. In *Proceedings of the Ninth Symposium of the International Society for Veterinary Epidemiology and Economics* (Breckenridge, Colorado, USA, 6–11 August 2000). Nairobi, Kenya: ISVEE (International Society for Veterinary Epidemiology and Economics), Paper 501.

Gitau, T., J. J. McDermott, D. Waltner-Toews, J. M. Gathuma, E. K. Kang'ethe, V. W. Kimani, J. K. Kilungo, R. K. Muni, J. M. Mwangi and G. O. Otieno. 2000b. Agroecosystem health: Principles and methods used in high potential tropical agroecosystems. In *Agroecosystems, Natural Resources Management and Human Health-Related Research in East Africa: Proceedings of an IDRC-ILRI International Workshop (Addis Ababa, Ethiopia, 11–15 May 1998)*, eds. M. A. Jabbar, D. G. Peden, M. A. Mohamed-Saleem and H. Li Pun. Nairobi, Kenya: ILRI (International Livestock Research Institute) pp. 55–62.

Gunderson, L. H. and C. S. Holling (eds.). 2002. *Panarchy: Understanding Transformations in Human and Natural Systems.* Washington, D. C.: Island Press.

Hannon, B. and M. Ruth, 1994. *Dynamic Modeling* (1st Edition). New York, Springer-Verlag.

Puccia, J. C. and R. Levins. 1985. *Qualitative Modeling of Complex Systems: An Introduction to Loop Analysis and Time Averaging.* Cambridge, Mass.: Harvard University Press.

Smit, B., D. Waltner-Toews, D., J. Rapport, E. Wall, G. Wichert, E. Gwyn and J. Wandel. 1998. *Agroecosystem Health: Analysis and Assessment,* Guelph, Canada. University of Guelph.

Waldegrave, C., K. Tamasese, F. Tuhaka and W. Campbell. 2003. *Just Therapy—A Journey.* Adelaide, Australia: Dulwich Centre.

Waltner-Toews, D., T. Murray, J. J. Kay, T. Gitau, E. Raez-Luna and J. J. McDermott. 2000. "One Assumption, Two Observations, Some Guiding Questions, and a Process for the Investigation and Practice of Agroecosystem Health." In *Agroecosystems, Natural Resources Management and Human Health-Related Research in East Africa: Proceedings of an IDRC-ILRI International Workshop (Addis Ababa, Ethiopia, 11–15 May 1998)*, eds. M. A. Jabbar, D. G. Peden, M. A. Mohamed-Saleem and H. Li Pun. Nairobi, Kenya: ILRI (International Livestock Research Institute) pp. 7–14.

White, M. 2003. Narrative practice and community assignments. *The International Journal of Narrative Therapy and Community Work* 2003(2):1.

Whyte, W. F. (ed.). 1991. *Participatory Action Research.* Newbury Park, Calif.: Sage.

13

Food, Floods, and Farming

An Ecosystem Approach to Human Health on the Peruvian Amazon Frontier

Tamsyn P. Murray, David Waltner-Toews,
José Sanchez-Choy, and Felix Sanchez-Zavala

Introduction

Beginning at about the same time as the Kenyan project described in the previous chapter, a team of Canadian and Peruvian researchers began working on an ecosystem approach to understanding agriculturally altered landscapes in the neotropics of South America. We began by developing a research process and conceptual framework that brought together the most recent understanding of ecosystems as complex systems with secondary data and exploratory field work in the Ucayali region of Peru (Rowley et al. 1997). Equipped with this process and framework, we revisited the region to specifically investigate the key determinants of, and linkages between, ecosystem and human health. The actual process followed was modified to fit the context, while retaining its essential features (fig. 13.1).

The Ucayali region is populated by 370,000 people and spans 100,000 square kilometers. In the 1940s a road connecting Pucallpa on the Ucayali River, a major Amazon tributary, with Lima, hastened settlement from the coast and the Andes. By the late 1990s, about 80 percent of its population lived either in Pucallpa or on the Lima road, creating agricultural production and food security challenges. Despite the natural diversity and fertility of this region, remote rural communities struggled to meet their basic needs and face a range of nutritional and health problems (Instituto Nacional de Estadisticas y Informaticas 1997). As a result of slash-and-burn agriculture, deforestation steadily increased and logging activities continued unregulated and with unknown ecological ramifications. The relationship between household production, income levels, and health in the Ucayali region was complex and poorly understood. Exploitation of local resources resulted in diverse seasonal combinations of farming, fishing, logging, and hunting and gathering activities. It was not known, at the time we started this study, how these different resource strategies affected household health or whether health and earnings were related—problems that have important implications for agricultural and technology development in the region. Ucayali therefore presented our research

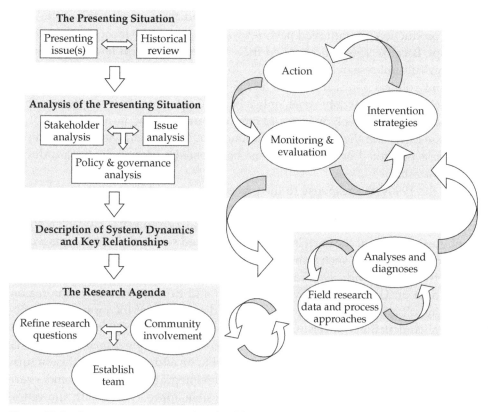

Figure 13.1   An ecosystem approach to health in the western Peruvian Amazon.

team with a very complex and dynamic set of interconnected issues. In addition, Ucayali was a benchmark site for the Consultative Group on International Agricultural Research's (CGIAR) Eco-Regional Program and the focus of coordinated research efforts whose findings would have potential application to other forest margins areas in the tropics. Therefore it provided a valuable challenge to determine whether the ecosystem approach was developing could synthesize different yet interdependent dimensions within the same region.

The Presenting Issues

When researchers or development workers enter a problematic situation such as that in Ucayali, both the opportunities to intervene and the way the problems are framed are based on how issues are presented to them. In the early 1970s, international and national research centers in Ucayali focused their efforts on livestock farming and deforestation. On the basis of their analysis of the system, deforestation was the result of declining land productivity and soil degradation that forced farmers to expand into forested areas. Research was directed toward technologies that would increase farm productivity and extend the economic life of the land. Cattle production on improved pastures was selected as the most efficient system to adopt. Because scientists confined their studies to the farm level and included only agricultural researchers, they failed to take into account important sociocultural and political issues at regional and national levels that affected farmers land use decisions. After twenty years, only a handful of farmers had adopted the technologies. Moreover, lack of access to credit meant that few farmers were able to purchase cattle and the resulting overabundance of *bracchiaria*, the improved pasture grass promoted by the researchers, has created a serious fire hazard during the dry season.

In the 1990s, the response to the presenting issue of deforestation was more expansive. Researchers recognized that farmers engaged in a multitude of natural resource activities including annual crops and agroforestry. However, efforts continued to focus at the farm level and overlooked sociopolitical issues as well as other resource sectors such as fishing and forestry.

Between 1996 and 1998, Canadian and Peruvian researchers gathered all existing data on Ucayali and developed a rich and detailed history of the region, including all ecological, social, and political dimensions at the farm, community, regional, and national levels. On the basis of the overall picture that emerged, we were able to identify relationships between sectors, key regional, and national constraints and information gaps that needed to be addressed. From this, in turn, emerged a consensus that human health was an integrating concern behind a variety of other presenting issues. Using an ecosystem approach to health allowed us to tease apart those key forces driving the system and therefore determine more effective methods of intervention than previously selected.

Issues and Actors: From Cattle to Complex Ecosystems

The evolution of understanding of this frontier region of the Peruvian Amazon demonstrates how the nature of the research approach influences the way in which issues and actors are identified and ultimately how the goals of the system and desired interventions are determined.

As described above, between the 1970s and 1990s, the key issues of concern in Ucayali have expanded from a narrow focus on cattle and pastures to natural resource management and ultimately to human health. To highlight the difference between past disciplinary approaches and the ecosystem approach, the issues, actors, and overall system descriptions are outlined in table 13.1.

The identification of issues and actors and the way the system was described were very different using the ecosystem approach from that used in previous research. Our understanding of the human activities within the ecosystem approach was predicated on three key principles, which, in turn, reflected our understanding of complex systems: (1) methodological pluralism and interdisciplinarity, (2) multilevel investigation, and (3) local participation and action research. With respect to actors and issues, this translated into a deliberate effort to ensure that all

TABLE 13.1	Comparing Disciplinary and Ecosystem Approaches in Pucallpa, Peru	
	Disciplinary and Traditional Approaches	Ecosystem Approach
Presenting issue	Increasing deforestation	Increasing human health problems
Actors	International and national research organizations Regional level Ministry of Agriculture	• International and national research organizations • Regional level Ministry of Agriculture, Ministry of Health, Ministry of Fisheries and Ministry of Education • Regional University • Nongovernmental groups (includes women and native groups) • Community groups (Village-level Mothers Club and Agricultural Committees) • School teachers
Beneficiaries	The main beneficiaries of research and development are cattle farms situated in the upland terraces along the Lima-Pucallpa road with easy access to markets. They include a relatively homogenous group of commercial farms fully engaged in the cash economy, owned and managed by male, Spanish-speaking mestizos, who have migrated from other parts of Peru. They have land titles and therefore access to government credit programs	Beneficiaries of the research on health include remote rural communities living in the floodplain and uplands, with special focus on the higher risk groups that include children and women of reproductive age. This is a very heterogeneous group that includes families and communities with different land use strategies, varying degrees of involvement in the market economy and settlements age ranging from 5 to 100 years. The majority of families barely surpass subsistence levels. Very few families have land titles. The group includes indigenous populations who have been living in the area for thousands of years.

interests were represented and the all the social, ecological, and political dimensions of issues, were also addressed. In contrast, in the 1970s and 1980s, the projects that focused on cattle production involved only a small set of actors, namely, agricultural scientists, the Ministry of Agriculture, and approximately forty to fifty large cattle farmers situated along the road to Lima. These farmers were unlike the majority of farmers in the region. They had land titles and year-round access to markets and derived little income from other resource sectors and off-farm sources. At the organizational level, there was no community-level involvement; relationships were set up between researchers and individual farmers. There were no women's groups, native groups, or other groups reflecting the diverse communities of this region included. Because the stakeholder group was so narrowly defined, issues raised were similarly narrow and confined to the interests of this small group. The issues they raised were mainly technical and economic in nature. They included market access, price variations, weed invasion, land degradation, and pest invasions.

Multiple Actors—Multiple Interests

As the authors in this book attest, the ecosystem approach demands a variety of forms of inquiry and multiple sources of evidence. Different methods are needed to address different forms of complexity and to answer different kinds of questions (Checkland and Scholes 1990; Midgley 1992; Holling 1995; Waltner-Toews and Wall 1997). In order to represent these different perspectives in Ucayali, we established a multidisciplinary team with expertise in nutrition, health, anthropology, agronomy, natural resources management, fisheries, forestry, ecology, rural planning, and economics. More importantly, the team included stakeholders from the region, local community leaders, as well as government professionals. The involvement of these key actors ensured that our efforts maintained their relevancy to local needs and ultimately that the information gathered was owned and used by local organizations.

The process of selecting researchers was instrumental in developing not only a competent team, but one that was representative of the different interests within Ucayali. The positions available were advertised through our partners, and we met personally with leaders of these organizations seeking their advice and suggestions. We included local representatives of these organizations in the project team. They had first-hand knowledge of the project and, more importantly, had input into its focus and direction. Because we selected eight communities that differed greatly in ecological, social, and economic factors, these representatives gained a significantly broader view of the diversity of problems in their region through their involvement in the fieldwork. This knowledge was then fed back into their organizations, thereby increasing their understanding of people's needs and the effectiveness of their programs.

Once health had been identified as an integrative presenting issue in the region, all our research questions were based on issues identified by local stakeholders. This entire phase of determining the project direction was done in close collaboration with our local partners. Notably, this phase was augmented with several focus group meetings with the only women's organization, Asociación de Mujeres Campesinas de Ucayali (AMUCAU), the main indigenous organization, Asociación Interétnica de Desarrollo de la Selva Peruana (AIDESEP) , and key informant interviews with leaders of other local organizations. Figures 13.2 and 13.3 show the different factors that the team identified as being critical to understanding child health: one set mediated through food intake (which identifies socioeconomic variables) and the other on food utilization (which focuses on biological

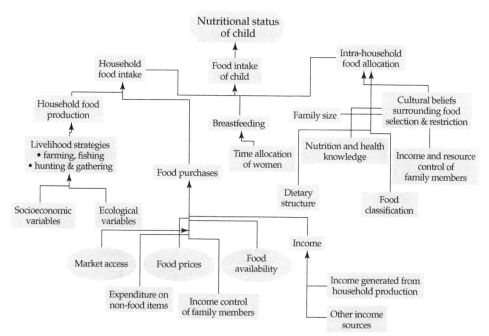

Figure 13.2 A web of causal influence on the nutritional status of children in Ucayali, Peru, mediated through food intake.

variables). There are clearly overlaps between the two. From there we were able to determine how each of these variables was to be measured. In addition, together we determined the selection criteria for the project sites, identified the eight communities that represented the diversity within the region and decided on the timing of visits necessary to capture seasonal fluctuations.

Once we had a rough plan that identified the communities and the main issues we were to address, we went back to the community groups and government agencies to ask for their comments and suggestions. In order to ensure that all the issues important to the individual communities themselves were included, we used a variety of participatory action research (PAR) methods. They included community mapping, seasonal calendars, focus groups, time line, life histories, pile sorting, and key informant interviews. At the community level, the research team adapted their investigation to meet priorities identified by the community. During the process of data collection, the team and the community began to assess different options for intervention. As results became available, they were used to guide local initiatives and provide feedback for ministry programs. Characteristic of multimethodological studies in general, and specific to systems-based research, there was an iterative cycle between research and action. We developed action plans that were then evaluated and the results were used to refine our hypotheses and future research agenda. The issues were not static; they evolved as we learned

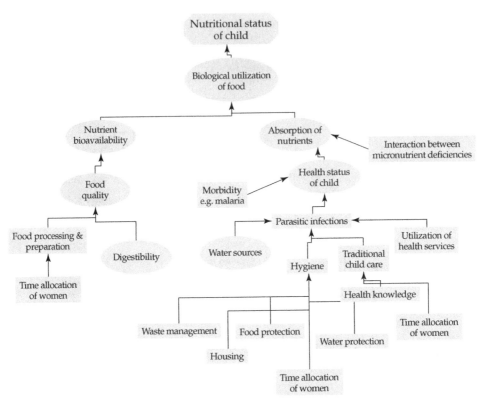

Figure 13.3   A web of causal influence on the nutritional status of children in Ucayali, Peru, mediated through biological utilization of food.

more and were better able to focus our efforts. There was not the usual lengthy delay between research, analysis, and intervention often carried out by different sets of actors.

The research was conducted in eight different communities spread across the region of Ucayali. From a methodological perspective, it was important to include sites that differentiated those factors affecting human health. To capture the region's heterogeneity, the following criteria guided the community selection process: (1) ecosystem type (floodplain versus upland forests), (2) ethnicity (native versus colonist), (3) access to markets and involvement in market economy, (4) time of settlement (early versus old frontier), and (5) dominant land use strategy (slash and burn agriculture, fishing, cattle ranching, and oil palm plantations).

Multiple-Issue—Multiple-Level Hierarchy

One of the basic principles of the ecosystem approach is that ecosystems exist within nested hierarchies (Allen and Hoekstra 1992; Checkland and Scholes 1990). They are comprised of smaller systems while at the same time being part of a larger whole. A household is therefore part of a community, while similarly being made up of different individuals. Recognizing that often the determinants of individual human health may occur at levels higher within the ecosystem hierarchy, we investigated variables at four spatial scales: the individual, family, community, and region/landscape. Figure 13.4 demonstrates the multilevel nature of the issues facing Ucayali as well as the different actors at these levels. All issues are linked to and have consequence for others higher and lower within the nested hierarchy. The ecosystem approach focuses investigation on the cross-scale interactions of key variables that explain the complexity and multidimensionality of health. For example, landscape spatial mapping determined the extent to which families depended on an area larger than their farm or community for food and income. The temporal scale was seasonally determined with three field visits that captured the driest period, the start of the rains and the height of wet season. Data on seasonal flooding levels were correlated with water quality and parasitic infections to investigate links between the hydrological cycle and disease periodicity.

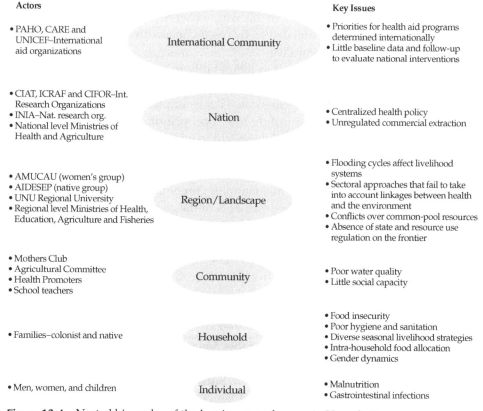

Figure 13.4   Nested hierarchy of the key issues and actors in Ucayali, Peru.

Multiple Decision Makers—Multiple Constraints

Understanding the governance structures in Ucayali was key to understanding how the situation could be changed. Centralized policy making at the national level severely undermined the ability of the regional ministries to respond appropriately to local problems. Moreover, powerful commercial interests that dictate resource use in Ucayali are based and regulated in Lima. Therefore all regional level efforts must take into account the constraints of these political and institutional dynamics.

Multiple Methods—Multiple Disciplines

Methodologically, we combined several quantitative and qualitative methods, implemented at four levels within the ecosystem: individual, family, community, and landscape or region. Tables 13.2 and 13.3 outline these methods. For data on health and nutritional status, we complemented medical diagnostic tests, household surveys, anthropometry, and food recall with ethnographic and participatory methods exploring local diagnoses and understanding of health. For data on ecosystem dynamics and natural resources management, we complemented landscape-level geographic information systems (GIS), spatial mapping, and soil tests with in-depth household surveys and community-level participatory methods detailing the livelihood systems of families. In this way, we were able to compare the results from the methods designed and driven by researchers with those that were led by the community members themselves.

Methods were complementary in their ability to verify results. The sequence of methods used allowed us to compensate for weaknesses in individual methods and to build knowledge systematically. For example, participatory techniques (tables 13.2 and 13.3) provided insight into local priorities and needs, yet extensive household surveys determined the extent to which these issues were common among all community members and which groups were most at risk. In addition, different methods were used to include different actors. For example, surveys involved the Ministry of Health, while mapping, timelines, and other participatory research techniques involved other members of the community. The data derived from each method were specifically targeted to the different end-users or decision makers. For example, GIS and the regional maps generated were directed to the regional government, medical testing results to the community health workers, and maps and drawings from PAR were used in community meetings to further discussion of the community action plans.

System Dynamics: Building an Understanding

The natural and human ecosystems of this Amazon region are in a state of constant change. These changes are the result of the annual 8–12 meter rise in river levels that occurs during the rainy season. Such annual floods affect every aspect of the local ecology as well as human settlement. Understanding of the dynamics of the rivers and its effects on the changing configuration of the floodplain and

Level	Indicators	Variables	Round 1 Jun–Jul 1999	Round 2 Oct–Nov 1999	Round 3 Mar–Apr 2000	Methods
Landscape	Ecosystem productivity	Soil fertility		X	X	Soil tests Existing information
Community	Access to food	Food prices	X	X	X	Observe
		Food availability	X	X	X	Observe
	Access to education	Education facilities available	X	X	X	Interview with leaders and teachers
	Access to health services	Health services available	X	X	X	Interview with health worker
Household	Socioeconomic status	Wealth index		X		Recall
		Nonfood expenditures			X	Recall
	Nutritional status Food Security	Energy and nutrient intake	X	X	X	24 hour recall
	Dietary Quality	Nutrient intake	X	X	X	Food frequency recall
	Dietary diversity	Food sources	X	X	X	Recall
	Income level; diversity	Income by source	X	X	X	Recall
	Production levels; diversity	Production and extraction outputs	X	X	X	Recall
	Food security	Crop storage		X	X	Recall
	Water access; contamination	Water quality	X	X	X	Water samples E. coli, pH, and turbidity
	Environmental health	Hygiene practices		X		Recall
Child	Growth and development	Anthropometric measurements	X	X	X	Actual measurement
	Nutritional status	Energy and nutrient intake	X	X	X	Recall

TABLE 13.2   Data Collection for the Household Surveys and Field Tests

(continued)

Level	Indicators	Variables	Round 1 Jun–Jul 1999	Round 2 Oct–Nov 1999	Round 3 Mar–Apr 2000	Methods
	Health status	Breastfeeding history	X			Recall by mother
		Diarrhea incidence and patterns	X	X	X	Recall by mother
		Incidence and patterns of respiratory infections	X	X	X	Recall by mother
		Morbidity patterns		X	X	Recall by mother
		Mortality		X		Recall by mother
		Iron status		X	X	Measurement of hemoglobin
		Parasitic infection		X	X	Stool samples
Women	Nutritional status	Energy and nutrient intake	X	X	X	Recall
	Female fertility	Reproductive history	X			Recall
	Health status	Morbidity patterns		X		Recall
	Labor demand	Time allocation	X	X	X	Recall
Men	Nutritional status	Energy and nutrient intake	X	X	X	Recall
	Health Status	Morbidity patterns		X		Recall
	Labor demand	Time allocation	X	X	X	Recall

TABLE 13.2   Data Collection for the Household Surveys and Field Tests (continued)

surrounding environment provided insight into the patterns of resource use of families and therefore ultimately, the determinants of health, food security, and nutrition. Figure 13.5 provides a basic understanding of the relationship between flooding and human health.

TABLE 13.3	Data Collection for Spatial Mapping and Ethnographic and Participatory Health Assessment		
Level	Variables	Methods	Participants*
Landscape	Spatial mapping of ecological diversity; lakes, swamps, forests, rivers, palm forests etc.	GPS/GIS	Farmers/fishermen/ hunters and gatherers
	Spatial mapping of resource use activities; farming, fishing, hunting, gathering and logging	GPS/GIS	Farmers/fishermen/ hunters and gatherers
Community	Sources of information	Observe/key informant interviews	Community leaders, teachers, students
	Community organization	Observe/key informant interviews	Community leaders, teachers
	Community facilities and resources	Community mapping Wealth ranking	Two groups of male comm. members approx. 20 people
	Hygiene and sanitation	Observation, Community mapping and health walk	Two groups of female comm. members approx. 20 people
	Disease periodicity	Seasonal calendar (focus group)	Two groups of female comm. members approx. 20 people
	Historical information	Timeline (focus groups)	Groups of 6–9 people • Elderly (1) • Fishermen (2) • Hunters (2) • Farmers (2) • Women (3)
Individual	Diarrhea management	Key informant interviews/case histories and decision models	Interviews with: 8–10 people knowledgeable of health issues 10 women with children with a recent diarrhea episode
	Nutritional ethnography (local classification)	Pile sort/food attributes/attribute rating	Small groups of 2–3 people include: • women with children • < 5 years (5) • women > 45 years (5) • men (2) • adolescents (2) • teachers (2)

(continued)

TABLE 13.3   Data Collection for Spatial Mapping and Ethnographic and Participatory Health Assessment *(continued)*

Level	Variables	Methods	Participants*
	Health ethnography	Focus groups Body and sexuality mapping	Focus groups include: women with children <ul><li>< 5 years (2)</li><li>women > 45 years (2)</li><li>men (1)</li><li>adolescents (1)</li></ul>
	Fertility (pregnancy, prenatal care, birth, diet, contraception etc.)	Interviews	Interviews with 10 women of different ages
	Risk Management (identification of risks, frequency, predictability and coping strategies)	Focus groups Life histories	Groups of 6–9 people <ul><li>Elderly (1)</li><li>Fishermen (2)</li><li>Hunters (2)</li><li>Farmers (2)</li><li>Women (3)</li></ul>

*Number in brackets indicates the number of groups.

In the natural ecosystem, changes in river levels dictate migratory patterns of fish and wild animals and the seasonal availability of forest foods. During the times of low water, the commercially important seed and fruit eating fish inhabit the river channels and lakes, fasting, yet avoiding predators. Once rivers rise, these fish disperse into the flooded forest seeking food and the protective cover of

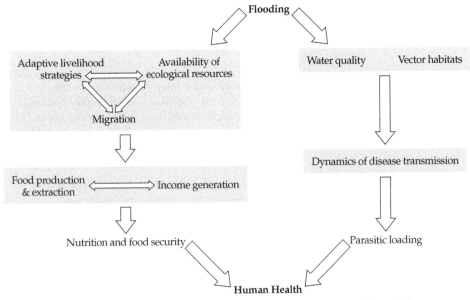

Figure 13.5   Relationships between human health and ecosystem in Ucayali, Peru.

the trees. This general pattern of fish migration results in periods of abundance, in the dry months, followed by scarcity, when the rains come and fish are difficult to catch. Terrestrial animals exhibit complementary migratory patterns. During the rainy season, they move into the floodplain to feed on the abundant supply of fruits, seeds, and nuts available in the forests. At this time, hunting is facilitated as the animals tend to concentrate on the higher "restingas" or levees.

In terms of the human ecosystem, all natural resource use decisions are dictated by the dynamics of the changing floodplain. The Ucayali riverbanks shift 100 meters each year and large floods every 7–10 years can change the river's entire course, wiping out communities and flooding arable land. Agricultural production cycles follow the rise and fall of the rivers and precipitation levels. During the dry season, as the river recedes, fertile alluvial banks are available for annual cropping, in particular rice, maize and plantain. Natural levees arc across the floodplain, the remnants of ancient river channels that have changed course. These areas provide valuable agricultural land that is inundated only during the larger, less frequent floods.

Moving from the floodplain to the upland forests, the river's impact is reduced and less direct. In these uplands, the driving force is the construction of roads and the ensuing access to natural resources as well as markets. As logging roads enter the frontier regions, colonists soon follow, building homes and clearing a couple of hectares for such crops as corn, rice, and cassava. These crops are then transported out to the markets in Pucallpa. Often farmers with existing farms along an older road will clear land near a logging road, taking advantage of the untouched and, at least temporarily, more fertile soils.

Rivers and Roads as Nexus of Organization

In complex systems terms, the river can be seen as a nexus around which a whole range of other activities and variables self-organize in the floodplain, resulting in the formation of a particular attractor; roads serve a similar function in the uplands. There are a number of important system variables that are part of the river-centered attractor. First, changes in food availability that result from flooding create a situation of annual yet predictable food insecurity. Second, health status exhibits a similar cyclical pattern. Disease and sanitary conditions change with the floods as rising water levels alter animal and insect habitats and affect water quality, vector incidence, and disease outbreak. At certain times of the year the combination of these two factors, malnutrition and ill health, result in a critical situation where some form of intervention is needed

Ecosystem Dynamics and Migration

Human migration is the result of the interaction between the river- and road-centered attractors. Human migration, similar to that seen in fish and animals, is cyclical and widespread. The rural population is constantly moving, synchronized to the changing rivers and lakes and the migratory paths of fish and animals. This continuous flow of people and resources sustains their livelihoods, as rarely can the area immediately surrounding them provide adequately for their needs. Families and individuals may leave their communities for several weeks at a time, engaging in a number of different activities in other parts of the region. As rivers rise and riverine villages are flooded, people travel to the uplands, using the roads to access markets as well as other farms in need of labor. The difference in agricultural cycles in the floodplain versus the upland areas allows for the labor pool to be shared between these different ecosystems. As labor shortage is one of the main constraints for farmers, this movement of farmers is critical to agriculture is the entire region.

Ecosystem Dynamics and Human Health

Determining the issues facing communities of the Amazon through an understanding of the dynamics of their ecosystem led to the discovery of a much larger set of complex interconnected problems that until this point had been overlooked. These new insights gained are as follows:

1. Patterns of natural resource use can only be understood at a fairly large, regional spatial scale, and using a variety of criteria [as defined by Allen and Hoekstra (1992)], particularly landscape and ecosystem. On-farm activities as well as those in the close surroundings do not capture the family livelihood strategies. Families make use of a great diversity of ecological resources located in different biotypes and parts of the region. We discovered that this large and diverse set of resource use activities interacts with each other. Often income from one resource use, for example, logging in the wet months, is used to pay for land preparation for crops in the subsequent dry months. Surplus harvest income is later used for equipment purchase and maintenance for logging and fishing. Thus resource uses are inextricably intertwined and interdependent and cannot be meaningfully analyzed separately.

2. Patterns of natural resource use can only be understood if examined at different times of the year. Seasonal resource use is determined by the flooding cycles that dictate the availability of arable land, wild foods, fish and animal migrations, and access to valuable forest resources. Figures 13.6 and 13.7 portray the different cycles of disease, food availability, and income level as perceived by the community members. The numbers indicate relative values: 1 is low, 3 is high, and 2 is medium. This information was gathered during a community meeting where the group, using colored markers, drew the graph on a large piece of paper. The figures demonstrate the difference between the cycles of the upland forests versus the floodplain. For example, in the floodplain, disease (in this case, diarrheal infection) is most prevalent in the dry months (June–August) when the water quality is at its worst. In the uplands, there are similar problems in the summer, though they also face malarial outbreaks in the wet season. In the floodplain, fish and wild animals provide sources of protein rich foods during the summer and winter months respectively. In January, little food is available as the flooded lands inhibit farming, hunting, gathering, and fish have dispersed into the flooded forests. In the uplands, food shortages occur in the winter months when fish prices are high and the

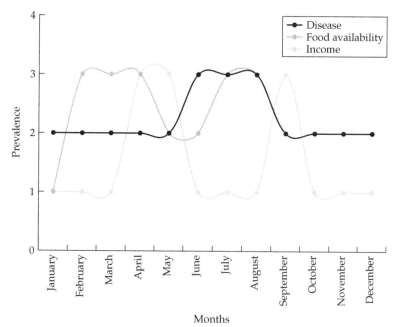

Figures 13.6 Cycles of disease, food availability and income level as perceived by the community members in the Ucayali floodplain (5) and upland forest (6). Numbers indicate relative values; 1 is low, 3 is high and 2 is medium.

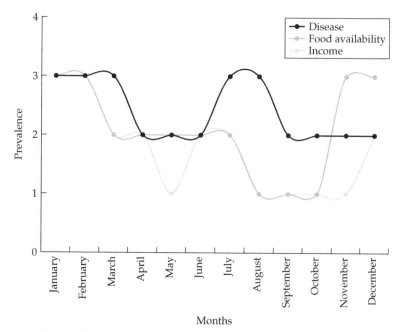

Figure 13.7 Same as fig. 13.6, but for upland forest.

agricultural harvest is still months away. Lastly, income in the floodplain is linked, first, to the sale of timber, usually sold in April when it has all been harvested and carried along the flooded rivers, and, second, to the sale of the agricultural harvest in September. In contrast, in the uplands, income is highest in January and February when their crops reach maturity and are sold in the Pucallpa market. There are few other sources of income.

3. Individual and family nutritional status can only be understood at a regional scale using landscape and ecosystem criteria. As food security and nutrition are determined primarily by the family's ability to produce, gather, or purchase food from a diversity of resources that extend across the landscape, an understanding of when and why there is low nutrient intake must take into account landscape level variables.

4. Individual health can only be understood if linked to seasonal change. Patterns of disease periodicity are linked to the environmental changes that occur with flooding. This affects vector habitats, dynamics of disease transmission, and water quality. Health is dynamic and therefore cannot be understood at only one point in time.

5. Population health can only be understood if linked to both national and international health policies and programs. All regional policies are designed in Lima and then implemented by the regional Ministry. Programs are not often appropriate, as they are not adapted to local conditions, nor are there baseline data to allow for an impact assessment. Health priorities are often influenced by the availability of resources from international aid agencies. Thus critical health issues facing communities in the Amazon are not effectively identified and addressed.

Development of a Common Vision

In any complex system, there are always multiple system goals and therefore decision choices that need to be negotiated among the actors or stakeholders, some of which are trade-offs. Emphasized by other case studies in this book, research into such complex systems requires legitimate involvement of the local communities who are most directly affected by decision choices and trade-offs. The affected communities may include government agencies, women's groups, nongovernmental organizations, or individual citizens. Together, professional researchers and local stakeholders define the problems to be examined, generate knowledge, analyze the findings, and take action. Identifying a common vision or goal is a critical step in the process from research to action.

In this project, there were two levels of goal-setting, or visioning. First, the research team developed a common vision. This vision evolved over the course of the project on the basis of input from our partners as well as our experiences in the field. Our vision dealt principally with issues and solutions at a regional level, bringing together information gathered in the different communities. This vision, developed in collaboration with the Ministry of Health, included two key aspects: (1) we recognized that, because the health problems of rural communities were so poorly understood, the Ministry's programs were not appropriate and therefore not effective; and (2) we highlighted the key role ecological factors play in determining health. For the first time, the Ministry considered the environment as a source of, as well as a solution to, health problems.

Second, we worked with communities to develop their own visions. In community meetings, as well as with smaller groups of women, men, and children, we asked people how they imagined their future and what, if anything, they would like to change. Then, in each community we developed an overall idea of what type of community people wanted to live in and how they thought they might arrive there. It would be overly optimistic to assert that each community had a common vision. However, the process of identifying goals allowed the communities to see where their common ground lay.

Interventions

Historically, interventions based on the narrow focus of cattle and deforestation were confined to improved plant varieties and animal species and more effective farm management strategies that extended the economic life of the land such as improved fallows and agroforestry systems. These programs were unsuccessful as farmers failed to adopt the proposed technologies. Such technologies were developed by researchers on research stations, using evaluation criteria developed in isolation from the affected communities. Simple issues, such as the fact that most farmers were without a land title and therefore ineligible for government credit schemes to purchase cattle meant that most technologies, in reality, held little promise for the majority of farmers.

Our intervention strategies were guided by the communities themselves as well as by the regional Ministry of Health. Although our focus was the family, our efforts were directed at the community and regional level. On the basis of our understanding of the system's dynamics these were determined to be the most effective channels for improving human health.

Regional-Level Intervention

At the regional level, our intervention strategies included policy evaluation, development, and training. On the basis of our experiences in the field and the results of our research, we worked with the Ministry in modifying their existing programs as well as exploring ways by which they could begin to incorporate the issues that linked flooding and ecosystem dynamics with human health.

There were several areas of intervention. These included the following:

1. Water quality and environmental health

The environmental health data collected by the project included parasite prevalence and re-infection rates, seasonal changes in water quality, basic hygiene conditions, and sanitation infrastructure. The Ministry of Health had an antiparasite program that contained neither baseline data nor impact analysis. The Department of Environment Health had several models of wells and latrines that are neither monitored nor maintained. There was no water quality testing in rural communities.

All the data on water quality and parasite levels in children were shared with the Department of Environment Health. This gave them substantial evidence of the severe water quality issues in rural communities. The project helped support further monitoring of these rural communities by donating over 1000 filters to Departamento de Salud Ambiental (DESA). We provided DESA with information on appropriate water treatment technologies for upland and floodplain ecosystems (solar and sand filters).

2. Nutrition

Our main data sets included nutrient deficiencies (micronutrients), seasonal food security, nutritional status (anthropometry), anemia levels, and diversity of foods consumed. As the above information had not been collected by the Ministry of Health in the rural communities, our results provided critical information that fed directly into the evaluation and design of their rural nutrition programs and the training of their health workers.

Using these data, we have begun to develop food composition tables appropriate for the Peruvian Amazon. The Peruvian food tables do not contain nutritional information on the majority of foods unique to the Amazon region. We also developed dietary guidelines based on our results of key nutrient deficiencies (e.g., vitamin A deficiencies) and sources of nutrients in regional foods that are being promoted through the Ministry's nutrition programs.

Through the World Health Organization (WHO), we were able to acquire several thousand field anemia test kits for the Ministry's Micronutrient Program. With these tests, the Ministry was able to monitor the impact of their new region-wide program in iron supplementation. This test is also be used by the Ministry's Malaria Program to monitor the recovery of malaria patients. Before this was implemented, we set up region-wide program to train all rural health workers in the use of the WHO anemia test.

3. Health: Training manual for community health promoters

Although most rural communities have a health outpost, with a nurse or health technician, these people are away from the community for as much as 30 percent of the time. During these times, communities are left with no means to treat illnesses. The Ministry of Health has a three-month training program available to community members who would like to become health promoters. As this training can be given by any of the Ministry of Health's professionals, the quality of the course has varied considerably. In addition, several key areas, such as environmental health, nutrition, and links between diet and natural resources management are missing, and the protocol needed to be adapted to the two main ecosystems: the uplands and the floodplain. Our project has modified and adapted the training protocol for health promoters on the basis of findings from the research. This has resulted in an extensive training manual now being used by the Ministry. In five of the eight communities the two nurses who are members of the researcher team spent one year working in the health outposts with the community health promoters. Using the new training manual, they ran several workshops with the health promoters, specifically targeting the health issues that were linked to ecosystem dynamics.

4. Nutrition, food security, and natural resources management strategies: Linking health and agricultural policy

The project team worked with professionals from Ministry of Health identifying areas where there needed to be coordination with Ministry of Agriculture. On the basis of the project's evidence of the linkages between natural resources management strategies and food security, it became clear that a link is needed between agricultural programs that increase food production with those health programs that address food insecurity and malnutrition. In each community we facilitated linkages with Ministry of Agriculture's extension agents and other groups working in agriculture that can integrate nutrition and food security goals with production and income generation. One of the team members is now working part-time for the Ministry's of Health food aid programs. Using our research findings, he has helped to direct their efforts at the specific times of year when seasonal food insecurity is at its greatest.

Community-Based Interventions

Intervention at the community level was confined mainly to education and some small initiatives based on the data collected on water quality and parasite levels. Each community was different in terms of the issues they faced and their available resources.

Although most interventions occurred after the research phase was complete, there were several key initiatives that took place in conjunction with our research activities. In all the communities we asked the women to identify areas of health and nutrition that were of most concern to them. Once we understood these issues, we set up small workshops where the women could learn some of the basic health and nutrition principles and ask questions. This provided us with a valuable opportunity to check the relevance of our work as well as providing the community with something in return for their time. Our experience demonstrated that variables such as water quality and parasite loading provided a very good starting point from which to link analysis and action. The community gained a better understanding of the significance of the research results and was, in turn, able to incorporate these results into their understanding. For example, stool and water testing initiated action in cleaning and controlling defecation along stream banks and reducing contamination of wells. In addition, two of our health workers presented a talk on hygiene practices. With the villagers as partners we identified common areas where hygiene could be improved.

Once the research phase was complete, we organized a series of educational workshops in five of the eight communities, emphasizing areas of water quality, parasites, hygiene, nutrition and basic health prevention measures. These occurred over a one-year period and drew upon the results from our research. The project formed the basis for a general assessment of health care delivery in Peru and how it might be improved (Goy and Waltner-Toews 2005).

Monitoring and Adaptive Management

Although the official research project has ended, continuing impacts are mediated primarily through the Ministry of Health's professionals who were involved first hand in our activities and through several of the project's researchers who now hold positions within the Ministry and other nongovernmental organizations.

The monitoring of basic health indicators set up by the project in remote rural communities continues through the Ministry of Health. This includes anemia and parasitic infections. Monitoring was made possible by the donation of resources from WHO and the University of Western Sydney.

Unfortunately, the highly centralized policy-making structures made it next to impossible to set up organizational structures for adaptive management at a regional scale, which underlines a major challenge for locally based, "bottom-up" initiatives such as this and the need for multiscale initiatives. Raez-Luna (chapter 18) addresses some of the challenges in doing this kind of work in parts of the world where there marked disparities in economic and political power.

Reflections

As expected, there were both successful and unsuccessful aspects to the project. We believe that the project's greatest success was with the Ministry of Health. We were able to bring a new, systemic perspective to the Ministry and the professionals with whom we worked. They have now incorporated many of our findings into their programs. We have drawn their attention to the links between health and ecology, and, as a result, they have redistributed their resources so that measures such as parasite infections and water quality are now being monitored. Further, the Ministry was able to adopt some of the methods used on the project. For example, two Ministry professionals were on the research team and therefore had first-hand experience with our methods and the skills to use them and train others in their application.

The inclusion of Ministry professionals on the research team also meant that the Ministry had a sense of ownership of the information. We worked with them in the initial analysis of the results, which were then immediately included in their reports. By including Ministry officials on the research team, we managed to avoid the perception that the research was conducted by outsiders and therefore not of particular relevance. All too often, when outsiders gather data, analyze it, and report their findings in isolation, the resulting information lies unused on government shelves, having little or no impact on decision making.

Our project helped to develop baseline data on remote rural communities that were previously beyond the Ministry's scope of impact, owing largely to their resource constraints. Our work provided critical information on the health and nutritional status of these communities and therefore on the nature and extent of their need for assistance.

Furthermore, the project linked members of the community with professionals in the Ministry of Health and the local university. Access to these decision makers was essential to providing community leaders with an avenue to voice their concerns. We helped them improve their understanding of the governance structures and how best to communicate their needs to policy makers.

The impact of the project on local community groups came largely from their direct involvement in the research activities. They received valuable training, and because of the large scope of the project, were exposed to many different parts of the region. As a result, they were able to return to their respective organizations with a greater understanding of the range of problems facing their members, and they were able to talk about these problems from first-hand experience. In addition, the training achieved during the project afforded these individuals important analytical and communicative skills that will continue to empower them in their capacity as community leaders.

With respect to the research team, the researchers gained a valuable and unique experience in working with different disciplines. We were all better able to see the limitations of the sectoral approaches taken by the government and other research organizations. Many members of the team are now working in other projects where they are able to apply the knowledge gained during this project.

The single most limiting aspect of the project was the short duration. Once the research was completed, there remained less than a year to implement the action plans. On the basis of our experience, we would recommend nothing less than a five-year period for projects using the ecosystem approach. This is especially true in a region where there is so little institutional capacity and social capital, which are seen to be essential for the implementation of the approach. Indeed, our experiences resonate with those of other authors in this book, and many share our conclusions. In particular, that "deep" or meaningful participation by, effective collaboration among, and fundamental empowerment of diverse constituencies, are essential achievements to navigating the complex cultural-natural ecologies in which we dwell. The work in Nepal, for instance (see chapter 15), unfolded over a period of ten years. There is no quick fix, no panacea: an ecosystem approach takes time, but the results are lasting.

References

Allen, T. H. F. and T. W. Hoekstra. 1992. *Toward a Unified Ecology*. New York: Columbia University Press.

Checkland, P. and P. Scholes. 1990. *Soft Systems Methodology in Action*. New York: Wiley.

Goy, J. and D. Waltner-Toews. 2005. Improving health in Ucayali, Peru: A multi-sector and multi-level analysis. *EcoHealth* 2:47–57.

Holling, C. S. 1995. What barriers? What bridges?. In *Barriers and Bridges to the Renewal of Ecosystems and Institutions*, eds. L. H. Gunderson and C. S. Light, pp. 3–34. New York: Columbia University Press.

Instituto Nacional de Estadisticas y Informaticas. 1997. *Población. Muher y Salud. Resultados de la Encuesta Demografica y de Salud Familiar*. Lima, Peru.

Midgely, G. 1992. Pluralism and legitimation of systems science. *Systems Practice* 5(2): 147–172.

Rowley, T., G. Gallopin, D. Waltner-Toews and E. Raez-Luna. 1997. Developing an integrated conceptual framework to guide research on agricultural sustainability for tropical agroecosystem: Centro Internacional de Agricultura Tropical—University of Guelph Project. *Ecosystem Health* 3:154–161.

Waltner-Toews, D. and E. Wall. 1997. Emergent perplexity: In search of post-normal questions for community and agroecosystem health. *Social Science and Medicine* 45:1741–1749.

PART III

Managing for Sustainability
Meeting the Challenges

In Part I of this book, we set out a theoretical framework for addressing some of the issues of complexity and uncertainty in managing for sustainability. These include matters of both substance (bio- and cultural diversity, for instance) and process (management and knowledge in multistakeholder systems). Part II introduced case studies at a variety of scales and in a variety of places. The intent was to determine if there was anything that could be generalized; if the current state of the world is entirely the outcome of geographically bounded social-ecological evolutionary forces, then can anything useful be generalized about ecological management for human well-being? Is all knowledge in this field local and contextual? If so, where does that leave scientists and scholars in general? In this section, we set out what was learned through those case studies, i.e., James Kay's thinking on the subject in the year before he died. In particular, we (James and his colleagues and graduate students) generated guiding questions (chapter 14) that we then applied to a particularly messy and interesting case study in Nepal (chapter 15). We then explore some of the questions that arose from the theory and its application.

14

Implementing the Ecosystem Approach
The Diamond, AMESH, and Their Siblings

David Waltner-Toews and James J. Kay

Basic Principles

Over the decade of the 1990s, we engaged in a wide-ranging series of discussions on complexity and sustainability with the authors in this book, as well as with several other distinguished members of the Dirk Gently Gang not represented here. With our colleagues in various parts of the world, we implemented practical strategies to achieve sustainable health, agricultural, and/or environmental management goals based on the best theory we had. Some of those case studies are described in previous chapters of this book. We then reviewed the case studies and searched for basic principles and guidelines that would carry us forward both practically and theoretically. We asked: What principles can we induce, relevant to an ecosystem approach, from the discussions so far of the necessary and complementary ecological and sociocultural characteristics of managing for sustainability?

Managing for sustainability, when viewed in a complex systems context, is itself a complex activity. There is no straightforward program or method of social or political organization that will take us from here to "there." Indeed, we cannot easily define what "there" is. What then are we left with? Without recapitulating in detail what has already been well-said in previous chapters, we can propose some principles drawn from the investigations undertaken so far that can guide us in learning our collective way into a convivial and sustainable world. They are presented in no particular order; some of them overlap, and we do not consider this list to be exhaustive. Nevertheless, they emphasize different essential aspects of what we would consider to be an ecosystem approach to understanding and managing for sustainability.

Social Principles

The process involves collaborative development of visions desirable futures. In any process designed to achieve the sustainability of convivial human communities, we are trying to set and achieve collective goals. Hence some kind of collective vision-ing process is essential; in order to be sustainable and accounting for socioeco-nomic power differences, these collective processes are at best collaborative and not merely collective.

The process explicitly recognizes and utilizes multiple perspectives on the situation. As Funtowicz and Ravetz assert (chapter 17), "there is no single perception pro-viding a comprehensive or adequate vision of the whole issue, nor any particular criteria of quality that can hegemonically exclude all the others." Berkes and Da-vidson-Hunt (chapter 7) underline this when they suggest that different knowl-edge systems "can be combined only in the sense of dialogue and mutual enrich-ment, sharing across systems of knowledge." From a management viewpoint, this means that the point is not to arrive at a consensus view of reality but to negotiate collective actions that can accommodate multiple views.

The process explicitly addresses issues of power and inequity. In any situation, there will be a variety of legitimate stakeholders who should be involved in ecosystem approach activities. Activities such as the framing of the issues and determining the type(s) of collective futures envisioned are influenced by who is sitting around the research/development/management table (the "legitimate stakeholders"). Some corollaries to this are as follows:

1. Who is sitting around the table often reflects both diversity and power imbalances.
2. At all scales, there are power imbalances related to social, political, and eco-nomic status. These are often reflected in power differentials based on such things as race, gender, and ethnic background.
3. The process must include a way to address and redress power imbalances.
4. Finally, given the inevitable contradictions and paradoxes inherent in the situations in which we find ourselves (and which are at least partly of our own creation), any process must find ways to accommodate change, loss, and tragedy.

Ownership of the process should rest with the stakeholders. The achievement of goals related to sustainability implies ownership of the process, as well as the power to act in relation to the resources. The human systems necessary to achieve this ideal are most easily achieved if we are working locally (in spatial terms). Not all problematic situations can be resolved locally however, particularly in a world increasingly interconnected by trade, war, communications networks, and the like. The need for assertive ethical stances coupled with sound political-economic analyses and related issues of solidarity and mutual respect seem to take on a magnified significance at these greater spatial scales.

The process should include mechanisms for conflict resolution. Because not all goals are (simultaneously) achievable,

1. There will be trade-offs and probably (given item 2 above) conflicts will occur.

2. Some understanding of how the various goals interact systemically, incorporating *both* social and ecological variables, is important to guide the social process.
3. Some mechanisms and rules for resolution of disputes must be a part of any process to manage for sustainability.

The definition and assessment of sustainability is dependent on cultural history and context as well as natural history and context. The process should therefore involve exploration and analysis of the cultural and historical context of the situation and multiple scales. Given the deep cultural diversity across a variety of scales, any process to understand and/or promote sustainability should incorporate mechanisms for cross-cultural communication and collaboration.

Adaptive ecosystem approach projects and programs should incorporate short-term as well as long-term goals. Most people will not undertake activities that only have long-term possibilities of achieving good things. Some short-term goals undermine the possibility of achieving long-term goals. This is related to the importance of linking hard and soft systems. Therefore some goals that are achievable within relatively short time frames but that are on the path to long-term sustainability goals need to be clearly identified.

Adaptive ecosystem approach projects and programs should be ongoing, iterative, and responsive to changing goals and new information. Because of the apparently changing nature of complex reality and our place in it (that is, not as external observers), our understanding will always be incomplete, and negotiated goals will never, in any final sense, be attained. This conditionality of knowledge and achievement should lead to humility, mutual respect, and trust between scientists and community stakeholders in the face on an intransigent and mysterious reality.

Ecological (Biocomplexity)

Models of ecosystems are observer-dependent, purpose-dependent abstractions specific to the problem they describe. Although we might assume that we are all embedded in a common, complex reality, there is no single model or perspective that can capture this complexity. Hence the ecosystems we describe are constructions designed to achieve particular ends.

Adaptive ecosystem approaches involve the identification of appropriate spatial and temporal scales to address a situation. System exploration will focus on this hierarchical level at least one level higher and lower than this. Ecosystems are usefully understood as comprising multiple, nested hierarchies. Other researchers have referred to this complex, nested context as holonocracy (Regier, Appendix A) and panarchy (Gunderson and Holling 2002). Although the exact language used to describe this nested complexity will vary, it is clear that issues related to sustainability can only be resolved with reference to this context.

It is crucial to understand the sets of feedback loops for the system(s) relevant to the situation. Linear causal models are inadequate to describe what goes on in ecosystems. Outcomes can become causes through series of loops and webs. These interactions are somewhat bounded within holons (at least to the extent that they operate by a particular set of rules within the holon), but interactions also take place across levels in a holonocracy. Therefore changes in a holon may change, stabilize, or destabilize a range of other holons within which it is nested.

It is crucial to understand the propensities for self-organization and the set of potential attractors for the system(s) relevant to the situation. Sets of interactions may self-organize around an attractor (operating state or domain of behavior), that is what we "see" on a landscape as being "natural." However, complex systems have multiple attractors and may shift or "flip" suddenly from any one of them (Holling 1986; Kay 1991; Regier and Kay 2001).

The construction of narratives is a key technique to understand complex systems. The dynamics of SOHO systems are described by narratives. The narrative description of a SOHO system is a description of its propensities, the mutual causality of the feedback loops and autocatalytic process that give the system its coherence as an entity. This set of propensities that define a holon is referred to as its "canon" (Kay 2000). While individual elements may consist of traditional scientific models and descriptions, synthesizing these elements together into a narrative transcends normal scientific descriptions and models. For complex systems, narratives are a cornerstone of the description of their dynamics (Allen et al. 2001).

Adaptive ecosystem approaches need to explicitly deal with uncertainty. Given the feedback loops, multiple attractor states, and the requirement for multiple perspectives, complex socioecological systems are characterized by inherent uncertainty and limited predictability. Interventions in a system may need to be large in order to notice a response. However, such changes may be irreversible. Management, while taking into account the canon and narrative, needs to be cautious, anticipatory, and adaptive.

Some of principles that emerge from consideration of social context are the same, or similar to, those that emerge from consideration of ecological context. This convergence, or overlap, is reassuring that we might be on the "right track." Having set out some basic, guiding principles, we can ask: What are the questions we need to ask to guide our work, what are the activities and tools we can use to answer those questions, and how can we assure quality of information and activity? If we think generically, we can identify a series of relevant components:

- the situation is brought to someone's attention, often because there is a perceived problem
- the "responders" try to understand the situation systemically
- identify systemically based alternative courses of action
- choose a course of action
- develop a plan that incorporates a collaborative learning system
- begin implementation
- ensure that governing, monitoring, and management coevolve with the changing situation

Assuming that a problematic situation has been presented, we are faced, as investigators, with trying to understand what is going on. How did this situation come to be? How did we get to where we are now?

Beyond Theory and Principles: What Do We Do Now?

The Diamond Diagram: A Basic Framing of the Process

At the core of an adaptive ecosystem approach to managing for sustainability is the premise that a sustainable society maintains itself in the context of the larger ecological system of which it is a part. Decision making for sustainability in this context comes to be understood for what it has always been, finding our way through partially undiscovered country rather than charting a scientifically determined course to a known end point. A framework for undertaking this is presented in fig. 14.1. This diagram, developed by James Kay together with his colleagues and students at the University of Waterloo, is sometimes referred to as the "diamond" diagram because of the centrality of the diamond in which social preference and ecological possibilities are integrated into a set of choices for human agents. This framing of the process has been used and adapted by a variety of scholars and practitioners in different parts of the world (see, for example, Boyle and Kay, chapter 16; Bunch, chapter 10; Charron, chapter 11; Gitau et al., chapter 12; Lister, chapter 6; Murray et al., chapter 12; Neudoerffer et al., chapter 15). Several alternative versions can be accessed through James Kay's Web site (http://www.jameskay.ca). This heuristic is fundamental to the ecosystem approach we are advocating.

In this post-normal approach to sustainability, the scientist's role in decision making shifts from inferring what will happen, that is making predictions that are the basis of decisions, to providing decisions makers and the community with an appreciation, through narrative descriptions (scenarios), of how the future might unfold. Through SOHO descriptions of the situation, science informs society about known ecological constraints and possibilities. (See the top left-hand box in fig. 14.1.) People provide an image of how they would like to see the landscape of human and natural ecosystems coevolve. (See the top right-hand box in fig. 14.1.) A dialogue must ensue (the diamond box in fig. 14.1) that explores the desired and the feasible and reconciles these in a vision of how to proceed. Scientists inform this dialogue by providing future narratives that will evolve as scientists partake, as equals, with other participants in the process of resolving the vision. Having resolved a community vision for the future, the next phase is to design an adaptive program for the realization of the vision.

This adaptive program consists of a plan and infrastructure for the triad of activities, governance, management, and monitoring. *Governance* refers to the continuing process of learning, revisioning, resolving trade-offs, and planning by the parties to adapt to the unfolding situation. This entails the ongoing evolution of governance arrangements. We see this activity happening all around the Great Lakes with the emergence of virtual governance, community-based initiatives that organize to focus on specific elements of the landscape such as watersheds or

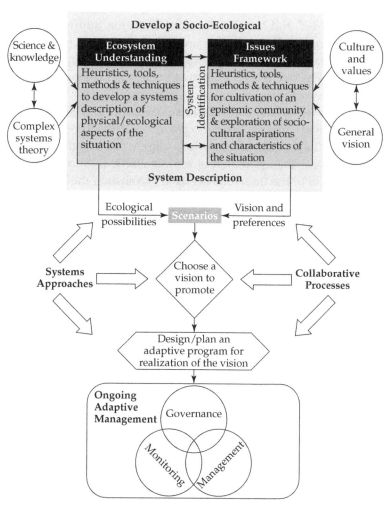

Figure 14.1   One version of the "diamond" heuristic of the ecosystem approach, showing how scenarios, visions and managing for sustainability emerge from the integration of ecological and sociocultural understanding. Other versions of this figure are available at www.jameskay.ca

bays. *Management* is the activity of translating the vision into reality. It involves the development and implementation of strategies to promote or discourage specific forms of self-organization in the context of the communal vision and plan. This means maintaining the context for the self-organizing complex (SOHO) systems, rather than intervening in the system in a mechanical way (Allen et al. 1991; Kay and Schneider 1994; Boyle and Kay, chapter 4).

Maintaining the context involves identification of external contextual changes, flows into and from the system, and feedback loops that are to be encouraged and discouraged. Generally speaking, management concentrates on the relationship between humans and natural ecosystems and guiding the human side of the relationship. *Monitoring* is the activity of observing the human and natural systems and synthesizing the observations together into a narrative of how the situation has actually unfolded and how it might unfold in the future (see chapter 16 for more on monitoring). This narrative is used as the basis for governance and management, that is for learning, revisioning, and adapting human activities as the human and natural ecosystems coevolve as a self-organizing entity.

In this adaptive ecosystem approach, monitoring, governance, and management make up a triad of activities that are carried out in the context of an issues framework of human concerns and an explicit conceptual model of the ecological-economic system. Taken together, the issues framework, and the conceptual model provide the focus for the discussion of sustainability. By furnishing a means for informed resolution of the trade-offs necessary to sustain the health and integrity of the ecological economic system, the activities of monitoring, governance, and management, carried out in concert, chart the course to sustainability.

While the diamond diagram has itself been used to frame and implement the ecosystem approach, some questions related to the context of an ecosystem approach remain outside its scope. For the most part, when working inside a culture, we assume certain kinds of stakeholders, participatory processes, and roles for scientists and nonscientists. When working across cultures, these assumptions need to be made explicit. Research and development interventions across cultures raise a series of issues that are not necessarily apparent when we are working "at home."

Adaptive methodology for ecosystem sustainability and health (AMESH) was developed through a series of research and intervention projects, coupled with in-depth reflection by several international teams of researchers (Murray et al. 2002; Rowley et al. 1997; Smit et al. 1998; Waltner-Toews et al. 2005). Many of the activities of AMESH have raised questions about stakeholder-governance-issue relationships, how these influence system definition, problem identification, and possible resolutions. That is, projects that have most strongly informed AMESH have focused on how the top part of the diamond diagram can be implemented and provide guiding questions for researchers and managers. Boyle and Kay (chapter 16) focus on understanding how the bottom section of the diamond (governance-management-monitoring) can be implemented.

The AMESH Family: A Brief History

In 1993, a major interdisciplinary project, Agroecosystem Health: Analysis and Assessment), was launched at the University of Guelph (Smit et al. 1998; Waltner-Toews 1994; see also Charron and Waltner-Toews, chapter 11). Drawing on a range of disciplines, from economics and geography to ecology and health sciences, a wide variety of models of both agroecosystems as systems and of the research processes to evaluate them were developed. The general research process drew heavily on the soft systems methodology of Peter Checkland, as well as the diamond diagram, and emphasized his insistence on systems approaches as being grounded in a vision of "never-ending learning" (Checkland 1981:285). We recognized that setting goals and objectives, establishment of a research team, and evaluation of results were all important yet left the processes by which these might be accomplished ambiguous.

In an academic research setting, the default option for research is that disciplinary experts of one sort or another determine the goals and evaluate results because we are sure that we know how to do them best. Even as the research progressed, some of us concluded that, if the purpose of the research was to foster better decision making by (and not just providing gratuitous advice to) farmers and agricultural policy makers, then this expert-driven approach was not very useful (Waltner-Toews et al. 2000). It was not only based on an unrealistic understanding of nature as merely complicated (not complex) and an outmoded notion of ecological management (prediction and control), but it also begged some fundamental questions of what we might mean by "better decisions" and who should be making those judgments.

Drawing on lessons learned from the Agroecosystem Health Project, David Waltner-Toews and Gilberto Gallopin initiated a project in South America with the cumbersome title "Development and application of an integrated conceptual framework to tropical agroecosystems based on complex systems theories: Centro Internacional de Agricultura Tropical-University of Guelph Project" (Rowley et al. 1997). The research area for the project was in the Ucayali region of the Peruvian Amazon, around the major city of Pucallpa. This project sought to deliberately root its research process in a complex systems understanding of ecosystems. As such, it drew heavily on a network of international scholars working in this field (the Dirk Gently Gang, described in the Postlude-tribute to James Kay, this book). The research process again drew heavily on Checkland's soft systems methodology.

In retrospect, it was still based on the usual (and naïve) scientific assumption that we could compare our models with some independent, objective version of reality to determine its correctness. However, we did specify and expand on several of the steps in explicit complex systems terms, listed out "guiding questions" for each of the steps, and made the process close in on itself, unlike most linear research hypotheses-testing methods. We also identified, for the first time, what roles expert researchers and local stakeholders might play and paid careful attention to the selection of local research partners. In the process, we discovered that the initial stakeholder-partner group we selected, which itself had been created through a previous (unrelated) research initiative, represented mainly academics and business people living outside of the communities being studied. This, along with our vague notion that we could compare our models to some "gold standard," turned out to be both a major failing and an important opportunity for learning.

A second research project in Pucallpa, Peru, led by Tamsyn Murray (chapter 13) focused on links between health, biodiversity, and natural resource use and was more adequately grounded in the realities of local stakeholders.

While the work was progressing in the Peruvian Amazon, two other projects, one in the Kenyan highlands (Gitau et al., chapter 12) and one in Nepal (Neudoerffer et al., chapter 15), were exploring the same methodological territory. One version of Gitau's research process appears in his chapter (fig. 12.1). What the Kenyan research team found was that mobilization of local stakeholders into the research process almost immediately generated a demand for action; it also became apparent that monitoring and assessment could be undertaken by local stakeholders using a set of indicators that were different from those used by researchers. Farmers and villagers worked on different time frames, with different resources and had different requirements than researchers. This underlines the fundamental insight that indicators can only be defined in relationship to goals (Boyle and Kay, chapter 16).

Gitau's process incorporated many practical concerns in implementing an ecosystem approach and highlights some features that need to be recognized in any version of an ecosystem approach to management. While AMESH has been visualized in several different ways, often reflecting different degrees of separation of roles for researchers and stakeholders, the version of AMESH that we present below and that has been described in several other publications, integrates researchers into the process as "just another" stakeholder group (Waltner-Toews 2004; Waltner-Toews et al. 2004). This recognizes that we are a stakeholder group with our own unique characteristics and that many of the questions we ask of all stakeholder groups we also need to ask of ourselves.

As a consequence of its post-normal and SOHO theoretical roots, AMESH does not prescribe a set of procedures and quality testing techniques. Rather, we draw on a set of "guiding principles," methodological processes are described in terms of sets of activities, and these are elaborated in terms of "guiding questions." The four guiding principles of AMESH are a distillation of the social and ecological principles set out at the beginning of this chapter: (1) Methodological pluralism and locally grounded multiple perspectives, (2) hierarchical and cross-scale interactions, (3) self-organization, and (4) unpredictability and uncertainty.

AMESH recognizes that complex socioecological systems can only be understood through the integration of many diverse and often differing perspectives. In order to gather and weave together these multiple perspectives, AMESH encourages the use of a range of qualitative and quantitative inquiry and analysis methods. Additionally, AMESH acknowledges the fundamentally important role local people play in any endeavor to address ecosystem sustainability and health and supports the full participation of local people and the inclusion of nonexpert perspectives to shape and inform the ecosystem understanding.

While the local perspective is critical to AMESH, the larger context in which the local is embedded is equally important. Hence AMESH describes complex socioecological systems as nested hierarchies. For example, in AMESH a household is understood to be embedded in a neighborhood, in a city, in a region, in a province, in a nation, etc. Several different types of such nested hierarchies may describe the same socioecological system, using, for example, various political or ecological criteria. The language used to describe these has been discussed in earlier chapters and will not be belabored here. We recognize that, given the multiauthor nature of this text, there is some variation in terminology.

In general terms and using Koestler's terminology, this characterization of the complex reality is important because the larger holons set constraints for the smaller holons and the system interactions across these different scales (from the local to the global or vice versa) may have important, but not easily recognized implications for the ecosystem. Self-organization is a defining characteristic of complex systems and occurs because of feedback loops within a particular set of constraints. AMESH guides researchers and practitioners to look for and understand system feedback and the emergence of self-organization. Finally, AMESH recognizes that complex socioecological systems are both unpredictable and uncertain because they hold the potential to dramatically reorganize at critical points of instability and may have multiple stable states.

Although there is no single prescribed course of action for AMESH, we can describe it in five broad categories of action (fig. 14.2). Each of these has associated with it a set of guiding questions. In the text, we have summarized the intent of the questions. Some specific questions that we have found useful are set out in the accompanying boxes.

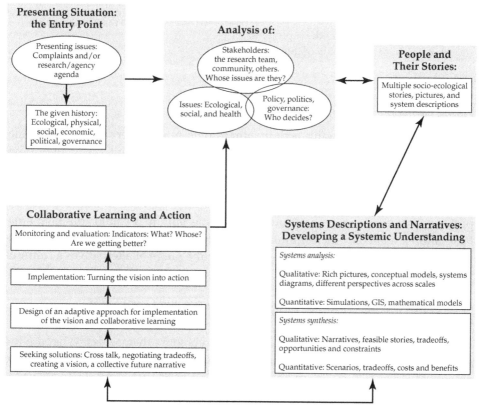

Figure 14.2   An adapative methodology for ecosystem sustainability and health (AMESH).

BOX 14.1   Guiding Questions for the Presenting Situation

General Questions
- What has been the overall historical development of the system?
- What have been the key ecological, economic, demographic, and social developments in the system? What kinds of rich interconnections have occurred?
- At what level have the changes taken place?
- What or who have been the agents of change?
- Which are the critical fast, medium, and slow variables within this overall development?
- Is there historical evidence suggesting sudden shifts or changes in the structure or behavior of the system?
- What has stabilized the system? What has tended to destabilize the system? Have there been surprising changes? Perverse stability?
- What are the positive and negative feedback loops?

Questions for Researchers
- What are the positive and negative feedback loops?
- What past projects have the research agency, or other groups, undertaken within the community?
- Who was involved in these projects?
- What was the outcome of these projects?

Documenting the Presenting Situation

The guiding questions in this activity aim to determine the ecological, economic, demographic, and social developments and how they have interacted over time. When projects are done at the behest of a research agency, we also need to ask about the history of the research agency in the community. This history will impact upon future work the research agency hopes to do. We need to understand past projects that other groups have attempted, especially in a similar problem area. Past failures will color the community's perception of the utility of working in a given area, while past successes might narrow the range of alternatives a community is willing to consider.

Analysis of Stakeholders, Issues, and Policy, Politics and Governance

This activity involves three subsets of questions.

1. One set of questions relates to the stakeholders themselves. What are the criteria for determining stakeholders? Who are they? What resources do they have access to? How do they group and interact? We need to ask questions about local leaders who are entry points and potential key informants for the research team. Without an understanding of local power and politics, it is too easy to unwittingly favor one power group over another. The researchers themselves need to be

BOX 14.2   Guiding Questions for Stakeholder Analysis

Stakeholders
- Who are the principle stakeholders?
- What is the process for identifying stakeholders?
- Under what and whose selection criteria will they be considered stakeholders? What are the critical discriminating features that allow us to differentiate the groups of stakeholders?
- What other groups, perhaps not directly related to the issue at hand, exist in the community?
- Which groups have been excluded?
- What are the factors that stakeholders perceive as being within and outside of their control?
- What are the power dynamics or relationships among and between stakeholder groups?
- What are the barriers and constraints to communication and collaboration between different stakeholder groups?
- What are the main resource coalitions?
- What goals/interests do these coalitions share?
- Which stakeholders are excluded from these coalitions and what are the consequences of this?
- What are the conflicts of interest? Which organizations conflict and why?
- What are the important gender relations among and between stakeholder groups? How do gender differences define people's rights, responsibilities and opportunities to participate?
- What are the assets and resources (skills, financial, material, connections, etc.) of each stakeholder group?

Researchers as Stakeholders
- What is the role of the researcher as stakeholder, and what power does the researcher wield to convene the others?
- How was the research team selected?
- Are they local or nonlocal?
- What is their level of experience with the research issue?
- What training have they received?
- What skills training does the team need?
- Does the research team have the necessary skills to conduct the AMESH analysis?
- What is the research agency's mandate?
- How might this mandate bias the AMESH process?
- What is the power relationship between the research team and the local stakeholders?
- What gives the research team legitimacy to work in the community?
- What power structures does the research represent to the community?

BOX 14.3   Guiding Questions for Issues and Governance Ecological, Social, and Health

Issues Analysis
- What are the critical issues?
- How do they relate to one another?
- At what spatial scale and temporal scale do they occur?
- For which stakeholder groups are these issues critical?
- How does the resolution of these issues come together in a vision for the future?

Policy, Politics, and Governance
- What are the key policies affecting the capacity of local stakeholders to sustainably manage their natural resources?
- At what level are these policies formulated and implemented?
- How do informal and formal governance structures affect local capacity to adapt to and deal with stresses?
- At what level are these governance structures formulated and implemented?

---

queried as stakeholders, and we now recommend a series of questions related to the power, role, and skills that the researchers bring to the process.

2. A second set of questions tries to unravel what the critical issues are, and at what scale, and to whom, they are important.

3. Finally, we are faced with questions related to governance, both formal and informal; in other words, the relationships between the stakeholders and the issues.

People and Their Stories

Although these narratives are central to our understanding of the basic science and other scholarly investigations, we have not (yet) given them their own guiding questions. The stories are a way of structuring different perspectives of the stakeholders, the relationships among the stakeholders, and their relationships to the various issues, over time. Thus the guiding questions of the previous steps also guide this part of the process, with a greater emphasis on temporal unfolding. Some of these are explored in a little more depth in chapter 15, a case study on Kathmandu, Nepal.

Systems Descriptions and Narratives: Developing a Systemic Understanding

This is the part of the process with which scholarly researchers feel the most comfortable because the tools used (especially for analysis) are those that have been traditionally valued in academic settings. Whereas cultural and ecological stories emerge from the first, participatory processes, systemic narratives, including the canon of possible states open to the system, are the outcomes of a more detailed systemic examination.

BOX 14.4   Guiding Questions for Systems Descriptions and Analysis

- What is the core purpose or essence of the system from each perspective?
- To whom are they of interest and why?
- How do we define and delimit the system?
- What are the spatial and temporal boundaries and scales of observation?
- What are the key ecological and social processes that define the system?
- What are the different subsystems of interest?
- What are the relevant subsystem behaviors?
- What are the relevant subsystem variables?
- What is the nested hierarchy in which the system is situated?
- What are the defining contextual relationships between the system and its subsystems and the larger system in which it is embedded?

In some cases, where very good historical data are available, one may even begin to frame the questions in explicitly complex systems terms and to build narratives about how the system behaves. For instance:

- What are the discrete organizational states (attractors) which may be accessible to the system?
- How does the system behave in the neighborhood of each discrete organizational state, potentially in terms of a quantitative simulation model?
- What are the positive and negative feedbacks and autocatalytic loops and associated gradients that organize the system about a discrete organizational state?
- What might enable and disable these loops and hence might promote or discourage the system from being in the neighborhood of a discrete organizational state?
- What might be likely to precipitate flips between discrete organizational states?

BOX 14.5   Guiding Questions for System Synthesis

- How do the different system models articulate with each other?
- How do the parts of the system that can be modeled (e.g., hard systems, historical components) articulate with those that cannot (e.g., soft systems, future-expectation models)?
- What are the possible future states of organization of the system?
- What is our understanding of conditions under which these states might occur?
- What are the trade-offs that the different states represent?
- What is the range of feasible and desirable management actions?
- What are appropriate schemes for ensuring the ability to adapt to different situations?
- What is the appropriate level of confidence that the narrative deserves? This is our degree of uncertainty.

Systems Analysis

Questions here focus on the definition of the systems involved, including the variables, processes, and behaviors. In some cases, more detailed quantitative questions may be pursued, relating to system dynamics, feedbacks, and flips.

BOX 14.6   Guiding Questions for Collaborative Learning and Action

Cross Talk and Seeking Solutions

- What are the areas of conflict and complementarity among stakeholders?
- What are the feasible interventions that account for the multiple perspective goals and multiple perspective impacts?
- How can the different perspectives be synthesized?
- How are the different perspectives and descriptions of reality reconciled?
- What are the necessary trade-offs?
- What are the desirable management options?
- What is the combined vision of the stakeholders?
- What are commonalities and differences within that vision?
- How can the stakeholders design a more sustainable future?

The visions and aspirations that the stakeholders hold for the future may be elicited by asking questions such as the following:

- How can the stakeholders design a more sustainable future?
- What story do you want your grandchildren to tell about their home? How would you like your community to look in twenty years? What changes would you like to see?
- What needs to happen to realize these goals?
- What barriers and constraints are stopping you from achieving these goals today?
- How might these be overcome?

Design of an Adaptive Approach for Implementation of the Vision and Collaborative Learning

- What are the management directives, goals, and objectives?
- What are the desirable and feasible changes in the systems, in order to reach the identified goals?
- What institutional arrangements are necessary to support these changes?
- How can they be implemented?

Implementation

- What are the steps, and in which order, which need to be part of the implementation process?
- Who will take responsibility for ensuring that each of those steps is implemented?
- How can they be implemented?
- How can people be motivated to adopt the suggested changes?
- How will implementation be sustained, both in terms of finances, and in terms of personnel and logistics?

Monitoring and Evaluation

- What are the relevant indicators that can measure performance in terms of the identified goals?
- Who "owns" the indicators; that is, who will use them for making decisions?
- Who will measure them?
- How will decisions be made on when to take action?
- Who will take action?

Systems Synthesis

Here we must ask questions about how the modeled and nonmodeled parts of the system articulate, what the range of feasible system states is, and what is our confidence level in various narratives.

Collaborative Learning and Action

Because of the complexity of ecosocial systems as described in this book, collaborative learning and action are at the core of the ecosystem approach. There are no definitive experiments, no final, unassailable truths. We are learning as we go. At this point, the emphasis of the process shifts from what has traditionally been seen as research to what has been viewed as management. In AMESH (as in the diamond framework), every policy initiative is a hypothesis, and every implementation or management program becomes a test of that hypothesis. Because of the complex nature of the issues with which we are dealing, with people inside the system, proposed solutions must now be negotiated based on the trade-offs identified in the systems investigations. Here, as researchers, we ask questions about how different perspectives conflict with, or complement, each other, and what kind of sustainable futures might be feasibly designed. Moving from visioning and designing to implementation, we can ask specific questions about the logistics of management.

Finally, we need to monitor changes, with a view to adapting our interventions in the light of changes in context and outcome. We need to ask questions about what the indicators actually measure and who "owns" them; that is, who thinks they are important and is likely to take action on them? If measuring certain indicators requires very expensive technical equipment and/or take a long time, then they are not likely to be useful for day-to-day decision making by local householders, farmers, or businesses. They may, however, be useful for detecting long-term trends and setting policies. At the local household and community levels we need indicators that are measured and understood with relative ease, so that people will "own" them and act on them. There is no one correct set of indicators, as these reflect both our understanding of the complex reality and our goals. Because the goals we choose are linked to the indicators and the indicators are what drive action, we discuss these in separate chapters that follow this one.

Ultimately, what we seek is a way of integrating our ways of knowing—not just what has been considered science in conventional terms but different kinds of knowledge, reflecting different epistemological stances—with our ways of doing. The ecosystem approach, as we have outlined it, is a coherent, self-reflective way for us as a species to learn our way into a convivial and sustainable future.

References

Allen, T., B. Bandurski and A. King. 1991. The ecosystem approach: Theory and ecosystem integrity. Report to the Great Lakes Advisory Board, Washington, D.C.: International Joint Commission.

Allen T. H. F., J. A. Tainter, J. C. Pires and T. W. Hoekstra. 2001. Dragnet ecology—"Just the Facts Ma'am": The privilege of science in a postmodern world. *Bioscience* 51:475–485.

Checkland, P. B. 1981. *Systems Thinking, Systems Practice*. New York: Wiley.

Gunderson, L. H. and C. S. Holling (eds.). 2002. *Panarchy: Understanding Transformations in Human and Natural Systems*. Washington, D. C.: Island Press.

Holling, C. S. 1986. The resilience of terrestrial ecosystems: Local surprise and global change. In *Sustainable Development in the Biosphere*, eds. W. M. Clark and R. E. Munn, pp. 292–320. Cambridge, UK: Cambridge University Press.

Kay, J. J. 1991. A non-equilibrium thermodynamic framework for discussing ecosystem integrity. *Environmental Management* 15(4):483–495.

Kay, J. J. 2000. Ecosystems as self-organizing holarchic open systems: Narratives and the second law of thermodynamics. In *Handbook of Ecosystem Theories and Management*, eds. S. Jorgensen and F. Muller, pp. 135–160. Boca Raton, Fla.: CRC Press-Lewis.

Kay, J. and E. D. Schneider. 1994. Embracing complexity, the challenge of the ecosystem approach. *Alternatives* 20(3):32–38.

Murray, T. P., J. J. Kay, D. Waltner-Toews and E. Raez-Luna. 2002. Linking human and ecosystem health on the Amazon frontier: An adaptive ecosystem approach. In *Conservation Medicine: Ecological Health in Practice*, eds. A. A. Aguirre et al., 297–305. New York: Oxford University Press.

Regier, H. and J. Kay. 2001. Phase shifts or flip-flops in complex systems. Monitoring in support of policy: An adaptive ecosystem approach. In *Encyclopedia of Global Environmental Change*, vol. 5, ed. T. Munn, pp. 422–429. New York: Wiley.

Rowley, T., G. Gallopin, D. Waltner-Toews and E. Raez-Luna. 1997. Development and application of an integrated conceptual framework to tropical agroecosystems based on complex systems theories: Centro Internacional de Agricultura Tropical-University of Guelph Project. *Ecosystem Health* 3:154–161.

Smit, B., D. Waltner-Toews, D. Rapport, E. Wall, G. Wichert, E. Gwyn and J. Wandel. 1998. Agroecosystem health: Analysis and assessment. Report. Ontario, Canada: Agroecosystem Health Project, Faculty of Environmental Sciences, University of Guelph.

Waltner-Toews, D. 1994. Eggheads on a chicken farm: The academic pursuit of agroecosystem health. In *Agroecosystem Health: Proceedings of an International Workshop*, ed. N. O. Nielsen, pp. 101–109. Ontario, Canada: University of Guelph.

Waltner-Toews, D. 2004. *Ecosystem Sustainability and Health: A Practical Approach*. Cambridge, UK: Cambridge University Press.

Waltner-Toews, D., T. Murray, J. Kay, T. Gitau, E. Raez-Luna and J. McDermott. 2000. One assumption, two observations and some guiding questions for the practice of agro-ecosystem health. In *Agro-ecosystems, Natural Resources Management and Human Health Related Research in East Africa*, eds. M. A. Jabbar et al. 7–14. Nairobi, Africa: International Livestock Research Institute.

Waltner-Toews, D., J. J. Kay, T. Murray and R. C. Neudoerffer. 2004. Adaptive methodology for ecosystem sustainability and health (AMESH): An introduction. In *Community Operational Research: OR and Systems Thinking for Community Development*, eds. G. Midgley and A. Ochoa-Arias, pp. 317–349. New York: Kluwer Academic.

Waltner-Toews, D., C. Neudoerffer, D. D. Joshi and M. S. Tamang. 2005. Agro-urban ecosystem health assessment in Kathmandu, Nepal: Epidemiology, systems, narratives. *Ecohealth* 2:1–11.

15

Return to Kathmandu

A Post Hoc Application of AMESH

R. Cynthia Neudoerffer, David Waltner-Toews, and James J. Kay

Introduction

The adaptive methodology for ecosystem sustainability and health (AMESH), described in chapter 14 and elsewhere (Waltner-Toews 2004; Waltner-Toews et al. 2004), emerged from ongoing practical and theoretical debates within an extended peer community. One particular series of projects, lasting over a decade, served as a source and a testing ground for many of its components.

Between 1991 and 2001, we collaborated with Nepalese researchers, first, to investigate a parasite that cycles between dogs, livestock, and people (Baronet et al. 1994), and, second, to generate a community-based initiative that used an ecosystem approach to human health (Waltner-Toews et al. 2005).

Because the Nepalese work spanned the full period of time during which AMESH was developed, it contained many elements of the approach and had a strong influence on its final "shape." As such, we felt that it would be useful to apply AMESH post hoc, with a view to clarifying the stages of such investigations, and the questions that need to be asked at each stage.

The characterization of the project in this chapter thus follows explicitly the various categories laid out in the AMESH process, question by question.

As set out in chapter 14, the five major activities that comprise AMESH are the following: (1) Documenting the presenting situation; (2) Analysis of stakeholders, issues, and policy, politics and governance; (3) People and their stories; (4) Systems descriptions and narratives: Developing a systemic understanding; and (5) Collaborative learning and action.

AMESH is a complex set of activities, and there is no one "correct way" to apply the process. The various activities of AMESH and the guiding questions that accompany them are intended to provide guideposts to researchers and practitioners, but they are not cast in stone. In practice, each activity will flow into and often overlap others; some will be revisited multiple times in the process as new knowledge is gathered that informs and shapes earlier decisions and understandings. Still, it is legitimate to ask, if we apply the framework of this process to a series of projects already completed, what more can we learn?

Presenting Situation: The Entry Point

The Presenting Situation

In 1991, Waltner-Toews at the University of Guelph and a Nepalese partner, Dr. D. D. Joshi, Director of National Zoonoses & Food Hygiene Research Centre (NZFHRC) initiated a collaborative research project to study a tapeworm infection of dogs (echinococcosis), which causes tumor-like cysts (hydatid disease) in people and other animals (Baronet et al. 1994). The study was in two wards of Kathmandu, wards being the smallest political administrative unit. The wards are composed of distinct neighborhoods, which in Nepali are called toles. Pilot studies by Dr. Joshi had shown that many of the animals being slaughtered along the banks of the Bishnumati River adjacent to Wards 19 and 20 in Kathmandu were infected with hydatid cysts. Furthermore, the open-air slaughtering, with dogs and pigs foraging freely, children playing, and the use of the river for waste disposal, drinking water, general laundry, and bathroom facilities, suggested that this was a high-risk situation for disease transmission. The research team initiated a series of epidemiological studies of the slaughtering practices and the health status and behavioral habits of stray dogs, households, and community members.

As the epidemiological studies progressed, what had initially appeared to be a relatively straightforward disease study turned into a complex tangle of ecological, social, health, and political issues. At the final workshop, researchers realized that previous interventions had not been very successful and that resolution of the health problems in these communities would require something more than scholarly documentation and public health education workshops and communiqués. As a consequence, after 3–4 years the project was transformed from a collection of epidemiological studies into an urban ecosystem health initiative, combining participatory action research techniques with normal and system sciences methodologies. In preparation for this, one of the researchers in the epidemiological study revisited the wards and summarized their histories and demographic statistics in the context of Nepalese development overall (Baronet 1996).

When the project transformed from an epidemiological study to a participatory ecosystem health project, the two initial partners, together with the International Development Research Centre (IDRC), which was funding the work, recognized the need for local expertise in social action and mobilization. Consequently, a local nongovernment organization, Social Action for Grassroots Unity and Networking (SAGUN) was invited to be a part of the research team, to complement the epidemiological and biomedical skills of the collaborators from the University of Guelph and the NZFHRC.

The goal of the project was to enhance the ability of the communities in Wards 19 and 20 in Kathmandu to set in place sustainable processes to identify, implement, and monitor factors that will improve the health of the socioecological systems in which they live and, ultimately, human health.

The specific objectives that emerged from an October 1997 Planning Workshop hosted by the two local nongovernment organizations (NGOs) and IDRC in Kathmandu, were the following:

1. To increase the technical understanding of Kathmandu's urban ecosystem (on the part of both researchers and community members).

2. To enhance the capacity of individuals and organizations to improve currently poor public health conditions, articulate demands, and lobby for change through the combination of the ecosystem approach and participatory action research.

The Given History

What has been the overall historical development of the system?   Nepal, a small, mountainous, land-locked nation, is nestled between India to the south and China to the north (fig. 15.1). Until the 1990s, Nepal was ruled by a hereditary monarchy. It is a traditional society and the world's only Hindu kingdom, a legacy of India's influence that defines its culture and its caste-structured society. The Nepalese legal system reflects its heritage, blending Hindu legal and English common law traditions.

Although the Nepalese kingdom can trace its roots back more than 1500 years, its current problematic situation emerged from the difficult transition from absolute monarchy to democracy. In 1962, King Mahendra Bir Bikram Shah Dev devised a centrally controlled, party-less council system of government called panchaya. This system, based on ancient village traditions in Nepal, was incorporated

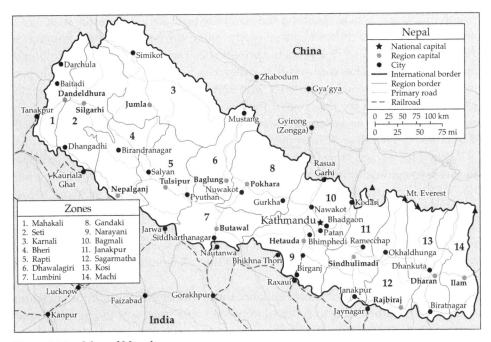

Figure 15.1   Map of Nepal.

in the constitution of the same year and was established at the village, district, and national level. This system of government by an absolute monarch behind a democratic façade lasted for thirty years (Library of Congress 1991).

The wave of international changes that rocked the global community in the late 1980s and early 1990s, including the fall of the Berlin Wall and the Soviet Union and the pro-democracy movements in Eastern Europe, also touched Nepal. This was a time of political turmoil, with the prodemocracy movement in full swing, holding demonstrations, rallies, and protests across the country, especially in the capital. On February 18, 1990, the Movement for the Restoration of Democracy was formally established. Ethnic groups agitated for official recognition of their cultural heritage and linguistic traditions and demonstrated against the monarchy. The goal of the prodemocracy movement, however, was to establish a more representative democracy and to end the *panchayat* system.

In 1990 the King approved and promulgated a new, more democratic constitution that vested sovereignty in the people. The *panchayat* system finally ended in May 1991, when general elections, deemed, "generally fair, free, and open" by international observers, were held. Sixty-five percent of eligible voters turned out for the election. The Nepali Congress Party won 110 of the 205 seats in the House of Representatives and the Communist Party of Nepal (United Marxist-Leninist) won 69. In the decade following these elections, Nepal saw no less than eleven different elected governments. In retrospect, this unrest in the country, nested in an optimistic global political climate, could be interpreted as a phase of "creative destruction" within Holling's four-box model (Holling 1986; Gunderson and Holling 2002) and thus an opportunity for innovative organizational change, without which the project would have been much less successful.

What have been the key ecological, economic, demographic, and social developments in the system? What kinds of rich interconnections have occurred? At what level have changes taken place? What or who have been the agents of change? Which are the critical fast, medium, and slow variables within this overall development?   The ecology of Nepal is dominated by its geographical position as a mountain nation. Easily eroded hills and mountains with steep to very steep slopes occupy about two thirds of the country. A mutually reinforcing cycle of erosion and deforestation has resulted in serious annual losses of Nepal's forest cover between 1950 and 2000 (ADB and ICIMOD 2006), as well as in migration to the cities, particularly Kathmandu. Deforestation is linked to a number of factors, including demand for grazing land, farmland, and fodder and energy needs dominated by fuelwood.

Despite only 17 percent of the land being arable, agriculture accounts for 41 percent of GDP and the livelihood of 80 percent of the population (CIA 2002). Industrial activity is dominated by processing of agricultural products, but textile and carpet production have increased, associated with water demands, and accounted for 80 percent of the foreign exchange earnings between 1998 and 2001. Tourism, associated with increased demand for meat, is also a significant industry. Political instability has hampered the government's ability to move forward with economic reforms and to increase essential government services. Nepal is highly dependent on foreign aid—the international community funds more than 60 percent of Nepal's development budget and 28 percent of its total government budgetary expenditures (CIA 2002).

Nepal is considered one of the poorest and least developed nations in the world. In 2001, the population of Nepal was estimated at over 25 million, and growing at 2.32 percent per year (CIA 2002), with an infant mortality rate of 74.4 deaths/1000 live births and average life expectancy of 59 years for men and 58 years for women. Of the population over 15 years, 40.9 percent can read and write, compared to 14 percent of women. Although the constitution officially guarantees equality, in practice, women, in general, occupy a secondary social position. The actual status of women varies by ethnic community and by caste. In addition to high morbidity and death rates, the absence of modern medical care, clean drinking water, and adequate sanitation result in a high prevalence of gastrointestinal diseases and malnutrition.

Although Nepal is the world's only official Hindu kingdom, the Tibetan influence is seen in its minority Buddhist population, and, in effect, most people practice a mix of Hinduism and Buddhism. Ethnic identity—Indo-Nepalese, Tibeto-Nepalese, and Indigenous Nepalese—colors many aspects of Nepali life, including the selection of spouse, friendships, and career and is evident in the social organization, occupation, and religious observances.

Broadly speaking, several factors weave together to create the ecological-economic-sociocultural fabric of Nepal. The mountains are both the defining and the controlling feature of the landscape. The fragile mountain ecology and monsoon-regulated rainfall drive many ecological processes. Travel is by road, and rail is difficult at best and near impossible in many parts of the country; meaning many mountain settlements are isolated and accessible only by foot. A national electrical power grid is also next to impossible. The physical location of the country is also important. India exerts significant power over Nepal owing to India's control over access to raw materials and to supply and trade routes. The combination of a traditional, caste-based society dominated by a landed, high-caste agricultural elite along with the hereditary monarchy combines with the fragile and natural disaster-prone ecology to create a challenging context for sustainable development.

Is there historical evidence suggesting sudden shifts or changes in the structure or behavior of the system? What has stabilized the system? What has tended to destabilize the system? Have there been surprising changes? Perverse stability? Within the hazard-prone mountainous environment, the strong traditional culture has served to stabilize the society and the agricultural practices. The imposition of Western ideals and standards of "development" have tended to destabilize the traditional socio-cultural-ecological system. From the outside, Western perspective, the dominance of the traditional social customs and norms, in conjunction with the control by the high-caste landed elite and a hereditary monarchy, has combined to create a perversely stable, but "underdeveloped" nation. By contrast, from the perspective of the traditional society, the imposition of Western ideals and standards of development have tended to destabilize the traditional social-cultural-ecological balance.

What are the positive and negative feedback loops?   Poverty and environmental degradation are a mutually reinforcing positive feedback loop. Another detrimental positive feedback loop is broadly formed by poverty, malnutrition, and ill-health. A third positive feedback loop is formed by environmental degradation, pollution, and ill-health. Finally, the traditional social-cultural system and control of economic and political activity by the traditional elites are mutually reinforcing.

Wards 19 and 20 reflect many of the characteristics of Nepali society in general. Residents of both wards had been active in the prodemocracy movement. The community hall of the butcher's association prominently features memorial photographs of two local martyrs, a young man and a young woman, who died fighting for the democratic cause. The sense of optimism and empowerment that accompanied the early period of promise at the birth of democracy quite possibly facilitated cooperation and the later success of the project. When the epidemiology project started in 1991, the level of ecological degradation, especially along the river, in both wards was overwhelming and extensive (project area photographs are available at www.nesh.ca). The outside perception of the futility of the situation was such that a well-meaning associate of Waltner-Toews suggested that Kathmandu was "beyond help and beyond hope." Clean air, clean water, and access to sanitation facilities seemed far-off dreams.

Specific Project History

The Community Researchers (one male and one female selected and trained per ward) used an Appreciative Inquiry Process (AIP) (www.appreciativeinquiry.cwru.edu) to explore the social-ecological history of the community in the two study wards. This AIP process emphasized the discovery of past success stories, present urban ecosystem health conditions, available opportunities, and existing strengths within the community. The list of past success stories for Ward 20 included such activities as construction of a community park, a closed underground drainage system, and a health center, vaccination and dog control programs, and safe water and earthquake awareness programs. These stories helped both the community and the researchers to understand what activities were attempted in the past, which ones were successful (or not) and why, and who might comprise a preexisting local "action community."

Analysis of Stakeholders, Issues, and Policy, Politics, and Governance

The second major step in AMESH comprises several overlapping activities with the overall objective of creating a "rich picture" of the local ecological-social system. One activity focuses on determining the identity, relationships among, and issues important to the project stakeholders. A second activity focuses on exploring the ecological, social, and health issues from a systemic perspective, and a third activity focuses on understanding the local governance structures, political systems, politics, and policies that influence ecosystem sustainability and health.

Stakeholder Analysis

Who are the principle stakeholders? What is the process for identifying stakeholders? Under what and whose selection criteria will they be considered stakeholders? What are the critical discriminating features that allow us to differentiate the groups of stakeholders? The researchers defined a stakeholder as someone who could gain or lose by project intervention as well as someone who was capable of affecting the outcome of the project. People do not immediately step forward to identify themselves as "stakeholders," rather, participants and interest groups, both strong and weak, gradually emerge out of discussions and interactions among the research team and the local community. Wards 19 and 20 are adjacent to one another in the oldest section of Kathmandu, between the famous Durbar Square and the Bishnumati River. The identification of water and waste management as important issues, as well as initial identification of key stakeholders, emerged from the surveys and epidemiological studies in the early 1990s.

SAGUN characterized stakeholders according to criteria of "importance" and "influence." *Importance* referred to the level in which the project objectives addressed the needs of a particular group. *Influence* referred to the level in which a particular group had the decision-making power to either promote or challenge the implementation of project activities. The stakeholder cross-classification that emerged from this process (table 15.1) was used to identify those who should be closely involved with the project (high importance, high influence), those whose participation should be ensured (high importance, low influence), those who could promote or challenge the project (low importance, high influence), and those with whom information should be shared (low importance, low influence).

A variety of critical discriminating features were used to differentiate major groups of stakeholders. The stakeholders fell into one of four general categories representing a group or major institution [Kathmandu City (KMC), Department of Drinking Water Supply Corporation (DDWSC), ward committee, local clubs, ward clinics, schools, Lumanthi, and ENAPHC (both NGOs); an occupation (butchers, sweepers, street vendors, hotel owners); status in the community (community leaders); and location of residence in the community (squatters)]. The low-caste groups were identified as stakeholders by their occupation—butchers, sweepers, or street vendors. Higher-caste people, however, were simply identified as the "community leaders" stakeholder group (that is, not occupationally related), and it was not clear whose specific interests they were intended to represent. Additionally, the squatters, similar to the community leaders, did not

TABLE 15.1   Summary of Problems and Concerns as Prioritized by the Main

Influence	Importance		
	High	Medium	Low
High	Community Leaders Butchers	Ward committee Local clubs	Kathmandu Municipality Corporation (KMC)
Medium		Ward clinic Schools Lumanthi (NGO) ENAPHC (NGO)	
Low	Sweepers Squatters (including some butcher employees)	Street vendors Hotels Restaurants	Department of Drinking Water Supply Corporation (DDWSC)

represent a singular occupation grouping but a different type of group based on their type of residence.

The classification used in table 15.1 represents only one possible description and does not indicate the cross-scale relationships among stakeholders nor does it identify how local interests interact with those beyond the ward itself (KMC and DDWSC, for instance). A stakeholder holarchy is instructive as it helps to establish the broader context in which people live out their lives. Different individuals or groups often take on different roles at different levels of the holonocracy (see table 15.1 and fig. 15.2). For example, individuals (men, women, and children) belong to castes and households, live in *toles*, belong to local clubs, and have different occupations (sweeper, butcher, etc.).

We identified seven biophysically defined holons, seven human activity-defined holons, and six governance holons (fig. 15.2). Because holons in the different holarchies did not directly overlap, we needed to include decision makers from several different holons when working on particular biophysically defined problems. In the final two columns in fig. 15.2, the actual project stakeholders for the water and waste management training activities are shown with respect to the governance holon. Because stakeholders are generally primarily concerned with issues that emerge most clearly at their holarchic level or scale, they often differ in their assessment of what is important.

What other groups, perhaps not directly related to the issue at hand, exist in the community?   Two local community clubs (comparable, e.g., to a Lions Club, but without the international connections) appear to have been initially overlooked. Their later, very active, involvement in implementing garbage management activi-

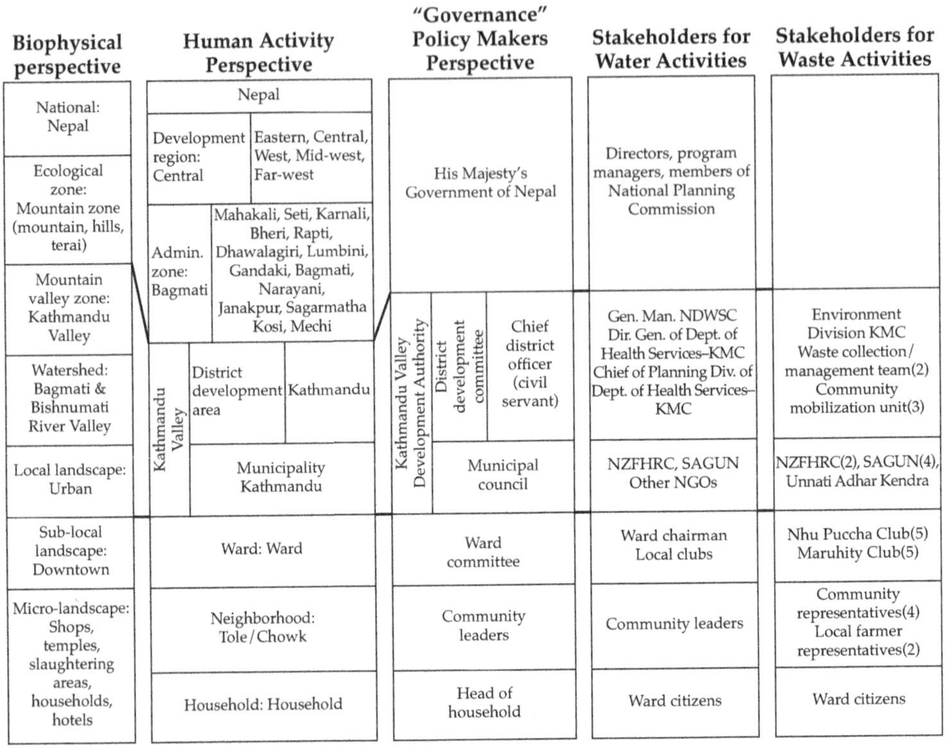
Biophysical perspective
Human Activity Perspective
"Governance" Policy Makers Perspective
Stakeholders for Water Activities
Stakeholders for Waste Activities
National: Nepal
Ecological zone: Mountain zone (mountain, hills, terai)
Mountain valley zone: Kathmandu Valley
Watershed: Bagmati & Bishnumati River Valley
Local landscape: Urban
Sub-local landscape: Downtown
Micro-landscape: Shops, temples, slaughtering areas, households, hotels
Nepal
Development region: Central
Eastern, Central, West, Mid-west, Far-west
Admin. zone: Bagmati
Mahakali, Seti, Karnali, Bheri, Rapti, Dhawalagiri, Lumbini, Gandaki, Bagmati, Narayani, Janakpur, Sagarmatha Kosi, Mechi
Kathmandu Valley
District development area
Kathmandu
Municipality Kathmandu
Ward: Ward
Neighborhood: Tole/Chowk
Household: Household
His Majesty's Government of Nepal
Kathmandu Valley Development Authority
District development committee
Chief district officer (civil servant)
Municipal council
Ward committee
Community leaders
Head of household
Directors, program managers, members of National Planning Commission
Gen. Man. NDWSC
Dir. Gen. of Dept. of Health Services–KMC
Chief of Planning Div. of Dept. of Health Services–KMC
NZFHRC, SAGUN
Other NGOs
Ward chairman
Local clubs
Community leaders
Ward citizens
Environment Division KMC
Waste collection/management team(2)
Community mobilization unit(3)
NZFHRC(2), SAGUN(4), Unnati Adhar Kendra
Nhu Puccha Club(5)
Maruhity Club(5)
Community representatives(4)
Local farmer representatives(2)
Ward citizens

Figure 15.2 Holonocracies and hierarchies in Nepal.

ties demonstrates the importance of searching out and seeking the involvement of preexisting community groups, instead of or in addition to trying to create new local institutions to achieve the project objectives.

Which groups have been excluded? Although some stakeholders belonged to more than one group (butchers, sweepers, and street vendors often being squatters), others did not appear to explicitly belong to any one specific stakeholder group.

The researchers themselves were not initially identified as stakeholders, even though the goals, objectives, and perceptions of each of these research groups certainly influenced the activities undertaken in the research project. NZFHRC and the University of Guelph had a history working in Wards 19 and 20, and the community already held certain assumptions regarding their interests, strengths and weaknesses. For example, because NZFHRC had been working hard to improve the sanitary conditions of the buffalo slaughterhouses and the retail meat shops, both butchers and meat retailers saw NZFHRC as an irritant. Consequently, NZFHRC and the rest of the project team had to work to overcome this notion and demonstrate to the butcher community that the team did not want to harm their business but rather work with them to improve both the conditions of their business and their interactions with the rest of the local community. SAGUN did not have a history of working in these communities. On the one hand, this meant that past biases and assumptions about roles and relationships were less at play between SAGUN and the community; on the other hand, SAGUN had to expend a greater effort at the outset of the project establishing a rapport with and gaining the trust of the local people.

Another influential group that was excluded from the stakeholder list were the community researchers and community volunteers. In the first year of the project (1999), one male and one female were recruited from each ward to work on the project as a community researcher. These volunteers, who acted as the core team for project implementation, were trained by SAGUN in a variety of participatory methods, including Participatory Urban Appraisal (PUA), community empowerment and training in village social mapping, seasonal analysis, and pairwise ranking (Pretty et al. 1995). Community researchers then carried out an initial PUA baseline survey in the two wards and subsequently identified 24 volunteers to serve as a Local Community Research Support Team. While not identified as a stakeholder group, these people were important participants in many of the project activities and exerted significant influence on the project process.

Although not excluded, DDWSC (and the Nepal Water Supply Corporation or NWSC) may have been undervalued. Because it was outside the immediate holon of interest, the organization was listed as a stakeholder with low influence and low importance. Accordingly, the project would only share information with them. However, in terms of decision making about water management the DDWSC/NWSC occupied an influential position. This highlights the importance of always paying attention to context.

Finally, the butcher employees appeared to be excluded from the stakeholder list. In table 15.1, the butchers who were identified as high influence, high importance stakeholders, were actually the owners of the slaughterhouses or the wholesalers. The researchers distinguished three groups within the "butcher" category: owners, retailers, and butcher employees. The latter group was identified as high importance, low influence, thus according to the stated definition the participation of this group should have been ensured. Later in the project, recognizing the conflict in power between the owners and butcher employees, the Nepalese researchers adopted the strategy of directly involving the owners and not the employees. This seemed to contradict the participation criteria laid out for selecting stakeholders (as defined with table 15.1).

What are the factors that stakeholders perceive as being within and outside of their control (exogenous and endogenous)?   This question was not asked until very late in the project. At the final workshop, held in Kathmandu on November 6, 2001, in a discussion among community members and representatives of the local municipal ward, the community members asked the ward officials why they failed to enforce the local waste management laws. The ward officials informed the community that they did not actually have the power to enforce the laws at the local level; this was the municipality's responsibility. Without this understanding, throughout the project, the ward office was believed to be negligent in their duties. Clearly, it is very important to understand what factors each stakeholder group controls and identify where decision-making power genuinely resides.

What are the power dynamics or relationships among and between stakeholder groups? What are the barriers and constraints to communication and collaboration between the different stakeholder groups?   Power and status hierarchies can act as significant barriers or constraints to communication and collaboration among different stakeholder groups. In Wards 19 and 20 the status hierarchy, which correlated well with power, appeared to relate to occupation and contain vestiges of the traditional caste hierarchy. These power and status differentials had implications for common community action. The street vendors and community leaders both complained about the street sweepers and felt that they did not do an effective job of keeping the streets clean. On the other hand, the street sweepers felt that the community ignored the rules for garbage disposal and thus made more work. Hence there was a significant amount of distrust between these groups.

Several of the primary stakeholder groups further subdivided into power/ status hierarchies, which, in turn, further inhibited collaboration and communication among stakeholders. The butchers were organized hierarchically as owners or wholesalers, retailers, and butcher employees. Because the owners employed the butchers, the latter were unwilling or afraid to criticize or speak out about their work conditions for fear of losing their jobs. The retailers, on the other hand, did not perceive any problems related to ecosystem health in the ward. When asked about the environment and public health situation the retailers replied, "Sab thik chha" or everything is okay. Consequently, the retailers did not appear to see a need to be involved in the project activities. The butchers were singled out within the community for their poor waste and water management practices. At the same time, however, they had no power to make decisions within their place of employment. Customers were specifically left out of the general "butcher" stakeholder group. However, customers, who were often also citizens of the two wards, played an important role in the system because their demands (for raw meat) or lack thereof (they did not complain about the quality of the meat or the fact that meat was not covered in the stores) were important driving factors in the system.

Power and status can be confounded by gender issues. The street sweepers/ cleaners in Wards 19 and 20 were always women, while all other positions in the sweeper hierarchy were occupied by men (see fig. 15.6). Furthermore, there were two different major groups of sweepers: those employed by the Ward Committee and those employed by KMC. It is unclear, however, how these two groups interrelated in terms of power and status.

Vendors were discriminated by whether they were local or nonlocal to the ward. These street vendors, called *doko* sellers, sold their wares either from a small cart or off a blanket spread out on the sidewalk and were an important feature of daily life in the wards. A local was someone who lived in the ward, while a nonlocal was someone who lived outside of the two wards and came to the area only to sell their vegetables several times a week. The local group was further discriminated, first, on the basis of whether they had an established place to sell their wares or if they were a temporary *doko* seller and, second, on the basis of whether they owned or rented their place of residence within the ward. Those who owned their places saw themselves as local, while the community perceived those street vendors who rented accommodation as nonlocal. Thus, for the street vendors, both perceived localness and competition for customers inhibited collaboration and communication.

Squatters were also differentiated on the basis of whether they were local or nonlocal. Squatters were considered local if their families had come to the wards no later than the 1960s. Local squatters may have had relatives in the ward who owned their homes on registered land. Nonlocal squatters were residents who moved into the ward after 1970 but who did not have strong kinship ties in the ward. For this reason, despite long residence for some squatters, they were considered to be nonlocal. This differentiation, between local and nonlocal, created a potential barrier for collaboration among squatters.

What are the main resource coalitions? Which stakeholders are excluded from these coalitions and what are the consequences of this? The butcher owners or wholesalers were members of the Small Meat Marketing Association. This group had access to financial resources and their level of organization was well evidenced by the fact that this group had launched a lawsuit against the police for harassment. The community leaders appeared to be well educated, financially secure, and organized. The squatters, encompassing some butcher employees, sweepers, and street vendors, were not well organized and did not have any sort of representative organization. This group was described as having low unity. The research team also described tension between the sweepers in Ward 20 and the butchers in Ward 19, both of whom were squatters.

The street vendors represented a potential resource coalition; however, they were identified as a group with very low unity because of the fact that they all competed for the same customers. Hence they were disinclined to work together. Similar to the street vendors, the hotel/restaurant owners represented a potential resource coalition group but for the similar reasons to the street vendors they had very low group unity.

What goals/interests do these coalitions share? What are the assets and resources (skills, financial, materials, connections, etc.) of each stakeholder group? Broadly, all stakeholder groups were concerned about access to clean water, water management, and garbage management in the wards. The butcher owners were interested in setting a fixed price for meat sold in Kathmandu, establishing clean public slaughterhouses, and providing training to all butcher employees regarding garbage management and hygienic meat preparation. The street vendors and community leaders identified the lack of efficient and regular sweeping by the street sweepers as an area of concern. The sweepers expressed a concern about the quantity of garbage produced by the slaughterhouses and the mismanagement of the same.

Sweepers were also concerned about the fact that ward residents threw garbage out of their windows as soon as they saw the tractor coming to remove the piles of garbage and residents ignored their requests to not carelessly throw garbage everywhere and to not throw fecal matter in plastic bags into the street. Sweepers were interested in raising awareness for all households in the wards regarding garbage management. The sweepers were also interested in receiving basic protective equipment (gloves, masks, and shoes) and hygiene training. The street vendors expected the sweepers to take care of their garbage. All of the street vendors were interested in forming a vendor association.

The assets and resources of each stakeholder group were not specifically identified, and this identifies a potential information gap. This type of information becomes very important when the community and the researchers are investigating alternative approaches and activities to address their social-ecological and health concerns. A lack of assets or resources will pose barriers to realizing some opportunities. The community needs to realistically identify what is possible with their current resources and what additional resources are needed to achieve their goals.

What are the conflicts of interest, which organizations are in conflict and why?   Although different perspectives among stakeholders were seen early, actual conflicts of interest clearly emerged later, as a result of the process of synthesizing system descriptions. When the concerns of the various stakeholders or actors regarding the water system and the food and waste system were linked together in one system diagram (one for water and one for food and waste), conflicts of interest or differing perspectives on the same issue clearly emerged.

Beyond a general acknowledgment of importance, stakeholders expressed different views on critical issues. The food and waste system exemplifies this well (fig. 15.3). Some stakeholders did not articulate specific concerns regarding food and waste, while others had strong, but unique, perspectives. For example, the wholesalers (slaughterhouse owners), "Feel that they receive bad media coverage," the retailers, "Do not perceive any environmental problems" and the community leaders felt that the, "Ward Committee does not properly enforce the rules and regulations" regarding garbage management. Differing perspectives on food, waste, and garbage management were evident between the sweepers and the community leaders and between the sweepers and the street vendors. While the street vendors felt that, "The sweepers do not sweep regularly enough" and the, "Sweepers should not get weekends off," the sweepers felt that they, "Do not have enough time to collect the garbage." At the same time, the community leaders felt that, "Lack of efficient and regular sweeping exacerbates garbage mismanagement" and, "Street sweepers are responsible for poor garbage management," and, "Sweepers only collect paper and vegetable waste and leave other garbage, e.g., dead animals." While the sweepers again felt that they did, "Not have enough time to collect garbage" and, "People throw garbage out their window as soon as they see the tractors," and "Citizens ignore requests not to carelessly throw garbage everywhere."

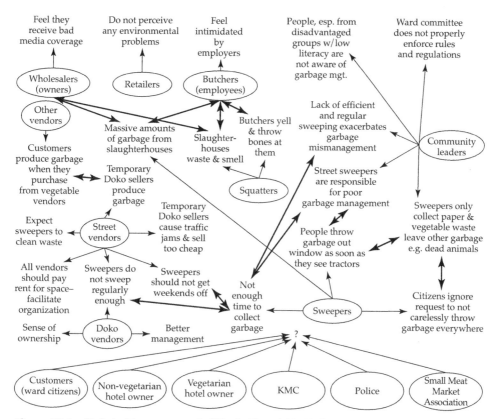

Figure 15.3   Stakeholder concerns in Wards 19 and 20, Kathmandu: Food and waste system.

By clearly identifying the various places where stakeholders held differing perspectives on the same issue, we highlighted points where the research team needed to negotiate understanding among stakeholders and key issues that needed to be addressed to get at the roots of the water or food and waste system problems.

What are the important gender relations among and between stakeholder groups? How do gender differences define people's rights, responsibilities, and opportunities to participate?   The research team took concrete steps to sensitize all stakeholders to gender concerns and to explicitly incorporate gender issues throughout the project. The two male and two female community researchers were given training in gender sensitivity and PUA gender techniques. Clearly, gender issues can be confounded with power and status roles both inside the home and, as in the case of the street sweepers, in the public domain. One result of this was that literacy training for underprivileged women in the wards became a project priority.

Ecological, Social, and Health Issues Analysis

The goal of this second activity is to identify the range of ecological and social issues that each stakeholder group (including researchers) feels is important and to reconcile differences in perception regarding the magnitude and importance of specific issues. All research teams are mandated to work on a predefined set of issues. The most important ecological and social issues for the community will not necessarily coincide with the mandate of the research team. The research team needs to be honest with the community regarding their own capabilities and limitations, so as to not set false expectations for the community.

What are the critical issues?   The two most critical issues for the communities of Wards 19 and 20 were *water quality and quantity* and *waste management*. Post facto, these two key issues were easy to identify. Although identified early, defining these problems substantively, however, and creating a commitment to deal with them only emerged through recurrent encounters among researchers and community members.

At the outset, drawing on both their own experiences in the community and previous surveys of residents, the *research team* identified the following as critical issues: water supply, quality, and access; microdrainage, sewage, and sanitation; waste and garbage management, especially in relation to open-air slaughterhouses, and recycling; cattle and dog control; and literacy.

At the first PUA Orientation workshop, held in 1999, the *research team* and *the local community participants* further identified the following common interests among the residents of the two wards: traditional vegetable farming, environmental concerns, cultural heritage preservation, community development, and political corruption. Seventeen key ecosystem health issues were selected from the identified critical issues and the research team used a pairwise ranking technique to determine the relative importance of these seventeen issues. Again, water quality and availability and garbage management ranked highest, with street vendors, slaughterhouse hygiene and air pollution close behind.

Collectively, at the Stakeholders' Intervention Action Planning Workshop, November 25–26, 1999, the project participants identified four key issue areas concerning ecosystem health and further elaborated on them in detail: air and noise pollution, water pollution, garbage, and others. In this meeting the stakeholders elaborated on the basic issues using specific examples to highlight the problems in their local communities.

How do they relate to one another?   Interaction among critical issues can be presented in an issues influence diagram (fig. 15.4). Such a diagram helped to illustrate the level of interconnection among the important ecosystem health issues. The more tightly interconnected an issue, the more "connecting influence" arrows the issue has with other issues. An issue that is not tightly connected has fewer arrows connecting it to other issues. In practice, this means that a "loosely coupled" issue, such as drug and alcohol abuse, may be (or may appear to be) relatively easy to address (compared to other issues in the ecosystem) because it is a relatively independent (i.e., not connected) issue. Water quality and garbage management, being influenced by and influencing many other issues, would be expected to be difficult to deal with.

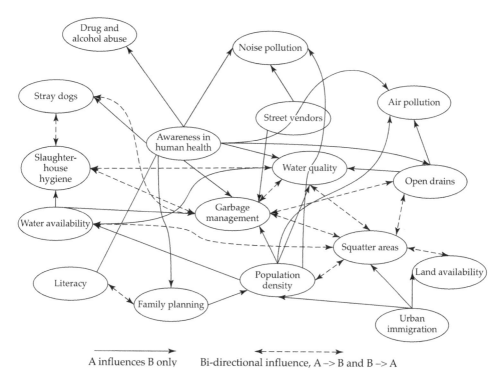

A influences B only   Bi-directional influence, A –> B and B –> A

Figure 15.4 Wards 19 and 20, Kathmandu: Issues influence diagram.

At what spatial and temporal scale do they occur?   The stakeholders specifically expressed concerns and issues that had an immediate effect on their daily lives in their local surroundings. Water concerns, for instance, were focused locally. However, their local water system was nested within the local landscape (Urban), the watershed (Bagmati and Bishnumati river valley), and the mountain valley zone (Kathmandu Valley), etc. Therefore the local level may not be have been the appropriate scale at which to address the roots of the water system problems.

In terms of temporal scale, water issues were ongoing, extending well into the community history and were expected to persist, especially if population growth and increased water demand continue in the area. Problems developed slowly over time, and because the problem existed at multiple scales in the system holarchy, any solution required time to implement.

Alternatively, while there were waste management problems at the KMC level and Kathmandu Valley levels, some of the problems such as disposal of garbage into the streets could be contained to a greater degree to the local scale, and it was easier to find and implement a local solution quickly.

How does the resolution of these issues come together in a vision for the future?
Visioning, while necessary for designing management plans later, is also important early in the process. This is what motivates people to participate and gives them energy. The visions of the "experts," NGOs, government agencies, and the local

community members, etc. all need to be reconciled and synthesized. Possible courses of action and future visions often emerge logically from this early visioning, and there is a natural tendency is to immediately jump to "action planning" without first developing a thorough systemic understanding. As we found in Kenya and Peru as well, residents demand action well before any substantive systemic understanding is achieved. The actions then become opportunities for learning.

While some researchers balk at undertaking a visioning process, it may be important or indeed essential for community engagement. Researchers should, at a minimum, not get in the way in such a process and, ideally, help to facilitate where conditions favor intervention. Visions can be revisited even after actions (as, in a sense, empirical tests of visionary hypotheses) are implemented.

The research team had a three-pronged vision for the project:

1. To foster a healthy environment that would include the development of well-planned roads, parks, trees, kitchen gardens, community toilets, well-maintained drainage systems, proper sanitation, water and waste management, river erosion prevention, well-managed temples and schools, and the maintenance of overall cleanliness.

2. To provide equal access of local resources and health facilities such as community clinics and government health posts to currently marginalized groups such as women and children, ethnic minorities, senior citizens, and the poor.

3. To establish an exemplary model of human and environmental health that can be subsequently replicated by adjoining wards of KMC.

At the Stakeholders' Intervention Action Planning Workshop the participants articulated the following vision for the future of Wards 19 and 20 (preserved in their own words). This vision was framed as a list of general indicators that would identify project successes.

1. No garbage piles and heaps will be seen on the streets, lanes, and along the riverside.

2. Around the houses (inside and outside of house) will be seen cleaned.

3. Garbage such as metal, leads, plastics, and other disposable matter will be separated.

4. Compost manure will be made of those disposable (organic wastes) and used in the gardens.

5. Drain water will not be seen mixing directly into the Bishnumati River.

6. Number of trees and plants will be seen in along the riverside, parks, and open land.

7. Traffic will be less in streets and lanes.

8. Community members will use boiled and filtered water.

9. Food items such as sweets will be seen free from nonedible colors.

10. Involvement of community members (from all stakeholders) will be increased on environment and public health program activities.

Policy and Governance Analysis

In many communities, official and informal governance systems are quite separate, and while on the surface key decisions appear to be made within the official governance structure, in actual fact the informal decision-making system drives the formal system.

Often community development work requires sanctioning by local leaders, who are then used as an entry point into the community and as a source of a list of potential key informants for the research team. The Ward Committees recommended potential candidates for the four Community Researcher positions. In this way the bias of the Ward Committee or local leaders, in terms of who they knew in the community entered the research process, and the status of the researchers in relation to the community was cast in a particular light. Realistically, using the local leaders to recommend key informants may be the only way to successfully gain entry into the community. Thus it becomes very important for the research team to understand the political dynamics of the community early on so that they can aim to contact as broad a cross section of the community as possible and ensure that they are not favoring one group over others or only listening to only one political perspective. Furthermore, an awareness of the governance holarchy is important as it influences the nature of the ecological holarchies considered (fig. 15.2).

For this project, the formal government could be defined—from the unelected King and elected members of the House of Representatives and National Assembly, to the elected District Development Committee, the Municipal Council [Mayor, Deputy-Mayor, member elected from each ward, and a secretary appointed by His Majesty's Government (HMG)] and the Ward Committee (Chairman and four other elected representatives from the ward).

The informal governance structure was much less well defined in these wards. Part of the process of identifying the project stakeholders included the definition of the numerous power and status hierarchies among the stakeholders. However, the relationship among power, status, and decision making was not clear. The traditional caste system in Nepal also added another layer of complexity onto the informal governance structure. Although the traditional caste system was officially abolished in 1963, many people still practice the occupation traditionally associated with their caste, so that the old system still casts a long shadow.

What are the key policies affecting the capacity of local stakeholders to sustainably manage their natural resources? At what level are these policies formulated and implemented? How do informal and formal governance structures affect local capacity to adapt to and deal with stresses? At what level are these governance structures formulated and implemented?   For several decades, Dr. Joshi had been advocating for a national meat hygiene act, so that all local butchers would operate under the same set of policies and regulations. Indeed, some international agencies refused to work to improve the riverside situation in these wards until such an act was promulgated. In light of this, the lack of formal policy and governance analysis represented a clear information gap in the project. The Ward Committee was identified as setting the rules and regulations for sweepers in their wards and the community leaders expressed the concern that these rules and regulations were not effectively enforced. According to the Chairman of Ward 20, the problem regarding the enforcement of rules for garbage management emerged from the fact that the authority for issuing fines rested at the municipal level, while the authority for making rules rested at the ward level. Hence the ward did not have adequate authority to enforce rules they may have made.

Local Capacity Building and Training

Meaningful participation may require skills that many members of the local community do not have. Hence local capacity building and training were key project activities, undertaken in parallel to the AMESH process and spearheaded by the SAGUN team members. These included training sessions on gender analysis, literacy, research, and Appreciative Inquiry.

Another training activity was the Water Quality and Management Training program, a two-week technical training course put on with the Nepal Water Supply Corporation. This course examined water supply, waste disposal, and sanitation. Trainees were selected from a broad coalition of groups including Nepal Water Supply Corporation (NWSC), Kathmandu Metropolitan City (KMC) Head Office, Ward 19, Ward 20, Lalitpur Municipality Office, Bhaktapur Municipality Office, Drinking Water Supply and Sewerage Department (DWSSD), NZFHRC, SAGUN, and others. Importantly (and, given the nature of the water problems, appropriately), the participants or stakeholders in this activity were drawn largely from holons beyond the wards, although there were participants from across several holons. The training also dealt with issues that crossed holarchic scales.

People and Their Stories

Emerging out of the participatory action research techniques, a fertile variety of descriptive narratives and stories form a rich picture of the local social-ecological system. The recounting and recording of these stories can be a very unifying and empowering step for any community and such stories give the research team invaluable insight into the community. If some people or group's stories are missing, the team needs to find ways to elicit them.

The community researchers and the local community stakeholders created narrative descriptions for each of the stakeholder groups through the combination of NZFHRC's biomedical study for understanding the local health situation and SAGUN's engagement with the local community using a variety of PUA techniques (Social Action for Grassroots Unity and Networking, National Zoonoses and Food Hygiene Research Centre 1999). Individual narratives were integrated into a comprehensive ecosystem story for each stakeholder group, i.e., one for the butchers, one for the sweepers, etc. These descriptive narratives were then used to create comprehensive system diagrams for system analysis and synthesis.

System Descriptions and Narratives: Developing a Systemic Understanding

System Analysis

The biophysical system   An analysis of the biophysical system establishes the ecological context within which our social, political, and economic systems are embedded. As described in the original context, Wards 19 and 20 are embedded in the urban landscape of the city of Kathmandu, within the Bagmati and Bishnumati River Valley watershed. This watershed is located in the Kathmandu Valley, in the Mountain Ecological Zone in the country of Nepal.

Nepal is a small, land-locked country dominated by the Himalaya Mountains, between two giants, India and China. The country is divided into three distinct ecological zones, each running the full east-to-west length of the country in three horizontal bands: the northern Mountains, the central Hills, and the *Terai* or flat river plains of the Ganges along the southern border with India. The rainfall distribution is extremely variable, and in Nepal the monsoon rains dominate the annual precipitation cycle.

The capital city of Kathmandu is located in the Kathmandu Valley. Thirty years ago, Kathmandu was a small city and the valley was a patchwork of lush green fields and forested areas. After several decades of intense population growth, the average growth rate in the 1980s was 5.2 percent (Schreier and Shah 1996), agricultural and industrial development, the entire landscape of the valley bears the mark of human activity and transformation. In addition to agriculture, carpet industries and brickmaking dominate the economy. The annual precipitation in Kathmandu (elevation 1350 m) is very seasonal with more than 70 percent of the annual precipitation occurring during a four-month monsoon season (mid-June to mid-September). Given the topographic conditions, this translates into massive runoff during the rainy season and unless a river originates high in the mountain glaciers, there is little water available during the dry season (Schreier and Shah 1996). Kathmandu relies heavily on water from the Bagmati River system as a main source of drinking water. However, both the Bagmati and Bishnumati rivers are extremely polluted because of, among other sources, the dumping of industrial and municipal wastes, e.g., dyes and chemicals from the carpet-making industries and open sewers emptying untreated into the River.

The physical and political boundaries of Wards 19 and 20 circumscribed the primary study area. The Bishnumati River, bounding both wards to the west, is a strong physical boundary, a primary source of water for many residents, and a magnet for a wide range of activities. Historic Durbar Square marks the northeastern boundary of the area. A dirt road separates the wards from the riverbank and despite being in the heart of downtown Kathmandu, a number of community garden plots dot the riverbank and are even found within the wards, along with small livestock such as chickens and goats. In between the river and the community gardens and the square is a densely packed mix of residential dwellings and commercial establishments. Several main streets, just wide enough for two small cars to pass (or more likely two rickshaws or bullock carts), wind uphill from the river to the main road leading to the square; jutting off these main streets at odd angles are a warren of small lanes and footpaths. Houses are squeezed together, stacked with multiple additions to accommodate growing and extended families, and giving the impression of a vast collection of teetering, multitiered wedding cakes. Along the main streets and lanes, various stores (butchers shops, tailors, general dry goods stores, etc.) occupy the bottom, street-level floor of many houses. Schools, health clinics, and the municipal ward offices also are found in the wards. The local residents span a broad socioeconomic spectrum, from low-caste street sweepers to high-caste doctors and lawyers.

These neighborhoods, called *toles,* are comprised of a bewildering (to an outsider) variety of courtyards, based on religion (Buddhist, Hindu), ethnic groups, family clans, castes, and whether they are private or public. None of this heterogeneity had been acknowledged in the original epidemiological studies but interacted profoundly with health, environmental, and literacy issues.

While many households count the Bishnumati River as their primary source of water, piped drinking water is available to a majority of residents in Ward 20 and a number of hand pumps are also found in the area. A unique feature in the community is a fourth major source of water, *dunge dharas,* or traditional stone water spouts. These *dunge dharas* are usually associated with a religious temple or shrine, the water runs continually, and, as many are several hundred years old, the location of the water source has been lost in history. These are an important source of drinking and washing water in the wards.

An example of system analysis   The identification and prioritization of the ecological and social issues of concern for the stakeholders set the context within which the team began to define multiple systems descriptions. The stakeholders in Wards 19 and 20 identified water availability and quality and improper garbage management as the two most critical issues relating to ecosystem health in the community. Hence the water and food and waste systems formed our team's entry point for understanding the local ecosystem.

In our retrospective framing of the situation using AMESH, the various descriptive stakeholder narratives were translated into multiple systems' perspectives by creating "system influence diagrams" (see also Neudoerffer et al. 2005). These diagrams contain five key elements: actors, activities, concerns, needs, and resource states (or indicators of ecosystem health). To create these diagrams, we began by first asking for each stakeholder group, "Who are the actors?" and "What do they do?" and by drawing influence diagrams to illustrate these actor/activity relationships. We enhanced our understanding by asking, "What are the concerns of the actors?" and "What needs arise out of these concerns?" We further built our collective understanding of the ecosystem by asking, "How do these activities and needs affect the indicators of ecosystem health (termed here 'resource states')?"

A rich picture of our ecosystem started to emerge when we were able to group actors and activities, not by stakeholder groups, but by ecosystem issues, such as water, food, and waste. We started to see the multiple perspectives of the stakeholders when we brought together all of the stakeholder needs and the indicators of ecosystem health (resource states). Multiple perspectives helped us to see where the stakeholders held divergent views of an issue and where we needed to negotiate solutions. A rich ecosystem picture is useful for thinking through the impact of potential solutions to ecosystem problems. For example, if we wanted to make changes to the food and waste system, we could use our rich ecosystem picture to try to understand how the existing system would be affected by any changes we might introduce, such as adding more garbage containers in the street. Finally, our ecosystem picture helped us to identify the activities and resources states that should be monitored to evaluate improvements in ecosystem health.

In each of the following diagrams, the actors appear in shaded ovals, activities in shaded boxes, concerns in italics, needs in bold text, and resource states (or indicators of ecosystem health) in shaded starbursts.

Butcher stakeholders   Figure 15.5 illustrates the ecosystem perspective of the butcher stakeholder group. The wholesalers group (contained in the black box in fig. 15.5) is made up of the wholesalers or the owners of the slaughtering houses, the butcher employees or the people who actually do the butchering of the animals, and the Small Meat Marketing Association, the local organization to which all wholesalers belong. The police were included as actors in this system perspec-

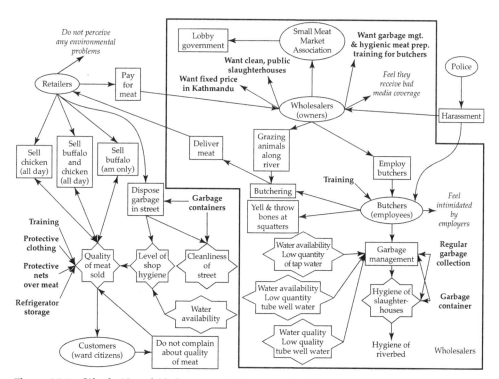

Figure 15.5   Wards 19 and 20, Issues and influences: Butchers' perspective.

tive, because both the wholesalers and butcher employees voiced concerns regarding harassment by the local police over the practice of slaughtering female buffaloes. It became apparent also that, although the wholesalers had the decision-making power over slaughterhouses, the butcher employees were responsible for garbage management and hygiene of the riverbed near the slaughterhouses. The tension between the wholesalers and the butchers' employees was captured by the concern of the latter that they "feel intimated by their employers." Although the stakeholders did not identify "customers" as an actor in this system, the research team added the customers, who are also citizens of the wards, as important actors because their demand for meat helped to drive the "butcher stakeholder" system. The important indicators of ecosystem health in this system were the hygiene of the slaughterhouses, riverbed, and butcher shop, the cleanliness of the street, the quality of the meat sold, and the availability and quality of water.

Sweeper stakeholders   The ecosystem from the perspective of the sweeper stakeholders is illustrated in fig. 15.6. In this system diagram we see the importance of

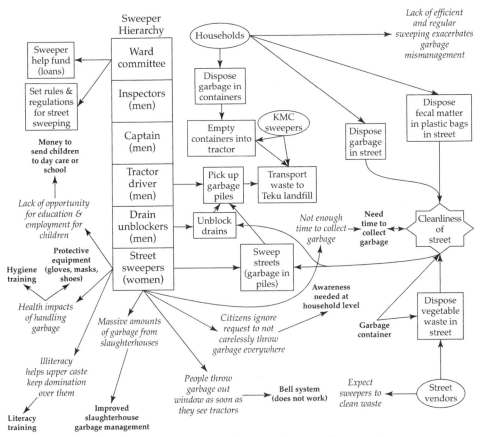

Figure 15.6   Wards 19 and 20, Issues and influences: Sweepers' perspective.

clearly delineating the specific actors in our system. While the stakeholder category was broadly termed "sweepers," the sweepers were actually differentiated based on a local power and gender hierarchy. Additionally, there were local sweepers, employed by the Ward Committee to clean the streets, and there were sweepers employed by KMC who were responsible for emptying the larger garbage containers in the ward and for sweeping the major streets that traverse the wards.

Any project activities that aimed to address garbage management needed to be cognizant of the different groups and roles of sweepers that worked and lived in the wards. The other key actors in the sweeper stakeholder system were the local households that were responsible for generating the majority of the garbage in the wards. The important indicator of ecosystem health in this system was the "Cleanliness of the streets." The street sweepers (primarily women) voiced a number of significant concerns, some related directly to garbage management and hygiene and others to broader social concerns over literacy and opportunities for their children. Their concerns and how they complemented or conflicted with those of others in the community were depicted in fig. 15.3.

Similar "issues and influences" diagrams were created from the perspectives of street vendors (internally differentiated by those that dwelt locally, or were from the outside, and/or had permanent or temporary stalls); hotel and restaurant owners (vegetarian and non-vegetarian owners, patrons); and two groups defined by quite different criteria—community leaders based on status in the community and the squatter stakeholders based on their location of residence.

Synthesizing the system descriptions

Each stakeholder-based system model brings together and highlights different issues. These can then be brought together into one diagram around a selected issue. Using the system influence diagrams, this may be done in one of several ways: all of the various actors and activities around one or several related resource states (or indicators of ecosystem health) may be brought together into one integrated diagram; all of the various stakeholder concerns around a particular issue or resource state (indicator of ecosystem health) may be synthesized; or all of the various needs may be synthesized around a particular resource state. Once again, a complementary quantitative synthesis of the multiple systems perspectives may be undertaken as well.

An example of system synthesis   The data available in this project were largely qualitative; thus the system synthesis was qualitative in nature. Bringing together actors and activities from different stakeholder groups around a common ecosystem health issue, the various multiple system descriptions were linked into more holistic ecosystem models.

The Production and Management of Street Waste system diagram (fig. 15.7) illustrates one such holistic ecosystem model. This system perspective highlighted the key points in the wards where garbage was generated and the primary processes for removing garbage from the streets and suggested points of intervention (and possible impacts of those interventions) for making changes to the local garbage

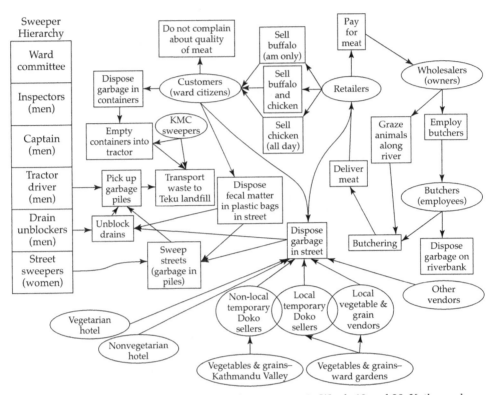

Figure 15.7   Production and management of street waste in Wards 19 and 20, Kathmandu.

management system. If we entered a new actor into the system, such as a mother's group or local clubs to help to manage garbage, we could have used this type of system diagram to determine the types of questions we needed to ask regarding what impacts these changes might have had on our system. The water system (not shown) was both more technical and tended to be at greater-than-ward scales.

Water quality and quantity are both serious issues of concern at the *national* level (Schreier and Shah 1996). A significant portion of Kathmandu residents and industries get their water from the Bagmati and Bishnumati rivers that pass through the city. The quality and quantity of water in these rivers is determined by factors largely exogenous to the sublocal landscape of the wards. While the people in Wards 19 and 20 could address some issues (cleaning around *dunge dharas*), their ability to address the larger issues was limited, and tackling these may have been inappropriate for the scale at which the research was undertaken.

Stakeholder needs and resource states: water, food, and waste   A final way of synthesizing the multiple stakeholder perspectives of our local ecosystem is illustrated in fig. 15.8. In this synthesis, we brought together all of the actors, indicators of ecosystem health (resource states) and needs for the ecosystem as a whole, with respect to water, food, and waste issues. This perspective highlights the needs of the different system actors and how these needs influence the state of the

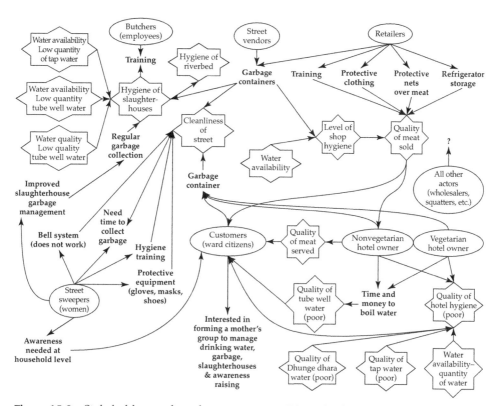

Figure 15.8   Stakeholder needs and resource states: Water, food, and waste.

indicators of the ecosystem health. The needs flag the various potential points of action or intervention in the ecosystem and the diagram is useful for asking questions such as: *How will the indicators of ecosystem health be affected by changes in the system that are made as a result of addressing different stakeholder needs?* This synthesis also illustrates where different stakeholders hold common needs, which may provide points of co-operation and collaboration in the community.

Collaborative Learning and Action

By now, a comprehensive picture of the current and historical state of the local ecosystem and a number of potential points or areas of action to improve ecosystem health and sustainability has emerged. The next phase is to move from understanding to action. Negotiation and compromise are critical at this step, as the stakeholders work together to create a common vision for the future, negotiate trade-offs, plan and implement activities and changes, and monitor and evaluate the same.

Cross talk and seeking solutions

Prior to taking action, the stakeholders need to revisit their common vision for the future and agree on a set of activities or an action plan to realize their local goals and objectives.

General visions for the future that were expressed earlier (see "How does the resolution of these issues come together in a vision for the future?") are refined to form the basis for action. Specific action plans were created to separately address the goals and visions of each stakeholder group: butchers, squatters, sweepers, hotel/restaurant owners, street vendors, community leaders, and an action plan of common activities of all stakeholders.

While this represented a positive step forward by not working together to address their collective problems, the stakeholders did not take advantage of potential systems synergies and in many cases merely avoided and often fueled conflict. The opportunity to integrate activities was lost when each stakeholder group was given a separate action plan. Each group worked separately on their issues and did not see or take advantage of interconnections among issues. From a system's perspective, when each group works independently, there is the potential that efforts will be at cross-purposes both at a social level (e.g., two different groups may attempt to address the same problem in two different ways, potentially leading to a conflict) and at an ecological level (e.g., two different groups working independently to address, for example, water quality or quantity may inadvertently negate the others improvements or compound existing problems). To avoid working at cross-purposes, it is important that systems analysis and synthesis are undertaken before attempting to define or create alternative intervention or action strategies.

As a next step, the goals of each stakeholder's group should be added to the system descriptions, both the multiple system perspective and to the synthesized system diagrams to illuminate where different stakeholder groups hold common and conflicting goals.

For example, the system diagrams of actors, activities, concerns, needs, and resource states identified different perspectives concerning who was "at fault" for solid waste management problems in the two wards. The street sweepers believed that the local citizens did not properly adhere to the garbage management rules and regulations and that they did not have sufficient time or adequate protective clothing to do their job effectively. The wealthy local citizens, in contrast, felt that the problem was one of poor garbage collection on the part of the street sweepers. One possible approach at this point would have been to bring the stakeholders together to discuss these different perspectives and attempt to find a collaborative solution that would address all of the stakeholder concerns.

While the stakeholders and the project team decided to focus on the garbage management issue as a primary action, the conflict among the various stakeholder groups was avoided by engaging a secondary stakeholder group as a primary actor in the garbage management activity. The community and the research team engaged two local service clubs, one in each ward, to facilitate garbage management around the *dhunge dharas*, as well as cleaning up and maintaining the areas around the garbage collection sites (garbage container sites). Given the strength of the traditional social structures in these communities, direct collaborative action among stakeholders to achieve garbage-related goals may not have been possible. In using a secondary group as a focal point for organization around a common issue, stakeholder groups that might not have otherwise worked together could collaborate on a shared action and goal.

Design of an adaptive approach for implementation of the vision and collaborative learning

In communities where there are strong traditional power structures, the creation of organizational arrangements that do not directly challenge those traditions but can effectively carry out important tasks may be both desirable and feasible. Thus, in these Kathmandu communities, in a country where Maoist forces, traditional royalists, and liberal democrats have fragmented the political landscape, the research team's quiet work with community clubs, volunteers, and Ward Committees appears to have accomplished a great deal more than would have been possible if more formal governance structures had been accessed. Formal government was often grid-locked, dysfunctional, or distracted with political negotiations, and, with a history of local democracy in these wards, there seemed to be a rich vein of local activism outside government channels.

Implementation

Because of the way this project unfolded, the implementation steps and responsibilities either stayed with Ward Committees or devolved to clubs and subgroups in the community, with researchers largely on the sidelines. If logistics and finances had permitted, it would have been informative for researchers to take part in the local activities, to help make them more transparent, serve as communicators between groups, and to ensure that larger issues such as gender, equity, and environmental sustainability were incorporated. On the other hand, having raised these issues in larger forums, it may be both necessary and desirable for researchers simply to "let go," remaining in the background as resources when called upon.

Monitoring and evaluation

The questions of what one should monitor and evaluate in complex social-ecological systems remains a matter of some debate. Indeed, the question of what to monitor for "learning" or "adaptation" remains implicit in some of the major recent tomes on complexity (see, for instance, Gunderson and Holling 2002). We do not yet know if the volumes of health and environmental indicators generated in the past twenty years have any value beyond project and planning evaluation. The underlying nature of the reality is often implicit, rather than explicit, for indicator lists, but many of them seem to assume linear realities with no feedbacks, holarchies, or complex nonlinear dynamics.

Much monitoring in international development projects such as this is focused on empirical and esthetic outcomes (no garbage in the streets, drinkable water, more trees; fig. 15.8) related to three-year project cycles. These are, indeed, useful for providing feedback to participants and reflect outcomes that are important to various stakeholders. However, whether they are important in terms of system dynamics remains to be seen. Linking measurements to both complex system dynamics and across scales remains a major challenge (Boyle 1998; Boyle et al. 2000; Kay and Boyle, chapter 4; Deutsch et al. 2003).

Monitoring and evaluation is not the end of the process. If AMESH has been truly successful at mobilizing the local community into action to address their own ecosystem sustainability and health issues, then a new set of concerns (or a next set of concerns) should emerge out of the evaluation of the recently completed activities. From these, the local community should be empowered to begin an AMESH process again, this time, without outside intervention.

In these wards, community members formulated qualitative indicators of success as part of their visioning process. See "How does the resolution of these issues come together in a vision for the future?" At least one outcome that could be measured quantitatively—water quality (in terms of coliform counts)—was also selected. For instance, the NZFHRC was able to obtain water quality testing kits and train local school classes to use them. What is apparent, however, is that these indicators do not necessarily reflect a systemic understanding of the community, the ecosystem of which it is a part, and its dynamics over time.

Conclusions

This post hoc application of AMESH was started several months prior to the final project meetings and a preliminary version of this work was presented to the complete project team, including approximately 50 local Nepali community members at the final project meeting in Kathmandu in November 2001. The team members from SAGUN and NZFHRC both found the system influence diagrams to be a very useful way to capture and share a large amount of information. The SAGUN team suggested that they were often guided by "gut feeling" during the project, and these diagrams helped to clarify to them why some of the various activities they had attempted to implement with the community made sense. Both SAGUN and NZFHRC felt that this evaluation and especially the understanding that emerged from the analysis of the systems diagrams would provide very strong justification for continuing the work in future funding proposals.

At the same final project meeting, two groups of community members, one from each ward, were invited to try creating an influence diagram for an issue of concern in their ward. This group activity demonstrated that the basic influence diagramming concept, using actors, activities, concerns, needs, and resource states, was sufficiently simple and flexible such that a small community group could start to use it to explore a local problem with only simple and straightforward explanation. Two vibrant discussions quickly emerged as the community members attempted to rethink the garbage management issue using these simple diagramming tools. It was during this discussion that the fact emerged that the Ward Committees do not have the authority to enforce the locally drafted garbage management rules. Prior to this point, other community members had been critical of the Ward Committee for failing to enforce the garbage management regulations.

Because AMESH was being developed as a process at the same time as the project unfolded, with theory and practice from Nepal, Peru, Kenya, Ontario

(Huron Natural Area), and other sites feeding into and out of each other, we were not able to fully utilize all of the insights generated. In particular, some important issues (such as governance and policy analysis) were not identified until too late to be useful for this project. Nevertheless, we felt sufficiently confident that, by the end of the decade of work, we were able to characterize the main features of an ecosystem approach to health.

Postscript

On June 1, 2001, just before the end of the project, King Birendra Bir Bikram Shah Dev died in a bloody shooting at the royal palace that also claimed the lives of most of the royal family. Crown Prince Dipendra is believed to have been responsible for the shootings before fatally wounding himself. Immediately following the shootings and while still clinging to life, Dipendra was crowned king; he died three days later and was succeeded by his uncle, the current King Gayendra Bir Bikram Shah. This royal tragedy rocked the nation and further increased the political instability. After this time, Maoist insurgencies, previously largely confined to the periphery in the remote, rural areas, moved into the capital city of Kathmandu. Bombs went off near Wards 19 and 20, and Dr. Joshi, after having being visited by two morose strangers, felt compelled to take down his sign reading "The National Zoonoses and Food Hygiene Research Centre," for fear that it might attract unwanted attention.

The recasting of the Maoist insurgency from an internal revolt against injustice to a terrorist war in 2003 and the suspension of democratic process by the king are as much nested into the international political context as the democratic movements of the 1990s. This makes clear, once again, that the promotion of an ecosystem approach to environmental and health issues has to be more politically astute (see Ráez-Luna, chapter 18) and directed across a wide range of scales, as well from a wide variety of perspectives, if it is to be robust and sustainable.

Finally, one of the greatest strengths of AMESH, we discovered, and perhaps the most difficult to capture in words or in any quantitative sense, was that the research process served as a kind of incubator of local self-awareness and self-reliance. As a consequence of the AMESH work in Wards 19 and 20, the local youth of the Khadgi (butcher) caste have injected new life into the local butcher's association and are aware both that a different life from that of their parents is possible for them and that they themselves have the power and ability to bring about transformative change in their lives and in their community. Since the Maoist War "heated up," Dr. Joshi and the members of Wards 19 and 20 have continued their work, and a new, IDRC-funded "Ecosystem Health" initiative on public education and outreach was launched and, as of December 2005, is ongoing and apparently successful. A test of the ultimate success of AMESH may rest in the enduring energy of this local group and provides those interested in local participatory development with an important lesson: the greatest success of any development initiative may not be that which is physically constructed but the ties, solidarities, and local strengths that are nurtured and develop through out this process.

References

Asian Development Bank (ADB) and International Centre for Integrated Mountain Development (ICIMOD). 2006. "Environmental Assessment of Nepal: Emerging Issues and Challenges?". Kathmandu, Nepal: ICIMOD. http://books.icimod.org.

Baronet, D., D. Waltner-Toews, D. D. Joshi and P. S. Craig. 1994. Echinococcus granulosus infection in dog populations in Kathmandu, Nepal. *Annals of Tropical Medicine and Hygiene* 88:485–492.

Baronet, D. 1996. Agroecosystem Health Project. Report. Ottawa, Canada: International Development Research Centre.

Boyle, M. 1998. An adaptive ecosystem approach to monitoring: Developing policy Performance indicators for Ontario Ministry of Natural Resources. Masters thesis. Waterloo, Ontario, Canada: Department of Environment and Resource Studies, University of Waterloo.

Boyle, M., J. J. Kay and B. Pond. 2000. Monitoring in support of policy: An adaptive ecosystem approach. In *Encyclopaedia of Global Environmental Change*, ed. R. Munn, 116–137. New York: Wiley.

Central Intelligence Agency. 2002. "Nepal Factbook." (Aug. 2002), at www.cia.gov/cia/publications/factbook/geos/np.html#Govt

Deutsch, L., C. Folke and K. Skånberg. 2003. The critical natural capital of ecosystem performance as insurance for human well-being. *Ecological Economics* 44:205–217.

Gunderson, L. H. and C. S. Holling (eds.). 2002. *Panarchy: Understanding Transformations in Human and Natural Systems*. Washington, D. C.: Island Press.

Holling, C. S. 1986. The resilience of terrestrial ecosystems: Local surprise and global change. In *Sustainable Development in the Biosphere*, eds. W. M. Clark and R. E. Munn, 292–320. Cambridge, UK: Cambridge University Press.

Library of Congress. 1991. "Nepal Country Study." (Aug. 2002), at http://lcweb2.loc.gov/frd/cs/nptoc.html.

Neudoerffer, R. C., D. Waltner-Toews, J. Kay, D. D. Joshi and M.S. Tamang. 2005. A diagrammatic approach to understanding complex eco-social interactions in Kathmandu, Nepal. Ecology and Society 10(2): 12. [online] URL: http://www.ecologyandsociety.org/vol10/iss2/art12/

Pretty, J. N., I. Guijt, I. Scoones and J. Thompson. 1995. *A Trainers' Guide to Participatory Learning and Action*. IIED Training Materials Series No. 1. London, UK: International Institute for Environment and Development.

Schreier, H. and P. B. Shah. 1996. Water dynamics and population pressure in the Nepalese Himalayas. *GeoJournal* 40:45–51.

Waltner-Toews, D. 2004. *Ecosystem Sustainability and Health: A Practical Approach*. Cambridge, UK: Cambridge University Press.

Waltner-Toews, D., J. J. Kay, T. Murray and R. C. Neudoerffer. 2004. Adaptive Methodology for Ecosystem Sustainability and Health (AMESH):
An introduction. In *Community Operational Research: OR and Systems Thinking for Community Development*, eds. G. Midgley and A. Ochoa-Arias, 317–349. New York: Kluwer Academic.

Waltner-Toews, D., C. Neudoerffer, D. D. Joshi and M. S. Tamang. 2005. Agro-urban ecosystem health assessment in Kathmandu, Nepal: Epidemiology, systems, narratives. *EcoHealth* 2:1–11.

16

Tools for Learning
Monitoring Design and Indicator Development

Michelle Boyle and James J. Kay

Introduction

The themes presented in chapter 15 of sustainable livelihoods, health, and integrity pose a challenge for monitoring and learning. A substantial portion of this book is devoted to determining exactly what these management objectives mean and how we get "there" from here. We raised the issue of monitoring in chapter 14, when we discussed the overall ecosystem approach process. Many of the case studies have had to wrestle with issues of who measures what and how to characterize and assess "progress." In many of these cases (Kenya, Peru, Nepal, India, etc.) the indicators were determined experientially; what did people value in their environments? This kind of monitoring is essential to keep stakeholders engaged in the process. However, understanding the ecological and social processes behind what we see, hear, and feel (garbage in the streets, dirty water, erosion, poverty) is essential for implementing effective management strategies and governance structures. Complex systems theory as described throughout this volume provides the theoretical basis for our understanding. Complex systems are extremely sensitive to initial conditions and often change rapidly and surprisingly. Our ability to predict is limited, and the future rarely unfolds exactly as it was envisioned. In adaptive management, these differences are used as opportunities for learning. Policies and programs must be implemented with the more realistic view that they are experiments. Therefore monitoring becomes the mechanism through which we test our theoretical understanding against what we can measure of the world that surrounds us. We incorporate learning through this iterative process of evaluation and adjustment.

There is a substantial body of literature on monitoring in general, covering topics such as frameworks and characteristics, indicator development and selection criteria, and existing sets of indicators. (See, for example: Environment Canada (2004) and U.S. Environmental Protection Agency (2004) for current initiatives; also see Boyle (1998), Maclaren (1995), Hodge (1994), Davies (1991), and Slocombe

(1990) for earlier comprehensive reviews on the subject.) Our intent here is only to highlight some distinctive features in the design of monitoring programs and processes within the ecosystem approach discussed in chapter 14.

In the past, monitoring frequently has been in the service of surveillance and compliance, error detection, or impact assessment. Monitoring within an adaptive framework, instead, plays a crucial and integrated role in decision making by providing information to generate narratives, advice to determine if alternative courses of action are required, and the feedback necessary for adaptive learning (the explicit assimilation of experience into decision making). It provides the key link between collaborative learning and action and our understanding of the system and future actions. This chapter presents an approach to monitoring design from this perspective.

What's Different About Adaptive Ecosystem Monitoring?

Despite the best intentions and conscientious work, normal science and available problem-solving tools are insufficient when dealing with complex systems. Issues of sustainability are persistently untidy and demand an examination of the wider ecological-societal system to resolve. Thus conventional nonsystemic monitoring methods applied to sustainability tend to fail. Adaptive ecosystem monitoring incorporates complexity and redefines the role and scope of monitoring in order to assist decision making and create the capacity for learning. Some characteristics of this approach that separate it from more traditional initiatives in the past and present are highlighted below.

- Monitoring must take into account the dynamic behavior of ecosystems and the relevant interrelationships between nested human and ecological systems.

- Monitoring must be well-grounded in the context of people and their concerns in order to ensure effective implementation, efficient resource use, and useful results.

- Stakeholders are involved from the beginning and participate in all aspects of design, development, and implementation. Expert opinion is in the service of facilitating this process rather than dominating it.

- Indicator development is not an isolated activity. Indicators are one aspect of a monitoring program that is meaningful only in the context of the interdependent activities of management, monitoring, and governance.

- There is no "correct" set of things to measure. The fact that each monitoring program is designed through participative methods around a particular issue for a particular system makes it necessarily unique.

- Much of the focus is on monitoring the context of the system. That is, indicators measure aspects outside the defined system boundaries or monitor the flows (of exergy, materials, and information) into the system. A change in context or inputs will likely cause the system to change. (Changes from within also can occur.)

- Emphasis is placed on monitoring the self-organizing processes that cause structures to emerge rather than measuring only what we can see. This provides early warning of changes about to happen.
- Because complex systems can behave in surprising ways, the system needs to be monitored for known threats to the system, and also for the general health or integrity of the system (Woodley et al. 1993).
- Monitoring is about learning. It helps us to assess the current state of the system, understand system dynamics, observe responses evoked by various management strategies, and determine how we can use this information to achieve our goals.
- The monitoring process is recursive and flexible in order to respond to changing circumstances, new knowledge, and experience. It is reflexive too, in that the monitoring program is periodically evaluated to determine if it is providing the required information and continuing to meet user needs.

These important features are described further in the remainder of the chapter as we discuss in detail the design of monitoring programs and processes within the adaptive ecosystem approach.

Monitoring Within an Adaptive Ecosystem Approach

In the ecosystem approach advocated in this book, governance, management, and monitoring are interdependent activities that form an integral part of collaborative learning and action (fig. 14.2). They are carried out in the context of an issues framework of human concerns and an explicit conceptual model of the ecological-societal system (fig. 14.1). Figure 16.1 shows monitoring in this larger picture.

The inputs to a monitoring program are the results of the analyses and participatory decision processes of the ecosystem approach described in chapter 14. It will help our discussion to briefly review some features of the ecosystem approach here. The issues framework in fig. 16.1 (see also the first few steps of the AMESH process) is designed to explicitly: (a) identify the actors and stakeholders, (b) describe and define the situation, (c) determine the important issues and concerns, and (d) disclose the values applied in making decisions and judgments. It hierarchically "maps" human concerns and preferences onto the ecosystem description (the conceptual model) to identify scenarios that are both ecologically possible and socially preferred. A chosen "vision for sustainability" thus defines the system(s) of focus, what aspects are important, and how it is hoped that these will change over time (i.e., a narrative of the future). This information defines what it is we want to know.

An output of the monitoring program is a narrative of how the system has actually changed and how it might unfold in the future. The results enable participants to evaluate changes that have occurred because of management intervention and to revise their understanding of ecosystem dynamics. This process is also meant to be reflexive so that methods of analysis and synthesis, the vision for sustainability,

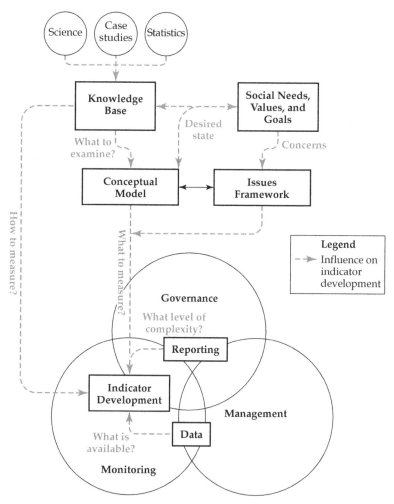

Figure 16.1 Monitoring in context. Within the adaptive ecosystem approach, monitoring, management and governance are interdependent activities. A comprehensive monitoring program cannot be established without describing the ecological-societal system of interest (conceptual model) and the societal goals and concerns (issues framework).

and the appropriateness of the monitoring system itself are reviewed periodically and adjusted (also see the AMESH diagram, fig. 14.2). The monitoring results feed back to inform management and governance as well as contribute to scientific knowledge and shape societal values. Table 16.1 sets out some definitions we use in this chapter related to monitoring within an ecosystem approach.

Elements of a Monitoring Program

If we "open up" the circle of monitoring in fig. 16.1, we find that it, too, is a multi-faceted activity. Monitoring encompasses various elements such as indicators, data, and its collection, and reporting (table 16.2). Developing these individual elements of a monitoring program should not be confused with the process of monitoring.

TABLE 16.1	Definitions of Monitoring-Related Terms in an Adaptive Ecosystem Approach
Term	**Definition**
Monitoring	Monitoring is the activity of observing the human and natural systems and synthesizing the observations into a narrative of how the situation has actually unfolded and how it might unfold in the future. This narrative is used as the basis for governance and management; that is, for learning, revisioning, and adapting human activities as the human and natural ecosystems coevolve as a self-organizing entity.
Indicators	Hodge (1994, p. 117) defines indicators as "measurable descriptors, quantitative or qualitative, of normative interest which facilitates assessment of the past, current, or future state or performance of the system constituent parts, controls, and feedback loops (as well as the system as a whole)." In applying this general definition to specific cases, we must always ask the questions: for what system?; and what is important in this situation? What an indicator is or means is shaped by the context, the "attributes" the complete set of indicators must possess, and the indicator "criteria" chosen by the users.
Data	Data are the quantitative or qualitative bits of raw information that are used to calculate an indicator. Data alone do not necessarily provide comprehensible information to assess progress toward goals. It is only used as input to the indicator which is in turn used for assessment.
Assessment	Assessment is used in the informal sense here (e.g., it does not refer to regulated environmental assessment processes). It simply means a synthesis of all information for a purpose. This can be done, for example, in the interest of determining the current state of the system, deciding whether we are on track to reaching articulated goals, or in evaluating the ability of the monitoring program to meet user needs.
Reporting	Reporting is the critical task of communicating all that we have learned (about the system, our actions, its responses, etc.) to those that have a stake or interest in the outcome. In systems terms, it is the feedback mechanism of the adaptive ecosystem approach. Reporting is intended for not only those directly involved in management and governance, but also the public, scientists, observers, and more. In each case the method, level of detail, and type and presentation of the information must be appropriate to the audience.

The former is about designing tools for the analysis of interacting ecological and social systems, and we discuss these in more detail later in this chapter. The latter is about designing a human activity system (fig. 16.2) to animate the elements.

The Monitoring Process

The development of monitoring program elements is only the first step in the process of monitoring (fig. 16.2). Collecting data and calculating indicators are activities traditionally associated with monitoring, and there is an abundant literature

TABLE 16.2   Elements of a Monitoring Program

Monitoring Program Element	Description
An issues framework	The impetus of a monitoring program is always to assess progress toward a set of human goals. Hence a clear articulation of the goals and users of the information is the foundation of any monitoring program.
A conceptual model of the world	The model represents how we look at the world in the context of the goals. It serves to delineate the system that should be monitored. It provides a framework that relates the indicators to each other in the context of the overall system being monitored.
A set of indicators	The indicators characterize the system being examined in a meaningful way for the users.
A methodology for data collection and storage	Carefully laid out procedures to address the practical and technical issues involved in data collection must be established to ensure accuracy, consistency, and statistical robustness. Equally important is the storage of the data, so that they are accessible and usable in the future.
A methodology for calculating indicator	The data collected will have to be manipulated in order to derive values for the indicators. Again, a method to do this accurately, consistently, and in a statistically appropriate manner is required.
A process for synthesis	Synthesizing the information that the indicators provide into an overall narrative of the status of the system is essential to completing the central task of the monitoring program, that is, to assess progress toward the human goals and aspirations which motivated the enterprise.
A methodology for reporting	The values of the indicators and the results of the system assessment (the narrative that results from the synthesis) must be reported to the intended audience or users of the information. A methodology for presenting it in a clear, purposeful, and timely manner for decision making is crucial to the utility and success of the monitoring program.

on how to complete these tasks. Synthesis is about generating the narrative regarding current and future status of the ecological-societal system and comparing it to the goals expressed by a sustainability vision or policy. The next stage is reporting. This is the communication of the results of monitoring and assessment to those involved in management and governance.

The report is a crucial interface between the functions of monitoring, management, and governance. It provides the opportunity to ask two questions essential to the learning process. The first involves the monitoring program itself: Is it meeting the needs of the people for whom it is undertaken? If not, adjustments must be made so that it better produces the information needed for decision making. The second question asks: What is the state of the system relative to our goals?

Given that ecological-societal systems are complex, it is reasonable to expect that the situation is not unfolding exactly according to the objectives articulated

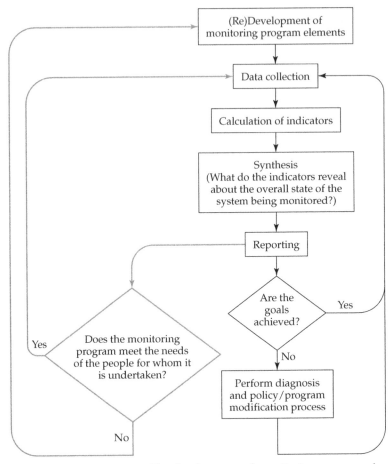

Figure 16.2 The monitoring process. The development of monitoring program elements and the results of indicator interpretation are used to develop a narrative of the current system status. If this differs from the sustainability vision, we must learn what this tells us about the system and modify our perception of the system and activities accordingly. The monitoring program must also be evaluated to ensure that it provides the necessary information to make sound decisions and meet user needs.

in the vision. Investigating these discrepancies, however, teaches us how the system behaves. Discrepancies should trigger both a reevaluation of our understanding and a reevaluation of the policy or program designed to achieve the vision. Does our description of the system, and what is deemed important, need to be revised? Is the vision still feasible? Are new management strategies or tactics required? (This reevaluation is represented on both the diamond and AMESH diagrams (figs. 14.1 and 14.2, respectively) as beginning a second iteration of the process from the top). Of course, any changes in these areas will nearly always require modifications to the monitoring program. Thus monitoring within an adaptive framework is dynamic and responsive, akin to the systems on which it focuses.

Attributes of a Monitoring Program

On the basis of a literature review of monitoring and indicators (Boyle 1998) and principles of complex systems theory and post-normal science, we have compiled a list of attributes that a comprehensive monitoring program should possess. This list is distinct from characteristics of the system of interest or from indicator selection criteria. Our research revealed that there often is confusion between attributes that apply to a monitoring program or to the set of indicators as a whole and criteria that serve to evaluate individual potential indicators. (The latter are discussed separately under the section "Developing Indicators.") The following list of attributes is helpful to keep in mind during the design of a monitoring program. It also is an important tool to evaluate whether the monitoring program provides the required information and meets the needs of users.

A comprehensive monitoring program should:

Be relevant and decision supportive. The monitoring program must be relevant to the issues of concern and to the users of the information. The information provided to decision makers must be understandable, timely, and as forward-looking (anticipatory) as possible, so that decisions may be made before the system is adversely affected.

Take into account considerations of scale and type (see Allen, chapter 3). The monitoring program should provide information at a range of spatial and temporal scales (hierarchically nested) that are appropriate to the vision, yet be sensitive to changes in time and across space or groups. The program must examine different types or perspectives of the system and consider abiotic, biotic, and cultural factors. A separate issue is that data must be collected and reported at appropriate but not necessarily the same scales.

Be based on a conceptual systems model that explicitly recognizes relationships between society and environment. The conceptual model of the system must be scientifically valid and should reflect the integration between society and the environment in several ways. It must be nested to deal with considerations of scale and type (as above) and be based on ecosystem and institutional boundaries. Human systems must be acknowledged as a subset of the ecological systems that support them, and linkages between the economy and the health of ecological systems must be made. It is also important to monitor the feedbacks between society and the environment (see figs. 4.2–4.4).

Allow for an overall integrated evaluation of the system. In order to evaluate the state of the system, the monitoring program should collect a variety of different types of measures. It is important not only to monitor the sensitivity to magnitude, direction, and duration of known stresses but also to monitor for ecological integrity (which could reveal unanticipated problems). An evaluation of the system should be able to assess its organization (structure and process), state, the quantity and quality of flows of energy and materials, and whether changes are reversible or controllable. Historical and baseline information can be used to identify natural ranges of variability and trends over time and perhaps the pathways of change between states. The monitoring program thus should reflect our knowledge of naturally occurring changes as part of normal system behavior and response. Cumulative effects should be measured and compared with threshold values for the system. Emergent properties of the system should also be detected and evaluated. Finally, all system assessments should rely on both scientific and objective measures as well as experiential and subjective ones.

Be adaptive and flexible. The program must be adaptive and flexible enough to deal with environmental (including catastrophic) changes; incorporate new information, technology, and scientific research; and adjust to changes in the political context or societal values. It is useful to have a monitoring scheme applicable to diverse operational situations at different scales and for different ecological systems, yet the program must also be tailored for specific applications.

Be practical. A program must be cost-effective to be useful. Employing existing expertise, data sets, and tools [e.g., geographic information systems (GIS) and modeling methodologies] can save time and money and incorporate the work of others.

Monitoring Program Development

The previous sections have provided an overview of adaptive ecosystem monitoring. We can now turn our attention to the design and development of the elements of a monitoring program. To refocus our discussion, refer back to table 16.2.

Strictly speaking, the issues framework and conceptual model should be developed as part of an overall adaptive ecosystem approach to sustainability and health. These elements would be produced as part of the system analysis (the right- and left-hand top boxes of fig. 16.1 and the Systems Description and Narratives box of AMESH) and provide the context for the monitoring program. However, it is often the case that the development of a monitoring program is not initiated in the context of a larger sustainability planning initiative. If this occurs, the monitoring program can drive the development of the governance and management activities necessary to complement monitoring.

The Conceptual Model: What to Monitor

The conceptual model explicitly identifies the environmental context and important scales, processes, structures, and feedbacks of the system of interest. As such, it provides a framework for deciding what to monitor (that is, the aspects of the system that should have indicators associated with them). These are summarized in table 16.3 and explained in more detail below. (A visual reference will assist the reader. Refer to figs. 4.1–4.4). In effect, the conceptual model defines different points where a "monitoring meter" should be "plugged into" the ecological-societal system. A monitoring program should include the following:

1. Measures of the state of well-being of each of the self-organizing entities, ecological and societal. (These are represented in the figures by process and structure boxes.)
2. Measures of the context for the societal system, that is, the biophysical surroundings and the flows into the societal system from the ecological system. (This is depicted in the figures as the societal system in the shadow of the ecological systems.)

3. Measures of the contextual and structural influences of the societal systems on the ecological. (These are represented in the figures by the arrows going from the societal system, up one level, and into the ecological structure.) These relationships demonstrate how societal influences, through changes they cause in the process and structure of ecological systems, can then alter the context for their own society. That is, societal influences on ecological systems are part of a feedback loop that affects societal systems.

Sustainability is about maintaining the well-being of the combined ecological-societal system. In systems terms, this requirement translates into ensuring that the self-organizing processes of the ecosystem be maintained. It happens naturally if we maintain the context for self-organization in ecological systems, which, in turn, will maintain the context for the continued well-being of the societal systems. This condition for sustainability is incorporated in the conceptual model.

Thus the conceptual model provides a framework for monitoring progress toward sustainability that is more elaborate than the conventional approaches. Traditionally, one would monitor external (usually negative) stresses influencing the system and its response. In this model, we emphasize the importance of

TABLE **16.3**   Aspects of the SOHO System Which Should Be Monitored

Category	System Aspect	Example Measures
State of the well-being of the societal and ecological entities	Self-organizing processes	• Evapotranspiration • Life expectancy at birth
	Self-organizing structure	• Biomass • Health care facilities
The context of the societal entities	Biophysical surroundings	• Average humidity • Land topography
	Flows into the system: - Exergy - Materials - Information	• Fossil fuels consumed • Aggregate consumption • Seed varieties
Influences of the societal systems on ecological systems	Structural influences	• Wetlands destroyed • Woodlots removed
	Contextual influences (biophysical surroundings)	• Average temperature
	Contextual influences (flows into the system): - Exergy - Materials - Information	• Phosphorous loading • Drainage flows • Hormone-like chemicals into the food chain

the hierarchical monitoring of both process and structure and their interrelationship in order to understand the current well-being of the system; the context of the ecological-societal system including the biophysical surroundings, and the flow of exergy, materials, and information into the system; and the contextual and structural influence of societal systems on ecological systems and ultimately on society itself.

By considering each of these aspects, we can decide where indicators may be placed to fulfill their purpose most effectively. For instance, if the intent of a monitoring program is to anticipate future changes in order to make wise management decisions in the present, indicators can be chosen to monitor any known factors that influence the processes and structure of the system. Measures focusing on the current state of the system will only reveal changes that have already occurred, likely too late to act to affect the outcome.

Anticipatory monitoring is particularly important in protecting certain areas or land uses, or in monitoring, e.g., to detect the early warnings of global change. Monitoring within the boundaries of the defined system is not sufficient to forecast change nor useful to learn about system dynamics and responses to external conditions. Therefore we must broaden our scope to examine the context that constrains or influences the behavior of the system, and do this at different scales to capture cascading effects (e.g., global systems affect regional systems which, in turn, affect local dynamics).

When using this model, it is important to bear in mind that self-organization inherently involves internal causality. A self-organizing system has the ability to maintain itself at an attractor despite changes in its environment. So the environment may change substantially without the system exhibiting major change. Self-organizing systems also can respond in a synergistic way to environmental changes from several sources. Another property of self-organizing systems is the capacity for new behavior to emerge independently of changes in the environment. These properties present a problem for policy and management evaluation because changes in the system can rarely be tied unequivocally to a specific cause. Rather, explanations of system behavior must be in terms of morphogenetic mutual causality (or feedback loops; see Kay and Regier 1999).

Developing Indicators

Indicators are often confused with a monitoring program. They are an important part of monitoring, but they must be generated, selected, and implemented within the context of the broader ecosystem approach. Indicators describe the status of different aspects of a system, but the conceptual model is required to provide a framework to integrate them. As well, data collection and calculation methodologies and a process for synthesizing all the information into a narrative of system behavior are required if the indicators are to be useful. Without these other elements of a monitoring program, the indicators cannot be used to describe and assess the overall status of the ecological-societal system in question.

TABLE 16.4   Indicator Development Activities

Term	Definition
Generate indicator selection criteria	Selection criteria that capture the issues and constraints regarding the desired set of indicators are used to screen potential indicators. They must be agreed upon by all those participating in the development of the monitoring program elements.
Generate potential indicators	A set of potential indicators consists of possible measures that emerge from consideration of the issues without regard for constraints. All potential indicators should be recorded and evaluated.
Select indicators	Not all potential indicators will meet all the criteria equally well and there will always be trade-offs when choosing some indicators over others. A method must be in place to assess the feasibility of potential indicators and to make a subjective judgement on their evaluation against the criteria.

Indicator development consists of three activities: choosing indicator selection criteria, generating potential indicators, and selecting indicators to be implemented for the monitoring program. These activities are summarized in table 16.4.

Figure 16.3 is schematic for organizing information while generating criteria and potential indicators. It delineates a set of considerations that directly or indirectly influence indicator selection, as well as the relationship between them. For each consideration (i.e., box in fig. 16.3), criteria for indicator selection and questions to guide the task of generating indicators are formulated. An explanation of the diagram follows, while compiled lists of selection criteria and guiding questions can be found in tables 16.5 and 16.7, respectively.

The use of indicators is motivated by the need to understand ecological-societal system dynamics and guide human activities. The articulated vision outlines the desired development of the system and identifies issues or concerns of import to the users of the indicators. This consideration defines what we want to know about. Changes in issues or concerns over time will be incorporated into the indicators through reiterations of the process.

The conceptual model defines the structural and functional components of the biophysical and societal systems and describes the linkages and feedbacks between them. It provides a reference frame that identifies the appropriate scales at which monitoring should occur and for which indicators are needed. It also provides a framework that shows the relationships between the indicators and some sense of how useful or timely the resulting information will be. The issues framework extracts specific management or user concerns and arranges them into a hierarchical framework to associate them with the appropriate scale and type. Thus the conceptual model and the issues framework are highly interdependent and both facilitate the decision of what to measure.

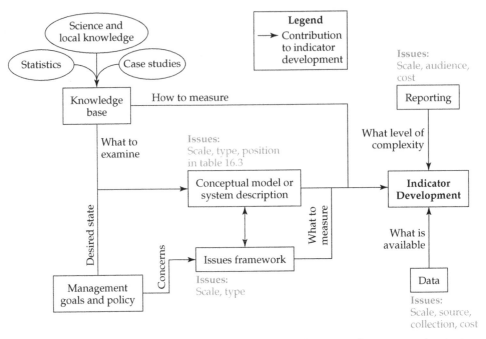

Figure 16.3   Considerations for indicator development. Arrows and accompanying text represent influences on the development of indicators. The main issues to be decided for particular categories are also listed.

Our knowledge base also determines what we elect to monitor. Science (from the perspectives of biology, ecology, geology, hydrology, etc.) aids in understanding the processes and structure of the system and the effects of specific influences or stressors. Complex systems theory adds valuable insight about how we perceive the system and what questions to ask when approaching sustainability issues. Statistics offer rules and standards for sampling and measurement as well as techniques for data analysis and transformation that ensure quality results. Case studies allow us to test chosen indicators and evaluate their utility and feasibility in a practical setting. This collection of information forms the body of knowledge that helps to judge what things are important to examine, how to measure, and how to calculate indicators accurately and consistently.

There are also many considerations attached to the data themselves that affect which indicators are chosen. Data availability encompasses considerations related to cost, collection, database maintenance, accessibility, and who is responsible for each task. The amount (i.e., historical extent) and the quality of data available are also important.

Finally, the development of indicators must be done with an awareness of the reporting requirements and the needs of the users. This category provides parameters for the appropriate level of detail and the number of indicators to use. Both determine how data will be aggregated and/or integrated to compose the indicators.

It should be noted that there are three independent considerations of scale that are relevant to developing indicators for a monitoring program. We must be concerned with (1) the scale and type (dimension) of focus, (2) the scale of data collection, and (3) the scale of reporting the results.

The scale and type of focus may be ascertained from specific issues of sustainability that are reflected in the issues framework. The scale of data collection needs not have anything to do with the level of focus for issues since information can be aggregated, disaggregated, or integrated to be meaningful for decision making. Likewise, the results may be reported at several different scales.

Indicator selection criteria   Using the above considerations as a framework and drawing upon the literature (Boyle 1998), we generated a list of criteria (table 16.5) to be used in evaluating potential indicators. These are properties each indicator should possess. Note that these are relatively generic criteria that apply to many different monitoring initiatives. For a particular project, however, project specific criteria would be added. Keep in mind that important characteristics of the moni-

TABLE **16.5**   Indicator Selection Criteria (Refer to Fig. 16.3).

Categories of Consideration	Indicator Selection Criteria
Conceptual model of the system	Indicators should: • clearly relate to specific societal or ecological elements of the system description
Issues framework	Indicators should be: • clearly relevant to specific articulated goals and objectives for sustainability
Knowledge base	Indicators should be: • scientifically valid • statistically and analytically sound • demonstrated by case studies to be practical
Data	Indicators should use data that are: • available and accessible (easy to obtain and maintain at a reasonable cost) • accurate • comparable over time • complete with historical information (to establish baseline conditions and thresholds)
Reporting	Indicators should provide information that is: • understandable to potential users • unambiguous • easy to use • at the appropriate scale for decision making

TABLE **16.6**   Indicator Categories

Indicator Category	Description
Level I	Indicators available for immediate use. All essential criteria are met.
Level II	Indicators with demonstrated potential for use. Most criteria are met, but, for example, data are not currently being collected, or there is a lack of historical data.
Level III	Possible indicators. There is evidence that these measures would be worthwhile indicators, but further scientific research and/or case studies are required to confirm their utility.

toring program (for instance, "anticipatory" or "monitor for specific threats") need not be a quality of every indicator. For a list of such overall characteristics refer back to the earlier discussion of "Attributes of a Monitoring Program."

Recognizing that trade-offs are inevitable; there are some criteria that, while desirable, are not essential. For example, an otherwise excellent indicator may require data that are expensive to obtain, or for which the necessary historical data are not available. Thus the list of indicator criteria may be used as a checklist to assist in indicator selection. Judgment and careful consideration of each indicator is still required. The use of indicator categories (table 16.6) is one way to accomplish this. This method of classifying potential indicators ensures that the most feasible are adopted immediately but that none are discarded as they may become more practical over time (as new research or funding becomes available for example).

Guiding questions to develop potential indicators   The considerations described in fig. 16.3 are synthesized in table 16.7 as a list of questions to help generate potential indicators. We have found such questions to be very useful in guiding the process of indicator development and encourage others to generate a similar set appropriate for their own circumstances.

Designing the Remaining Elements

Below, we discuss the design of the remaining elements of the monitoring program (see table 16.2) briefly as they are either case specific or are the subject of much literature in their own right.

A methodology for data collection and storage   Carefully laid out procedures that address the practical and technical issues involved in data collection must be established to ensure accuracy, consistency, and statistical robustness. This stipulation is especially important when different people will be collecting the data over time. An emerging design concern is the technical expertise of those

TABLE 16.7	Guiding Questions for Generating Potential Indicators
Rank	Guiding Question
1	What is the sustainability goal or objective?
2	What are the potential indicators?
3	Which aspect of the conceptual model does each indicator measure?
4	At what scale(s) will the indicators be reported?
5	What is the rationale for each of these indicators (i.e. what does it tell us about the sustainability goals and issues)?
6	Are there any confounding factors or problems that may limit the use of the indicator?
7	Data sets: What measurements are required? At what scale will the data be collected? Who is responsible for data analysis and reporting?
8	Data sources: Who has the data set? Who is responsible for its collection?
9	Which category does the indicator fall into (Level I, II, or III)?

collecting data. Increasingly, community groups and volunteers are engaged in data collection. Thus the reliability of the data depends on a methodology that allows the keen, but technically unsophisticated, to collect replicable data. In any case, the design of the methodology must reflect the capabilities of those using it, if the resulting data are to be useful.

Storing data so that they are accessible and usable in the future is equally important. Much has been written on this subject, but, in essence, there must be a way of recording the information that allows someone in the future (typically a decade) to understand what measurements were taken in the field and to be able to assess their reliability and to repeat them. Failure to do this has often resulted in data being rejected simply because its reliability is unknown, effectively wasting the time, effort, and money spent in collecting the data in the first place (Wiersma 1995).

A methodology for calculating indicators   The data collected will have to be manipulated in order to derive values for the indicators. Again, a methodology to do this accurately, consistently, and in a statistically appropriate manner is required. The methodology also must be easy to apply by those entrusted with the responsibility.

A process for synthesis   Synthesizing the information that the indicators provide into an overall narrative of the status of the system is essential to completing the central task of the monitoring program, that is, to assess progress toward the human goals and aspirations that motivated the enterprise. The synthesis should, in effect, bring the conceptual model to life, an art form akin to storytelling. As such, the narrative is dependent on the audience and case specific.

A methodology for reporting   The values of the indicators and the results of the system assessment must be reported to the intended audience or users of the information. The importance of defining a methodology for presenting the results in a clear, purposeful, and timely manner for decision making should not be underestimated. This task is crucial to the utility and success of the monitoring program. Many a monitoring report has collected dust on a shelf because the authors paid no attention to the issue of packaging and marketing it. In fact, creating the report should be recognized as a marketing exercise and should include different "reporting" products for different audiences. With the advent of the World Wide Web and browsers with capabilities to access geographic information systems and computer-aided drafting software in quite transparent ways, some exciting possibilities for reporting the results of monitoring programs are emerging.

In Closing

Ecosystems have the ability to generate novel behavior and to self-organize in response to change (often in a way that we cannot foresee). The reality of irreducible uncertainty associated with complex situations limits our ability to predict outcomes and hence where the path to sustainability lies. Thus any successful approach to sustainability must not only employ traditional anticipatory methods but also strategies for adaptation.

In an adaptive ecosystem approach, governance is seen as the ongoing process of resolving the trade-offs and charting a course for sustainability. Management is about operationalizing sustainability. Monitoring is a crucial and interdependent component that provides feedback on how the situation is actually unfolding and where it appears to be headed. It is the source for learning about system dynamics and responses to human activities.

In this context, a monitoring program must be capable of providing a narrative description of how the ecological-societal system is developing and the implications for the identified path to sustainability. The design of the monitoring program must collect information clearly tied to the conceptual model and the issues framework and report this information in a manner that informs governance and management. In this way, we successively gain a better understanding of the interrelationships between the natural and human systems, the meaning of sustainability itself, and appropriate goals and activities that allow us to move forward. Indicator development must be undertaken in this broader system context and not as a purely scientific exercise. Otherwise, it is more likely than not to produce a set of measures that satisfy disciplinary curiosity but provide little input to the process of resolving the ecological-societal challenges confronting us.

References

Boyle, M. 1998. An adaptive ecosystem approach to monitoring. M.E.S. thesis. Waterloo: Environment and Resource Studies, University of Waterloo.

Davies, K. 1991. Monitoring for cumulative environmental effects: A review of seven environmental monitoring programs. Report prepared for The Federal Environmental Assessment Review Office and Environment Canada. Ottawa, Canada.

Environment Canada. 2004. "Ecological Monitoring and Assessment Network." At www.eman-rese.ca/eman. Accessed October 12, 2004.

Hodge, R. A. 1994. Reporting on sustainability. Ph.D. thesis. Montreal, Canada: School of Urban Planning, Faculty of Engineering, McGill University.

Kay, J. and H. Regier. 1999. An ecosystem approach to Erie's ecology. In *The State of Lake Erie (SOLE)—Past, Present and Future. A Tribute to Drs. Joe Leach & Henry Regier*, eds. M. Munawar et al., 511–533. Leiden, Netherlands: Backhuys Academic.

Maclaren, V. 1995. *Developing Indicators of Urban Sustainability: A Focus on the Canadian Experience.* Toronto, Canada: Intergovernmental Committee on Urban and Regional Research.

Slocombe, S. 1990. Environmental monitoring for protected areas: Review and prospect. *Environmental Monitoring and Assessment* 21:49–78.

U.S. Environmental Protection Agency. 2004. "Environmental Monitoring and Assessment Program," at www.epa.gov/emap. Accessed October 12, 2004.

Wiersma, B. 1995. *Looking for the Forest Among the Trees.* Washington, D.C.: National Academy of Science.

Woodley, S. J., J. Kay and G. Francis (eds.). 1993. *Ecological Integrity and the Management of Ecosystems.* Ottawa, Canada: St. Lucie Press.

PART IV

Where to from Here?
Some Challenges for a New
Science in an Uncertain World

Although this book represents some of James Kay's last thoughts about the ecosystem approach, it is clearly not the last word on the subject nor was it intended to be. As we reviewed the material before he died, James agreed that a theory and its application to complexity must always remain an approximation that would require many lifetimes of continual intellectual and practical engagements with the world. This is reflected in some of the SOHO diagrams that James Kay and Julian van Mossel-Forrester were developing in the last months of Kay's life (See chapters 4 and 16.) Hence we agreed that the final section of the book would need to be focused on unresolved issues, conundrums, and challenges for those who wished to investigate and manage for, a sustainable, convivial planet.

17

Beyond Complex Systems
Emergent Complexity and Social Solidarity

Silvio Funtowicz and Jerry Ravetz[1,2]

Introduction

In response to the new leading problems for science, characterized by high levels of uncertainty (both measurable and unmeasurable), epistemological conflicts (whose knowledge counts), arguments over facts (which facts are relevant) and values (which goals are legitimate), and a sense of urgency that decisions with possibly major consequences need to be made quickly (e.g., climate change), complex systems have become the focus of important innovative research and application in many areas (Costanza et al. 1993). This development, described in some detail with regard to ecosystems studies in various chapters in this book, reflects the progressive displacement of classical physics as the exemplar science of our time and the emergence of the systemic sciences associated with sustainability.

Concepts of simpleness, complicatedness, and complexity and their systemic representations have already been introduced by Kay, Boyle, and Allen in Part I of this book and have also been elucidated by other authors (Casti 1986). We wish to explore this further and examine in some greater depth the implications of doing so. We find it useful to further refine "complexity" into "ordinary" and "emergent." These types are characterized by two different patterns of structure and relationships. In ordinary complexity, the most common pattern is a complementary one of competition and cooperation, with a diversity of elements and subsystems. By contrast, an emergent complex system may oscillate between hegemony (or domination) and fragmentation, in which there is an unresolved struggle for control among subsystems. Degenerate system states, such as "ancien régime" or "autolysis," are not found in ordinary complexity. The state complementarity and diversity is desirable. In ordinary complexity it occurs without intention; while in emergent complexity, this diversity requires a special commitment and awareness for its achievement and preservation.

We show that a full analysis of emergent complexity requires dialectical thinking, with "contradiction" as a key concept. In this way we can integrate apparently paradoxical concepts such as "creative destruction" (Schumpeter 1942) into a general framework. With these ideas we can also provide a philosophical foundation for "post-normal science" (Funtowicz and Ravetz 1991).

Ordinary and Emergent Complexity

In recent years the theory of "systems" has been developed and enriched by a number of approaches in which dynamical properties have been grafted onto what was originally a rather static concept; among these is complexity, which is now seen as manifesting in many scientific contexts. These new systems ideas, developed in conjunction with new concepts of structure, growth, qualitative change, and chaos, have provided powerful tools of analysis and have guided practice in many fields. As the concepts have expanded in their application from the abstract fields of their origin to the study of phenomena in the biological and social worlds, the problems of their relation to external realities have needed to be addressed. The ascription of some degree of reality to any intellectual construct involves many factors including culturally conditioned metaphysics, occupational or social group practice, and personal commitment. Thus we generally consider that some "things" and "causes" are good reflections of reality, while others are understood to be more artifactual. For us, we do not need to invest any system with a strongly self-subsistent reality; it is enough that it is a powerful heuristic concept. Others may disagree; such arguments are never completely resolved, but in a dialectical fashion they will evolve or decay in subsequent history.

Emergent complexity can be distinguished from other states of organization that have been studied. The simplest state is that which can be described by the tools of classical mathematical physics; this has functioned as the standard for generations of natural and social scientists. More recently, "complication" has been discovered, characterized by the number of variables along with the nonlinearity of its processes; beyond that lies ordinary complexity, which involves structure and self-organization (implying some teleology). Whereas complication has no teleology (although there can be unidirection, as in Fourier's theory of the dissipation of heat and classical thermodynamics), ordinary complexity has a simple teleology. The boundaries between the classes are not distinct; thus the dissipative systems studied by Prigogine are at the lower end of complexity. We can contrast ordinary and emergent complex systems in terms of their patterns of stability and change. Keeping biological species in mind as examples, we can list some relevant properties of ordinarily complex systems. Much of their behavior can be explained in terms of mechanisms enriched with a functional teleology with simple systems goals such as reproduction and survival. The normal state for such systems is one of diversity of elements, coexisting in (what *we* see as) a complementarity of competition and cooperation. The ordinary complex systems tend to maintain a dynamic stability against perturbations until they are overwhelmed. (For some purposes it is useful to enlarge the boundaries of the system to include such occasional extreme events.) The new ideas of chaos and its "edges" enable simulations and analyses of processes of extraordinarily articulation, variability, and apparent design (Kauffman 1993).

Emergent complex systems, by contrast, cannot be fully explained mechanistically and functionally. In them, some at least of the elements of the system possess individuality, along with some degree of intentionality, consciousness, foresight, purpose, symbolic representations and morality. Attempts to reduce human society completely to ordinary complexity can result either in unrealistic theories [as those of B. F. Skinner (1971)] or catastrophic policies (such as those of Pol Pot's Khmer Rouge regime in Cambodia). Another difference between ordinary and emergent complexity relates to novelty. In ordinary complex systems, although numerical properties of subsystems (population size and density) can vary strongly, genuine novelty among the elements (a true *Origin of Species* as opposed to the formation of varieties) is very rare and still not easy to explain in mechanistic systems terms. On the other hand, continuous novelty may be considered as a characteristic property of emergent complexity. "Emergence" can manifest in different degrees; thus we refer to "traditional societies," where the genuinely emergent properties of the society (and of its members) were considered as not so fully developed as in our own. Among us, the pace of novelty constantly increases in all spheres of life, including the symbolic realm and consciousness. The phenomenon of postmodernity reflects this flux. For its adherents, there is no stable interpersonal reality out there at all; everything is relative to consciousness and fashion. [For a discussion of postmodernity in relation to science-based technology, see Funtowicz and Ravetz (1992)].

We have mentioned that emergent complex systems can sometimes be studied and successfully managed as if they were ordinarily complex. Indeed, because we are natural as well as social beings, the emergent aspects of our social and technical systems will always be, as it were, the tip of an iceberg of which the greater part is ordinarily complex. We should also expect a significant border zone where the two sorts of criteria are present in varying degrees. This zone will tend to become the focus of scientific and ideological debate, for it is crucial in the determination of our own identity. Thus the higher mammals, possessing some cognitive and social skills, tease our conception of humanity in various ways. Pets are friends, primates can reason, and cetaceans are said by some to reeducate us in feeling. Species apparently lower on the evolutionary scale, such as ants, can exhibit highly organized hegemonic behavior that never ceases to fascinate and perhaps also to frighten. A completely different sort of category of systems based on computers presents challenges along different lines. Artificial intelligence raises questions about our definition of "rationality," and now virtual reality throws open questions of being and existence, both personal and social. As the technology of cyberspace develops, we can imagine closer integration and symbiosis of humans and machines, and at that stage the border zone between ordinary and emergent complexity might well require reexamination (Gibson 1984). It could be argued that in some respects, there are no longer any cases of pure, ordinarily complex systems. Any natural system that is of interest to us has properties that affect our welfare. These properties will be salient for us; our sociotechnical system will interact with that other one, and so both will be part of a larger system. Thus our culturally laden descriptions of systems and relations like "competition" (to say nothing of "selfishness") structures our perceptions, concepts, and research activities alike.

The *Ancien Régime* Syndrome

In emergent complexity, technique is complementary to consciousness; by its means, one species can influence all related systems for its own benefit through domination and control. This factor may explain the phenomenon of large-scale and long-lived hegemonies within the human species. By hegemony we understand a systems state where the goals of one element or subsystem are totally dominant to the point where all others are either annihilated or survive only on the margins. This state alternates (structurally and temporally) with fragmentation, which is a conflict among plural attempted hegemonies. The mixture of these polar opposite forms of relationships will depend strongly on the context, but the dangers of collapse of hegemonic societies into fragmentation are greater than we had previously imagined. In its way, modern intensive agriculture and bacteriological medicine can be considered as forms of species hegemony. They need to be supported by ever more sophisticated technologies and are obviously and increasingly unstable in their relations with their ecosystem context.

In the hegemonic state, the internal contradictions of the system are not resolved but are suppressed. We can speak of an "ancien régime" syndrome, characterized by underperformance in key attributes because of prohibition of diversity and prevention of novelty. The regime refuses to deal with or to recognize its problems even when they become obvious to everyone else. In traditional societies, there was a tendency to a cycle, analyzed by the great medieval Arab scholar Ibn Khaldun (1967), where after some generations a dynasty becomes totally corrupt and indifferent to the elementary requirements of governing. Instead of an adaptation to challenge and change (as in preventing or ameliorating techno-natural disasters such as floods or droughts), the system fails to respond. Then when there is a threat to the regime, there are no reserves of loyalty to call on; a relatively minor external challenge can topple it. This may explain the unlikely sequence of events described in the traditional aphorism, "For want of a nail the shoe was lost, and so through horse, message, battle, war and empire." Clearly, a system with so little resilience against the uncertainties of battle and warfare was in the ancien régime syndrome. More mundane examples of the same phenomenon in the political realm are the Chernobyl accident in the Soviet Union, and the earthquake in Managua, Nicaragua. In both cases, public confidence in the authorities was fatally weakened by the mismanagement of the event and its aftermath.

A similar phenomenon has been observed in the case of natural systems. Thus many pine and spruce trees can appear in stands of very old Krumholtz (stunted) trees, with a high density and no understory, a sort of hegemonic "biotic desert," which in spite of underperformance can persist for a long time, resisting collapse, until an external force or a broader-scale phenomenon finally destroys it. Such ancien régime states might be imagined to occur in isolated ecosystems as on islands; perhaps Australia could be counted as one. The vulnerability of emergent complex systems can be even subtler. Thus the collapse of the Communist bloc (a contemporary paradigm case of ancien régime) precipitated radical changes in several other countries, which had no obvious links with it. But their political systems, some in an ancien régime state, had been buttressed ideologically by a response to a presumed Communist threat. As that became officially recognized as nonexistent, many different sorts of conservative or repressive policies lost their justification. Thus the release and rehabilitation of Nelson Mandela in South Africa could be seen as a remote but real consequence of the fall of the Berlin Wall.

Emergent complex systems can also exhibit a more extreme version of the collapse of hegemony. This occurs even before the external threat is presented; the system simply grinds down and approaches paralysis in many of its functions. This happened in post-Czarist Russia, when the Bolsheviks needed only to capture the post and telecommunications center in the capital. By contrast, the Irish insurgents who the previous year had captured the post office in Dublin were speedily defeated by the English and then hanged for their efforts. We can therefore speak of an "autolytic" property, distinguishing emergent complex systems from those of ordinary complexity. This and the ancien régime syndrome can be seen to apply not only to political systems but also to particular technologies and indeed in some ways to our industrial culture as a whole. For this, E. M. Forster's prophetic short story "The Machine Stopped" might be considered as an example; it exhibits an ancien régime passing into its autolytic state (Forster 1954). This state can be fruitfully contrasted to the "autopoesis," which has been identified as characterizing genuine complex systems (Maturana and Varela 1980).

Contradiction

We have mentioned contradiction, and, because it is a key concept for this analysis, we should provide some clarification. It expresses a very general heuristic, a way of looking at the world, which encompasses complexity, change, and conflict as natural and essential. In one way it has affinities with Asian philosophy, in which the Yang and the Yin are complementary aspects of all things and processes, but there are also connections with Western science. Thus Newton's third law of motion, that for every action there is an equal and opposite reaction, is an expression of the presence of contradictory forces inherent in many systems.

Contradiction, as part of dialectics, emphasizes the coexistence of antagonistic forces. It provides a perspective that prevents oversimplified analyses of situations and problems. Within this style, one cannot envisage a beneficial progress without looking for its costs, the growth of knowledge without its interaction with ignorance, or the achievement of good without some production of evil. Had we possessed that enriched understanding, we might have been spared the naiveté and subsequent disillusion of many of the social and environmental crusades of the last half century, such as for the global eradication of hunger, disease, and war. In current systems theory, there are explanatory schemes that express the dynamic, contradictory aspects of ecosystems very well. Thus Holling has a four-phase cycle of change that includes creative destruction as an essential part of the process of renewal (Holling 1992). When applied to emergent complex systems, such a concept is highly charged ethically. In one sense, creative destruction is always with us, as in the free elections that periodically unsettle the careers of politicians, but if used to justify political change brought about by violence, creative destruction can be labeled "extremist." However, if a society is already in an ancien régime or especially an autolytic state, its creative destruction is less subject to simple ethical judgment.

For our purposes, we can consider contradictions as being of several sorts. One is of complementarity, where the opposed elements are kept in dynamic balance. Another is of destructive conflict, where the struggle results in the collapse of the system in which they coexist. Finally, there is creative tension, in which the resolution is achieved by the qualitative transformation of the system; this is the well-known Hegelian sense. In natural, ordinarily complex systems the contradictions are found among competing subsystems or elements, typically members of species whose relations are more of competition than cooperation. In such systems, it is most common for the competing populations to oscillate between limits, none ever completely displacing any of the others. In those cases, the contradictions are of complementarity; if (as in the standard mathematical models) the predators should eat all the prey, they would then starve! Thus the active and the passive subsystems are bound together. In a sense, each needs the other in spite of the inequity (anthropomorphically viewed) of the situation. By contrast, in emergent complex ecosystems the contradictions can all too easily get out of hand and become simply destructive. Fisheries are a notable example, where the livelihood of fishers and their communities from week to week depends on an exploitation of the resource that regularly leads to damage or even destruction of the stock of fish (Clark 1990). Managing such contradictions involves a complex adjustment of local, regional, national, and transnational relations on the economic, political, and sociocultural planes, as well as attempting to control a natural resource whose behavior is very imperfectly understood.

In technological systems, a design exercise can be understood as including the management of contradictions (although the same can be said of architectural or urban design processes, for example). This is because incompatible design specifications are produced by the various competing interests. The different prospective purchasers of, for example, an aircraft will have their special requirements in price, operating cost, and performance characteristics of various sorts, and they will also contend with conflicting interests among the makers. A design synthesis can bring a creative solution to the problem at the price of leaving some interests unsatisfied. However, on occasion the design process fails: the competing demands cannot be reconciled, and designs and prototypes are simply scrapped. In our terms, the contradiction becomes a destructive conflict. The design process, as a management of contradictions, can also apply to an ongoing system. Thus the design criteria for a society can include the optimization of internal stability over other attributes such as flexibility of response to challenges. This is said to have happened in Tokugawa, Japan. Interestingly, such a phenomenon can occasionally be seen among ordinary complex systems, as a feature of stable climax cultures of ecosystems. Their fragility against external assault can be analyzed in terms of depleted stocks of available energy. Alternatively, in design terms it can be interpreted as an optimization of the management of internal contradictions at the expense of defense against external threats

This last example reminds us that systems, as we imagine them to encompass real natural and social phenomena, are strongly articulated internally and have multiple relations externally. Thus an individual copy of a commercial aircraft is an element of a system of production and use of a particular aircraft design. This system overlaps with other systems realized in various organizations, and it is also part of a broader system of air transport and is further involved with systems of employment, training, resource use, defense, etc. In many respects, such systems are ordinarily complex, particularly in cases where the actions of individuals are mainly significant in their aggregate. Indeed, "the market" (as idealized in conventional economic theory) could be considered as an example of a most effective use of an ordinarily complex system to organize transactions among those emergent complex systems of desires and needs that are called individual people. However, because any successful market requires external regulations, as for the prevention of dishonest or criminal activities of various sorts, and in many cases also requires the establishment of trust between key actors, the impression of ordinary complexity is only partly true and perhaps conceals more than it reveals. Also, the rapid replacement of central planning by market economies has in some contexts led, not to the hoped for diversity but only to increased fragmentation—a subtle but important distinction.

Post-Normal Science

The introduction of the notion of emergent complexity enables the development of new conceptions of scientific practice, involving its epistemology, methodology, and power relationships. Traditional science assumed nature to be simple and capable of reductionist mathematical explanations, which are based on observations by a detached observer. This epistemological naiveté was matched by a lack of awareness of its societal power relations and by an arrogant dominance over all other ways of knowing. This complacency could not be sustained as science developed and changed through the twentieth century. First, physics and the social sciences seemed to justify relativism and later the growing self-awareness of science as a social activity further eroded the experiential basis of naive realism. Kuhn's notion of "incommensurability" raised the specter of fragmentation in science (Kuhn 1962), and, subsequently, the postmodern movement extended fragmentation to all of knowledge (Feyerabend 1975).

The social and intellectual contexts of scientific work have been transformed by the new problems of risks, the environment, and public suspicion of the works of science. There have been many attempts to achieve more sophisticated versions of reductionist science, employing a variety of mathematical techniques, ranging from games theory to Bayesian statistics and catastrophe theory, also including systems analysis at one stage. The recent growth in the appreciation of complex systems indicates a change in attitude and direction. Appreciation of diversity, which is not at all the same as relativism, can lead to a new practice of science in emergent complex systems. This is what we call post-normal science.

Emergent complexity provides a theoretical justification for post-normal science, in which the peer group for quality assurance is extended beyond the certified experts to include all those with a stake in the issue. This concept helps us to appreciate that there is no single perception providing a comprehensive or adequate vision of the whole issue nor any particular criteria of quality that can hegemonically exclude all the others. Casti has expressed the point of a plurality of legitimate alternative perspectives by equating degree of complexity with the number of nonequivalent descriptions of a system (Casti 1986). Atlan (1991 in Funtowicz and Ravetz 1994) has made a similar point: "plus un phenomène est complexe et singulier, plus toute théorie suceptible d'en rendre compte est sous-determinée, donc incertaine" (p. 578). In relation to policy dialogues, this translates to the principle that no particular partial view can capture the whole. It is therefore necessary and legitimate for the dialogue on such issues to include persons representing all the different interests, which may also include concerns for children, nonhuman species, and ethical values. This point reflects the growing practice in the resolution of global environmental issues, where earlier attempts to privilege one perspective corresponding to a hegemonism of Western culture, have proved inadequate at all levels including the practical, economic, political, and ethical (Davis 1993).

The conceptual apparatus of emergent complexity also provides a way of understanding uncertainty including the sort that has hitherto not been amenable to a structural analysis. Many attempts have been made to quantify or formalize uncertainty for the purposes of decision making, but as the uncertainties become more remote from classical probabilities, the methodological difficulties in such programs become more severe (Funtowicz and Ravetz 1990). We can appreciate irreducible uncertainty as a systemic property of emergent complexity. Because no perspective encompasses the whole, then all perspectives are partial, and each one is affected by ignorance of some aspects of the situation. This is partly overcome by dialogue and learning to appreciate each others' point of view. Each participant, however, must first learn that the others cannot simply think within precisely the same framework as them and then learn to respect those irreducible differences. The ideal of a total agreement based on a perfectly shared perspective, which has characterized the technocrat vision of policy, is now seen to be utopian.

We can understand the challenge that is met by post-normal science in systems terms, by invoking the property of complex systems of flourishing "at the edge of chaos" (Lewin 1992). This has a strict mathematical meaning for simulated complex systems; in the case of emergent complexity, we can translate the tendency to chaos as fragmentation and the complementary tendency to organization as diversity. Post-normal science provides concepts for managing debates on policy issues lying "at the edge of chaos." The contradictions that appear as differences of perception and value, normally involving debate and even tendencies to conflict, can be contained and can also be made the occasion for mutual learning and respect. Such a respect can survive even in deadly conflict, as is familiar from such diverse contexts as medieval chivalry and the Japanese samurai. A great task of environmental politics is to foster such respect in the ordinary business of decision making.

For an example of this plurality of perspectives, we may imagine a group of people gazing at a hillside . The perceptions they might have include the following: just a hillside, a pleasant expanse of green, a case study in geomorphology, an example of ecological succession, an archaeologically interesting site, an area of recreational potential, a prospective housing site, a center for earth energies, and a launching point for departing souls. Each uses their training to evaluate what they see in relation to their tasks. Their perceptions are conditioned by a variety of structures, cognitive and institutional, with both explicit and tacit elements. In a policy process, their separate visions may well come into conflict, and some stakeholders may even deny the legitimacy of the commitments and the validity of the perceptions of others. The challenge is to see those partial systems from a broader perspective and to find or create some overlap among them all, so that there can be agreement or at least acquiescence in a policy. For those who have this integrating task, it is important to understand that this diversity and the possible or even likely conflict that emerges from it is not an unfortunate accident that could be eliminated by better natural or social science. Rather, it is inherent to the character of the complex system that is realized in that particular hillside.

These two key properties of complex systems, radical uncertainty and plurality of legitimate perspectives, help to define the program of managing for sustainability. They show why policy cannot be shaped around the idealized linear path of the gathering and then the application of scientific knowledge. Rather, the formation of policy is itself embedded as a subsystem in the total complex system, of which the problem as defined is just another element. Let us return for a moment to the example of the people looking at a hillside. As we stated, the various perceptions of the observers overlap in some respects but might also involve differences, or even conflicts, over values and realities. Some perspectives may claim to be "true" or at least valid in some privileged way over all the others. Some of them are projections, but *other* relevant perspectives are admitted only if they can be interpreted in their own terms. The classic scientific perspective, involving reductionist quantification, attempted to legitimize itself in a logically closed way. It claimed to be both "rational" and "neutral," thereby claiming a privileged status while denying that it was doing so. The recent reaction against scientific hegemonism has produced some of the contemporary tendencies toward skepticism and fragmentation, including postmodernism.

Post-normal science enables us to avert the nihilistic implications of postmodernism by observing that there really is a hillside there, even though no one (including ourselves) can see it as a whole, in all its possible meanings and interpretations. The relations among the different perspectives, or projections, can vary widely. The participants may appreciate the complementarity among them or they may have debate or even conflict over the issues that are reflected in the various perspectives and commitments of stakeholders. These alternatives may be seen as the polar opposites on a continuum: with complementarity we have diversity, and with attempted hegemony we have fragmentation. Complementarity requires awareness by each stakeholder that their own perception is partial and (in terms of the phase-space metaphor) a projection of the whole configuration into their particular partial subspace. In the case of attempted hegemony, such an awareness of legitimacy of the other's perception (and with it, their values and rights) is either discounted or (in the extreme case) denied altogether. This is fragmentation, after which life becomes, as in Hobbes' description, nasty, brutish, solitary, and short.

Emergent Complexity, Sustainability, and Ethics

In systems with emergent complexity, symbolic representations and ethical judgments interact through contradictions to produce some of the greatest achievements. Thus ethical contradictions became the basis of classical Greek tragedy as an art form, and conceptual contradictions led to philosophical inquiry and creative mathematics. Thus human civilization, unlike an ordinary complex system, is constantly transforming itself, a process involving loss and forgetfulness as well as conservation and growth.

To what extent are such considerations relevant when we think about the general instabilities that threaten our planetary existence? First, it would be wrong to think of emergent complex systems with all their refined pathos as restricted to the cultural realm while the real business of life and survival can proceed at the level of ordinary complexity. We are realizing that the complexity of the ecosystem as a whole is not ordinary but emergent. Our actions, resulting from our lifestyles and visions of "the good life" have created new natures that have transformed all existing ecosystems. Many of these (including agriculture and landscapes) have the paradoxical property of being "unnatural" in that they did not and could not exist in our absence. There are no pristine habitats, just as there are no truly aboriginal peoples.

We can distinguish between those aspects of complex systems that are emergent and those that are not. Thus a cornfield is devoted to a crop that is the product of human invention, where even individual plants need technology in order to grow and survive, but where natural processes, devoid of conscious intention, are the driving forces. On the other hand, in the system of private passenger transport, the dimensions of individual choice, socially constrained aggregated actions, systems of law and regulation, physical technologies, and resource and pollution problems are all closely bound together. In this latter case, it is not so easy to sort out relatively independent subspaces lying in their own discrete dimensions.

The idea of contradiction becomes very powerful when we consider the total emergent complex system including humanity and the biosphere. The characteristic contradiction of this system is the incompatibility between the individual drive for material comfort, convenience and safety, and the ecological consequences of this being achieved even for a significant minority of humanity. This is a truly emergent system for such drives in individuals are strongly conditioned by conceptions of "the good life" that are derived from a very special cultural milieu. In Western industrialized nations our own pollution is transforming the environment quickly enough, even without the help of those nations that are still "developing." Hence we cannot allow them to join in our self-destructive affluence. This situation produces the leading ideological contradiction of our special civilization because it justifies itself on the humanitarian ideal of equality for all humankind. And this is quite impossible; just consider the ecological consequences of, say, four billion private automobiles and as many domestic air conditioners.

There is a tendency for both of these contradictions to be considered as mere "dilemmas" and to be masked by the title "sustainable development." As yet, only a few independent thinkers can imagine a "development" that means anything other than the achievement of a global affluent consumer society, which is universally known to be unsustainable on ecological grounds. To encourage the world's poor to consider developing along less destructive lines than ourselves combines two further contradictions: first, the physical impossibility of an environmentally benign consumerist society, and, second, the sin of the rich preaching the virtues of poverty to the poor. The resolution of these contradictions will not be accomplished at the level of ordinary complexity alone. To do that, we would need to imagine "sustainability," which entails the characteristically human qualities, being reduced to survival in the lower subspace of ordinarily complex systems. This latter concept of survival applies to fox-and-rabbit computer models or artificial-life strings, operating through dynamic stability but with no real novelty. In ordinary complexity, ethics is also mapped down to the single goal of group survival, and in the process it loses its meaning.

The assumption that personal survival is the *only* thing that counts simply does not hold true, even in extreme situations like the concentration camps. The characteristically human qualities reassert themselves even there, so that death is not the worst fate for communities or persons. The limited public response to the earlier "doom mongers" of the 1960s may have been due to a revulsion from their reduced vision. They gave the impression of not feeling the difference between the human race and a Petri dish culture. Their early proposals for the application of "triage" to nations, abandoning some because of their feckless fecundity, came very close to blaming and punishing the victims for a situation in which "our" share of the blame was fully commensurate with "theirs." In this way the logic of a mechanistic scientific worldview has taken it beyond ordinary morality, as had already occurred in connection with eugenics. This logic eventually becomes no less counterintuitive than the vision of death and even disaster as symbolic exchanges with a meaningful nature (O'Connor 1993).

Sustainability is therefore not a matter of mere survival; indeed, a global strategy that focuses only on survival would be very likely to encounter crippling social and ideological contradictions, and ultimately, to fail. Emergent complexity requires its special solidarity to maintain its own sort of dynamic stability. In general, in the absence of solidarity, a human society would degenerate not into ordinary complexity but into a horror. For this, history and fiction (as *Lord of the Flies*) provide us with no shortage of examples (Golding 1954). The concept of "coevolution" as developed by Norgaard (1993) offers the possibility of a synthesis with emergent complexity. By offering an enriched conception of sustainability, coevolution provides a full articulation of themes that we have sketched in connection with post-normal science. Authors in the following chapters of this part explore somewhat further what that solidarity might mean in terms of ecosystem approach sustainability.

What we have called "emergent complexity" is a heuristic device for asserting and explicating, in the technical context of systems theory, what is human about humanity. In these terms, survival is a mere shadow of sustainability. There is such a thing as a life worth living and also (as Socrates explained a long time ago) a life not worth living. The pursuit of a particular one-sided ideal of a life worth living as if we were elements of an ordinarily complex system has brought us to our present threatened state. The comforting simplicities of theoretical systems that assume ordinary complexity are achieved at a price: a denial of human reality with all its contradictions, the destructive and the creative alike.

Conclusion

The exploratory analysis developed here is complementary to those conducted with a more formal, mathematical, or computational approach. In those, the properties of what we call ordinary complexity are being developed, and results of great importance and power are being derived. Our concern is to articulate what lies on the other side of that somewhat indistinct divide, the conceptual space we call emergent complexity. One possible use of this present discussion could be to inhibit any further sterile debates about whether machines, or computers, can be fully "human" in some essential aspect. There is enough exciting and creative work to be done on ordinary complex systems without needing to claim more for them than it is justifiable or useful.

Our primary purpose is to begin the work of applying concepts taken from other areas of philosophy that will go into the construction of a systems theory that provides explanatory power for the specifically human aspects of human societies and human creations. In this way, there could be a very fruitful interaction and synthesis between the enriched conceptions of science now being forged in studies of complexity and a philosophical enquiry in which the perennial problems are recast in the light of the new realities that humanity is now creating.

Notes

[1] The views expressed are those of the authors and do not represent necessarily those of the European Commission.

[2] This chapter is based on our earlier paper: Funtowicz, S. and J. Ravetz. "Emergent complex systems." *Futures* 26(6):568–582 (1994).

References

Atlan, H. 1991. *Tout non peut-être*. Paris, France: Éditions du Seuil.

Costanza, R., L. Wainger, C. Folke and K. G. Mäler. 1993. Modeling complex ecological economic systems. *BioScience* 43:545–555.

Casti, J. L. 1986. On system complexity: Identification, measurement, and management. In *Complexity, Language and Life: Mathematical Approaches*, eds. J. L. Casti and A. Karlquist, 146–173. New York: Springer-Verlag.

Clark, C. W. 1990. Mathematical Bioeconomics: The Optimal Management of Renewable Resources. New York: Wiley.

Davis, S. 1993. The World Bank and indigenous peoples. Memorandum. Washington, D.C.: The World Bank Environment Department.

Feyerabend, P. 1975. *Against Method*. London, UK: New Left Books.

Forster, E. M. 1954. The machine stopped. In *Collected Short Stories*. London, UK: Penguin.

Funtowicz, S. O. and J. R. Ravetz. 1990. *Uncertainty and Quality in Science for Policy* Dordrecht, Netherlands: Kluwer

Funtowicz, S. O. and J. R. Ravetz. 1991. Three Types of Risk Assessment and the Emergence of Post-normal Science. In *Social Theories of Risk*, eds. D. Golding and S. Krimsky, 251–273. New York: Greenwood.

Funtowicz, S. O. and J. R. Ravetz. 1992. The good, the true and the post-modern. *Futures* 24:963–976

Funtowicz, S. O. and J. R. Ravetz. 1994. Emergent complex systems. *Futures* 26(6):568–582.

Gibson, W. 1984. *Neuromancer*. New York: Berkley Publishing Group.

Golding, G. 1954. *Lord of the Flies*. London, UK: Faber and Faber.

Holling, C. S. 1992. Cross-scale morphology, geometry, and dynamics of ecosystems. *Ecological Monographs* 62(4):447–502.

Kauffman, S. A. 1993 The Origins of Order: Self-Organization and Selection in Evolution Oxford, UK: Oxford University Press.

Khaldun, I. 1967. *The Muqaddimah (An Introduction to History)*. Princeton, N. J.: Princeton University Press.

Kuhn, T. 1962. *The Structure of Scientific Revolutions*. Chicago, Ill.: University of Chicago Press.

Lewin, R. 1992. *Life at the Edge of Chaos*. New York: Macmillan.

Maturana H. R. and F. Varela. 1980. *Autopoiesis and Cognition: The Realization of the Living* Dordrecht, Netherlands: D. Reidel.

Norgaard, R. 1993. *Development Betrayed*. London, UK: Routledge.

O'Connor, M. 1993. On steady state: A valediction. In *Entropy and Bioeconomics*, eds. J. C. Dragan et al., 414–457. Milan, Italy: Nagard.

Schumpeter, J. A. 1975. *Capitalism, Socialism and Democracy*. New York: Harper.

Skinner, B. F. 1971. *Beyond Freedom and Dignity*. New York: Knopf.

18

Third World Inequity, Critical Political Economy, and the Ecosystem Approach

Ernesto F. Ráez-Luna

Introduction

Funtowicz and Ravetz (chapter 17) have argued that "emergent complexity requires something like solidarity to maintain its own sort of dynamic stability." I wish to pursue this discussion further to determine what solidarity might mean in the realm of everyday inequities and political conflicts within which most people in the world struggle to make a living and what the implications are for the ecosystem approach proposed in this book.

If we agree that one key condition of sustainable development is intergenerational equity, it logically follows that intragenerational equity is a key accompanying condition. We cannot deny our present fellows the treatment that we grant to our future fellows. Still, this is the greatest challenge to sustainable development. In our times we witness a steady increase of overt, extreme, and seemingly rigid differences in political power and economic wealth among people, particularly in the third world. This perhaps constitutes the most important theoretical and practical challenge facing the ecosystem approach. It is one thing to develop rich pictures and discuss alternative management scenarios among unevenly powered citizens; it is a whole different thing to break barriers between subjects and masters or between ethnic enemies. When in Canada or Denmark, we expect government officers to act as public servants; in Peru or Bangladesh they often act as authoritarian administrators. When in Canada or Denmark, it is a given that anybody (in principle) can sit at a bargaining table; in Peru or Bangladesh some people may be shocked by the very idea of sitting with inferiors and listening to them. When in Canada or Denmark, we count on extensive middle classes, interconnected human settlements, and relatively open opportunities for volunteer organizations to structure and articulate the voices of civil society, while the third world is dominated by people living in absolute poverty, living in inaccessible places, and facing daily struggles for subsistence, leaving them little room for volunteering or strategic thinking. Centuries of domination and resistance have shaped every minute interaction among unevenly powered social actors in the third world and have often institutionalized miscooperation. The problem surpasses the approach's principle of respect to different perspectives because the problem is not just one of navigating among conflicting interests, discourses, and narratives but navigating among overt, dominant narratives while painfully uncovering covert, marginalized ones. In these circumstances, the researcher's rich narratives and scenarios will face enormous difficulties to avoid reflecting a political choice (particularly in the eyes of the stakeholders); while any attempt to integrate conflicting narratives or to obtain a balanced account may threaten the theoretical coherence and practical value of the product.

A related difficulty arises from the politicization and historicity of reason. Reasoning in the third world is often heavily layered, relational, and diachronic, as opposed to the task-oriented, focused, and atemporal ("universal") rationality of Western modernity. "Here and now" have very different meanings in the first and the third world. Still, the difference is not that of reductionist versus holistic thinking but rather that of apolitical versus political thinking. The key challenge resides in the historicity of power-laden divergences in the third world, which turns the here and now into a virtual reality of sorts, as people may be responding not only to the issues at hand but also to a nuanced and tangled web of historical, structural, unequal exchanges. The foreign researcher—still fighting with language and cultural barriers—may completely miss those tangles and nuances (or misinterpret them as dumbness); while the national researcher, acting within the web, will often be blind to them.[1]

In this chapter I will summarily (1) discuss some complexities of third world inequity; (2) review the concepts related to SOHO systems, as applied in the ecosystem approach, to account for an unequal and constructed world; (3) discuss the differences between ecosystem and sociosystem narratives; (4) suggest a draft code of conduct for the ecosystem approach practitioner in the third world; and (5) suggest a few paths for strengthening the ecosystem approach at the politico-economic dimension. This chapter has been strongly influenced by my readings of two Canadian scholars, James J. Kay and Robert W. Cox, and by twenty-something years of passionate journeying among the peoples and landscapes of tropical Latin America.

Third World Inequity

Inequity in the third world manifests in three politico-economic forms: inequity of access to the means of production, inequity of access to exchange, and inequity of political access. These inequities unfold in the material/instrumental as well as in the ideological/symbolic dimension: Access to production involves not just land, capital, and labor but the means of production and reproduction of one's group's culture and identity; exchange refers not just to gadgets and coins but to ideas; political access refers not just to universal and free elections of government officers and policy makers but to the enjoyment of human rights and the active experience of citizenship. (Evidently, the multiple inequities and limitations of human society are not exclusive to the third world. It is my contention that I am talking about global issues, even if, for the purpose of this chapter, my standing point is located in the south.)

The ominous rope of third world inequity is braided out of a number of historical strands: gender and age discrimination supported by native and Western patriarchal traditions, pre-Colonial traditions of unequal exchange and geopolitical tensions between different native cultures, Colonial institutions of racial discrimination and segregation between the descendants of the conquered and the descendants of the conquerors (geopolitically expressed in the centralist vocation of third world nation states), and the global-local work of modernism and globalization determining hierarchical dichotomies between Western and non-Western, universal and ethnic, modern and pre-modern, industrial and pre-industrial, capitalist and pre-capitalist ("free" and "distorted"), urban and rural, developed and underdeveloped. All these strands twist around each other, interpenetrate, reinforce, and reproduce each other. Out of their interaction emerge authoritarian regimes that oscillate between populism, paternalism, and autocracy and between demagogic persuasion and violent coercion. Regimes supported by heterogeneous masses of subject citizens who are (often in the same individual) nonwhite, nonaristocratic, nonmale, nonurban, nonauthorities, noncorporate, and too young. It is this complex and strongly historical inequity that sustainable development and the ecosystem approach face in the third world.

Sociosystems as Self-Organizing Holons

"Third world" has no meaning in itself. The term refers to a relational identity. There is a third world because there is a first world, and vice versa. Therefore, in order to discuss the third world I have to discuss world order. Here I find a first difficulty. In spite of its embrace of social issues and values, the ecosystem approach still contains a quite transparent biophysical bias. Human organizations are envisioned and pictured within the biosphere, depending on its functioning. Still, nature contains as much as it is contained within humanity, and its functioning depends of the fate of human affairs. There is a material and ideational interpenetration of humanity and nature that still tends to escape the theory of the ecosystem approach.[2]

In practice, the biophysical biases of the ecosystem approach have led to a tendency to refer to sociosystems as if they could be mapped in basically the same way than ecosystems or as components of ecosystems. They cannot. This is not to deny that ecosystems include humans, but humans cannot be understood just as part of ecosystems. In sociosystems, for every material flow, there are a number of ideational flows. And every material flow in a sociosystem is channelled through concrete human groups, distributed in concrete human ways, justified in exclusively human terms. Human and nonhuman nature interpenetrate each other, but they are not the same. Other than the obvious fact that sociosystems refer to just

TABLE **18.1**   Two Different Kinds of SOHO Systems

Feature	Largely non-human	Largely human
Self-organization	Power non-existent.	Power as key organizing driving force.
Holarchy	Nested holons correspond to land and time. Trophic webs and pyramids.	Nested holons correspond to territories (land under material or ideal control) and history (which is narrated time). Power webs and pyramids.
Complexity, synergies, emergence, non-determination, unpredictability, rapid change.	Response to sudden change as adaptation or extinction. Figure eight roughly corresponds to "r" and "K" strategies.	Surprise. Catastrophes. Response as magic and technology. Figure eight corresponds to hegemonic and non-hegemonic historical structures. Cycles can be locked.
Attractors	No attractor is better than other. Their nature can be abstracted from their value.	Attractors are never neutral. Their nature cannot be abstracted from their value.
Resilience (negative feedback loops)	In general, desirable.	Often undesirable. Locked systems preclude creative change, enhancement of human spirit. Some resilient states may be based upon generalized human suffering (perverse resilience).

one biological species, while ecosystems are ensembles of many species, or the also obvious fact that sociosystems are formed by the only known self-reflecting species, there is one dimension of sociosystems that is completely absent in non-human nature. This dimension is power. How does power affect our reading of the world order as a SOHO system? Table 18.1 attempts some draft answers: Most of the table should be self-explanatory, but the heavy-bordered cell introduces some terms and ideas that deserve clarification.

Hegemony

This is a concept coined by Antonio Gramsci. Here I will use it as Robert Cox has reinterpreted it.[3] Funtowicz and Ravetz (chapter 17) have discussed this is terms of emergent complexity. I wish to discuss this in terms of the mess of history where we live. In short, in any given modern society, there may eventually appear a ruling or leading group or class. This group rules over the rest of society by means of persuasion, concessions, and coercion (supported by a privileged access to the means of violence). If persuasion prevails over coercion, the ruling group has achieved hegemony, a more or less peaceful period of certainty, stability, and accumulation ensues, the ideology of the ruling group dominates social conscience, and we are in front of a hegemonic historical structure.[4] Of course, there will always be counterhegemonic groups, competing for power against the rulers, and supported by their own counterhegemonic ideologies. Still, they will be significantly weak and disorganized relative to the rulers. This would be the case of the first world and most Western democracies.

Still, hegemony may not happen, and the society will then become the constant battlefield of different groups fighting for power. A nonhegemonic historical structure ensues, where coercion prevails over persuasion, and each group maintains their own strong self-identity. This is the scenario of caudillos, dictators, and uncertainty. This would be the case of the third world.

However, the first and the third world are part of the same world order. They are linked together through institutions of unequal exchange that simultaneously reinforce first world hegemony and third world nonhegemony. Unequal exchange locks the first world in a strong cycle of material accumulation and concentration (that we have learned to perceive as development), while the third world is also locked in opportunistic, unstable, and short-term cycles of exploitive accumulation and sudden reorganization (that we have learned to perceive as underdevelopment). The hegemonic/nonhegemonic world order can be nicely described by a simple modification of Holling's lazy-eight (figs. 18.1A and 18.1B). Within the nonhegemonic (third world) locked cycle we can identify inner (regional) levels of unequal exchange, relative accumulation/concentration and relative underdevelopment.[5]

Narratives

The differences between ecosystems and sociosystems also determine deep differences in the nature and function of the narratives that can be developed from those systems. The ecosystem approach recognizes the multiplicity of ecological, politico-economic, and cultural perspectives around each ecosystem. However, examples often refer to multiple ecological perspectives such as the description of an ecosystem from multiple scales or in terms of multiple attractors. However rich these pictures may be, we are still within the realm of *one* social perspective: that of the biophysical researcher (which is, of course, a strength as much as it is a weakness of the approach).

On the other hand, as soon as we intend to describe the multiple scales and attractors of a sociosystem, we find that they are largely determined by power and by the groups that struggle to increase and concentrate their power (including the epistemic communities). Thus, for instance, as discussed above, a center-periphery dynamics largely determines the spatial politico-economic (and, in the end, ecological) cross-scale structure of the biosphere. Different politico-economic attractors will often correspond to different ruling groups in society. Therefore

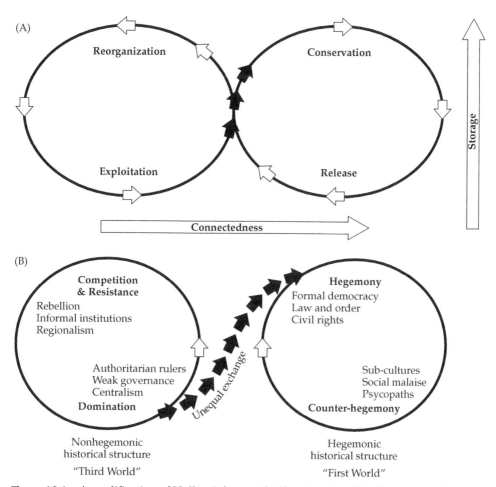

Figure 18.1   A modification of Holling's lazy-eight (four-box) model of ecosystem change (20.1A) to describe hegemonic/non-hegemonic world order (20.1B)

different scenarios (even different "strictly ecological" scenarios) will be necessarily linked to shifts of power. If power structures are too rigid (allegedly one of the problems in the third world), scenarios will be strongly conflictive. Because the theory and practice of the ecosystem approach prefer institutional change to end-of-the-pipe solutions and given that institutional change cannot happen without shifts of power, exactly those scenarios that are of more interest in the ecosystem approach might become conflictive. Under these circumstances, the ecosystem approach practitioner may be forced between two hard choices: to become a more proactive political actor (to openly walk the fine line between sound science and political action) or to surrender ground to business as usual.

A second problem stems from the conditions under which human life takes place in the third world (and among the first world's dispossessed). The urgencies of material survival often preclude the development of long-term (institutional) solutions. End-of-the-pipe solutions and a problem-fixing attitude tend to dominate the scenarios and the imagination of poor peoples and third world decision makers alike.[6] Under these circumstances, claiming a detachment to problem fixing would not only be political suicide but also immoral.

Therefore the ecosystem approach practitioner is challenged to imagine scenarios that coherently articulate short-term solutions and long-term changes. As a matter of fact, the only (political and ethical) way of legitimating the discussion of long-term scenarios would be through the approach's early success at fixing severe self-perceived problems. I suggest that the power of the ecosystem approach to develop rapid cross-scale intersectoral diagnostics and early warnings should be strengthened and given a more visible presence in our methodology. Also, it would not hurt to pay more attention to the work of participatory rural development practitioners such as World Neighbours. When we insist in our epistemological non-problem-fixing attitude, we must constantly make sure that we do not misinterpret it as detachment (indeed, this would be a distortion of the social role of the ecosystem approach researcher), we must avoid throwing the baby out with the bathwater.

In summary, every narrative or rich picture should evolve from a synchronic moment to a diachronic moment. In the first moment, a given reality is taken under its own terms. Here the advantages of the ecosystem approach at crossing scale, sectoral, and disciplinary barriers provide a powerful tool to identify "deep" quick fixes. In the second moment, the diachronic moment, novel elements and alternative long-term scenarios are brought to the table, but their discussion is already legitimized by the former phase. Along the way, we must steer toward systemic understanding and solutions and away from reductionist frameworks.[7]

Role of the Researcher

The above paragraphs bring us to a crucial question: What should be, in political terms, the role of the ecosystem approach researcher in the third world? At least until the dream of a network of local schools of the ecosystem approach becomes true, the ecosystem approach researcher suffers the dominating image of every scientific researcher in the third world: He/she will be perceived as Westerner, first-worlder, modern, patriarchal, rich, knowledgeable, and powerful. Who would ever want to look otherwise? Of course, there is a lot of truth in that image. Still, it is our moral and epistemic duty to use it to the political advantage of our approach (rather than to personal advantage). In any case, it becomes a crucial issue of self-reflection and scientific rigor to what degree the narratives and scenarios that we develop are biased by power. We must explicitly identify our political perspective (not to be confused with a political position).

In practice, this would help to solve one of the most common problems in the application of the ecosystem approach: that of identifying a point of attack among the multitude of stakeholders/issues. For example, for the last few years I have been applying a political ecology perspective to my research. Political ecology looks into ecological conflicts as conflicts over access to nature among unevenly powered social actors. These agents relationally reproduce (but also transform) inequity, responding both to their structural (historic, institutional) context, which determines the sources of uneven power or inequity, and to the political constraints and opportunities brought about by ecological transformation itself. The perspective does not affect the principles of the ecosystem approach; it simply provides a (politico-economic) fulcrum for their application.

The role of the politically aware researcher will also be to record the different narratives, discourses, and perspectives of stakeholders in terms of their political functionality (are they strategy or livelihood driven? are they overt or repressed or occluded?) and bring them to the open. Beyond learning the local language, talking to everybody and eating with everybody, budgeting and spending sensibly, paying earnest attention to practical self-perceived needs and seeking to empower the powerless without alienating the powerful, a challenging, vigilant (self-critical) attitude is required from the ecosystem approach researcher. Her role of facilitation should be shaped by a political understanding that promotes transparency and brings to the table invisible (often uncomfortable) dimensions, issues, and stakeholders that have been marginalized by power.

Strengthening the Ecosystem Approach

In conclusion, here I advocate for strengthening a set of theoretico-practical features in the ecosystem approach.[8] The key task would be to empower the approach to take a closer and deeper look into human society. This involves the following:

1. A more explicit consideration of history as narrated time (but rejecting neo-Darwinian conceptions of evolutionary development).
2. A greater attention to inequity and repressed/occluded/excluded narratives.
3. An explicit model of world order (such as the hegemony model above suggested) that should include (1) a more elaborate theoretical understanding of the interpenetration of nature and humans and of ideas and material reality [in the sense that nature is within (and depends on) society as much as society is within (and depends on) nature]; and (2) a deeper exploration of the theory of hierarchies: complex systems are not only holons but also webs/pyramids; and the inclusion of power as a key organizing force.
4. Expand the understanding of "different perspectives/multiple narratives" to different unevenly powered rationalities: not just "what I see/want," but chiefly "what makes sense" differs.
5. Strive to improve in the approach's theory and practice the articulation of reductionist and holistic research; problem solving and structural understanding; and short-term, linear, deterministic and long-term, nonlinear, uncertain institutional transformation.

I would like to finish with a short elaboration of the last item in point 5. Perhaps one useful concept in trying to articulate short-term, predictable change and long-term, uncertain transformation is pre-adaptation. Besides being adaptive, monitoring, management, and governance institutions should be pre-adaptive. They should be designed in such a way that they end up containing in themselves the seeds of long-term transformation. This is a complicated way of saying that a truly participatory ecosystem management scheme prepares the ground for a truly participatory society. The role of the researcher is to keep track of those seeds and to push their odds of germination, maturation, and propagation. We must be careful gardeners, even if we do not know exactly what kind of flowers will eventually bloom.

We also need to consider uncertainty more closely. What do we exactly understand by "uncertainty"? Certainly not likelihood. For many years now, I have been angered by the misuse of the term "uncertainty" for perfectly predictable ecological-economic phenomena such as extractive boom-and-boost dynamics (be the unfortunate exploited resources anchovies, mahogany, or what have you). These phenomena follow exactly the same phases. Even their graphs look the same. In a similar way, long-term (deep) institutional transformation and specific paths for such transformation are uncertain but predictable. They may become increasingly or decreasingly likely: we can measure this likelihood, and we can predictably act to increase it, even if the specific features, the date and time of the final outcome, remain uncertain.[9] For instance, by thickening support networks and by multiplying nodes and links, long-term transformation will become more likely (more socially necessary), if not more certain.[10]

I would offer a real-life example: A Peruvian nongovernment organization (NGO) working in the Amazon (which is inhabited by a myriad of ethnic groups) for several years now has developed an excellent program to train rural school teachers in intercultural environmental education. The problem is that rural teachers are often and suddenly relocated. Once relocated, keeping in touch with them and monitoring their maturation and their use of the program's concepts and materials become impractical: neither the NGO nor the teachers have enough resources. It is not difficult to predict that under these circumstances, in spite of the value of the initiative, no long-term transformation will occur in rural education. However, if networking and location stability could be significantly increased (which would take a relatively mild, predictable institutional change), then the likelihood of a long-term transformation of rural education would also increase. No need to say that greater teacher networking and location stability serves also a number of immediate needs and improves the overall efficiency of the rural education system.

Another way of dealing with the problem of articulating short- and long-term institutional transformation is by shifting the frames of reference from uncritical to critical. For instance, a bitter conflict has ensued in Peru around the legal regime of access to the Amazon timberlands. The de facto frame of reference for all competing stakeholders is that timber is a key resource for the development of the backward Amazon region of Peru. Under this frame, every institutional transformation being designed responds to better management criteria (better organization of production, better training of government officials, better policing, better rules of access, etc.). An alternative frame is that Amazon development is not a function of any economic activity in itself but of the social destiny given to the products of economic activity. By shifting the burden of proof from timber to the national society, the necessity of long-term institutional transformation enters the picture, not displacing but reshaping the criteria for better management-oriented institutions.

There are not clear-cut answers to the problem of applying the ecosystem approach in the third world. However, there are clear-cut requirements to the ecosystem approach. This is a fascinating challenge, indeed.

Notes

[1] Are these real differences? I suspect that in certain sense they are not. I suggest that both first and third world people reason in the same way, that is, locally, historically, and politically (or, in general, relationally). Universal and apolitical reason would be a relatively recent Western invention, serving the purpose of guiding and legitimating world domination. (And what could be more parochial and more egocentric than such an idea?) Many of us have learned to think within the Western tradition (in particular, the North American version), where theory is supposed to be universal ex ovum and therefore atemporal (ahistorical), arelational (apolitical), and legitimately applicable to all humanity. This thinking supports Catholicism as much as it supports biotechnology. The very fact that the benchmarks used to introduce the issues of this chapter were Canada and Denmark (not India and Brazil) betrays that even among the practitioners of post-normal science and the ecosystem approach (myself included) persists a certain Eurocentrism, a subtle political bias tainted by dreams of world domination.

[2] Alternatively, the ecosystem approach's ecological perspective (humans within nature) could be complemented by a "sociosystem" approach still to be developed, with a sociological perspective (nature within humans), also inspired in complex systems and, perhaps, dialectics.

[3] Cox, R. 1983. Gramsci, hegemony, and international relations: An essay in method. Millenium 12(2):162–175.

[4] Historical structures are dialectic configurations of politico-economic forces, perceived as pressures and constraints influencing (not determining) social behavior. At their base, historical structures are determined by the different modes of producing material life in a society (which includes the production of ideas). Cox (1983) identifies three interacting components of historical structures: material capabilities (labor, technology, accumulated wealth, and natural resources), institutions (which may act to stabilize/perpetuate the current order or may become the battlegrounds of competing social forces), and ideas (which are intersubjective meanings, supporting stability, and rival collective ideas of social order, supporting change). We may see historical structures as a special kind of dissipative structure.

[5] According to Cox, this would manifest in three basic scales or "spheres of the social world": the world order (affecting the whole biosphere), forms of state (roughly corresponding to national territories), and the level of the social forces

that obtain their identity from specific modes of production (roughly corresponding to regional landscapes). Within each sphere we can recognize the three components of a historical structure. See Cox, R. 1986. Social forces, states, and world order. Review of International Studies 18:161–198.

[6] In fact, this is the same story all over the world.

[7] By employing a technique that a friend of mine likes to call "facipulation."

[8] Or, alternatively, building a complementary "sociosystem" approach.

[9] Of course, this also applies to ecosystem management.

[10] Is this not the mantra of globalization? Is it not, by any chance, working?

19

An Ecosystem Approach for Sustaining Ecological Integrity—But Which Ecological Integrity?

David Manuel-Navarrete, Dan Dolderman, and James J. Kay

Ecological integrity (EI) has become a central concept in environmental manage-ment and conservation policy in North America and other parts of the world (Manuel-Navarrete et al. 2007). The ecosystem approach has been invoked as a tool for sustaining EI, but EI can be defined and interpreted in different ways. This chapter discusses the implications for society and for individuals stemming from three discourses to EI identified elsewhere (Manuel-Navarrete et al. 2004): (1) normative, (2) ecosystemic-pluralistic, and (3) transpersonal-collaborative. Each discourse implies a different interpretation of the concept and has deep im-plications for the design and application of an ecosystem approach. The norma-tive discourses describe human-ecosystem relationships in terms of separateness (being apart from ecosystems). The ecosystem-pluralistic considers humans as part of ecosystems. The transpersonal-collaborative emphasizes the construction of personal meanings of EI in the context of ecosystems being one with humans. Each discourse embraces different beliefs and worldviews, which, in turn, pro-mote specific social practices and, consequently, influence the way individuals engage in the world. These practices range from command and control strategies, to adaptive/participatory management, and to collaborative learning. Regardless, the commitment to any of the discourses, we consider very useful the awareness of, at least, these four strategies to address the transition toward sustainability from an ecosystem approach.

Normative Discourse

The normative discourse divides the planet into natural areas and zones of human occupancy. The traditional anthropocentric human-nature dualism is replaced in favor of a biocentric dualism (i.e., from managing ecosystems for the benefit of humans to controlling humans for the benefit of all living systems).[1] Nevertheless, by maintaining a dualistic stance, this discourse remains consistent with dominant scientific languages and provides an opportunity to address the environmental crisis without challenging the type of social practices and forms of personal engagement that are dominant in Western cultures. The emphasis is on control of "techno-humans" (Westra 2001: 273), while liberating natural systems to enable them to remain pristine or to follow their own evolutionary trajectories (apart from humans).

The normative discourse promotes (1) management understood as command and control, (2) coercion to enforce laws, and (3) objective scientific research (De Leo and Levin 1997). These three practices demand a "professional" kind of engagement (e.g., expert, bureaucrat, or manager), which discourages personal meanings, creativity, and emotionality so as to maintain "objectivity." Professionalism is informed by the sets of norms that define what it is to be competent and powerful. It implies a distinctive status and conveys an image of certainty (Grey 1999). This image may require the hiding of both uncertainties and errors (Michael 1995).

Those who do not belong to the managerial, legal, or scientific communities, namely, citizens, are supposed to engage in the participatory mechanisms of Western representative democracies (e.g., lobbying, voting, and protesting).[2]

Human agency is depicted as actions causally connected to mental intentions. These intentions are bounded to innermost values that are not directly accessible to empirical observation and therefore can only be "known" when brought outward through rational action [this is consistent with Šunde's (chapter 20) characterization of a "dualistic mind"].

How are these self-contained, isolated psyches at all related with the integrity of ecosystems? The answer, within this discourse, lies in the domain of ethics and morality. Individuals are supposed to engage in the world as moral beings (i.e., observing moral principles of what is the right thing to do). This assumes the existence of morally right choices based on deontological principles[3] determined by detached, disembodied logic (Zack 2002). As a corollary, a coercive hierarchy is necessary in order to guarantee the priority of universal moral arguments over personal unenlightened self-interest or emotions.

Ecosystemic-Pluralistic Discourse

The ecosystemic-pluralistic discourse to EI introduces many of the elements of the ecosystem approach presented in this book. First, a formal process of negotiation is promoted, where a large number of stakeholders' values are included and made explicit. This breaks up the monopoly of moral principles and experts in influencing managerial decisions and embraces a pluralistic approach. This discourse acknowledges the possibility of different but ecologically equivalent regimes in a given ecosystem. There is not a unique ecologically "correct" ecosystem to be preserved (Kay and Regier 2000). Thus, according to this discourse, EI must be discussed in political and social arenas as well as in legal or ethical ones. Scientific and managerial arenas are extended toward society through formal participation (for instance, Diamond and AMESH-like processes as depicted in previous chapters). Thus the ethical problem becomes making explicit and integrating the existing values in society rather than seeking arguments for the sorts of norms, goals, or standards that people ought to hold in relation to ecosystems.

Accordingly, this discourse suggests a social practice based on negotiation, consensus, and compromise in order to connect the realms of the innermost subjective values and the objective complex world. Management and scientific practices are democratized (Michaels et al. 2001). Adaptive and participatory management is proposed in order to involve everybody in a social practice that would normally be associated with a particular group (i.e., managers) and perceived as separated from that which is managed (Grey 1999; Ludwig 2001). "Local knowledge" can become part of management. However, local knowledge is often understood as pseudoscientific information, which is locally generated, versus that which is produced by official research institutions (Allen et al. 2001). Thus local knowledge would not include, for instance, personal meanings and experiences.

Citizens (i.e., nonprofessionals) engage in public participation and comanagement as stakeholder and partners. This usually implies the endorsement of social roles characterized by self-interest. For instance, in conservation conflicts, governmental officials, industry representatives, and environmental advocates are typically involved. These "stakeholders" are defined by the goals, objectives, roles, and resources they bring to the negotiation process. Thus individuals are characterized by their roles and the norms associated with those roles rather than by the person as an integrated individual. They act according to their place within the shared practices of the community. They understand themselves as agents acting in a public role and conduct themselves according to standardized performance criteria. Šunde (chapter 20) describes this as a "dialectical framework" that "blocks the possibility for dialogue."

Underlying this discourse, there is an oversimplification of economic and social dynamics, as well as of the individual's inner worlds. Generally speaking, it does not account for (1) changes in power[4] relationships existing in society (which at the same time determine the values of every stakeholder) or (2) the issues related to personal integrity or inner experience. Individuals are still considered as "black boxes" that interact with the outer world, without taking into account the intrapersonal changes provoked by such interaction.

Transpersonal-Collaborative Discourse

The transpersonal-collaborative discourse is predicated on the observation that integrity, in the end, is about people and their binding relationship with the biophysical world. This discourse claims that only by changing fundamental attitudes about our place in the larger scheme of things we will attain EI. Consequently, EI becomes an issue of consciousness transformation and meaningful understanding of human-environment relationships rather than one of law enforcement and the implementation of policies and managerial plans, regardless of how democratically they have been decided.

According to this discourse, the association with efficiency and control entailed by "management" severely constrains the ability of humans to successfully resolve the environmental crisis (Bavington 2002). The social practices described in the previous discourses are perhaps necessary to enable us to buy time and deal with short-term issues, but at the same time, their hegemony is digging us deeper into our current rut. We must promote other social practices that get at "the heart of the matter," so to speak, rather than simply dealing with surface symptoms. The deeper issue is one of inner change, resulting in cultural transformation (Erickson 2002). Because modern culture encourages us to seek meaning in social relations that are becoming increasingly commoditized and material-based, we have developed dysfunctional coping strategies of investing in short-term solutions that, while they help us in the immediate, are not sustainable over the long term.

The alternative (or complement) to managerial control is a collaborative learning practice (Daniels and Walker 1996). Šunde (chapter 20) describes this practice as a "dialogical opening." A learning community of practice includes learning as the very core of its enterprise (Wenger 1998). This implies a dialogical process of identity construction occurring within a community of practice. Because learning transforms who we are and what we can do, it is an experience of identity. Through experience, individuals internalize the ongoing dialogue from the world around them. Ideally, each voice incorporates its own evaluative position at the same time as it remains open to the potential truth of "the other." This self-reflective dialogue allows the emergence of new, temporary, and open-ended identities (Richardson et al. 1998). New and ever-changing identities or accretion points are experienced as conclusions that seem to be locally valid or right (i.e., concerned with the quality of our motivations, the worth of the ends we seek, or the good or right life for us in our time and place), even if we are aware that from our limited perspectives, we can never attain anything that is final or absolutely certain. In this perspective, to be ethically sensitive and responsible is not only to have knowledge of abstract moral principles or to defend someone's own position in a coherent way. It has more to do with participating from one's unique position in a dialogue about what is important for each participant (Duane 1997; Hester et al. 2000).

The main outcome of collaborative learning practice consists of a better understanding of the reasons and meanings guiding people's interactions with their biophysical environment. It seeks to engage individuals in local communities of practice dealing with planetary issues (i.e., the global environmental crisis) (Richards 2001). The agency for change rests on individuals through acknowledging that the whole planet is part of each one's identity (i.e., locally situated individuals with a deep sense of ultimately belonging to planet Earth). Because individuals integrate ecosystems (including social structures), any change in the former will scale up to the later. This bottom-up radical perspective places a huge trust on human evolutionary capacities and requires at the same time that the constraining structures of society work for, and not against, personal development (although it is not the point of this discourse to propose dogmatic methods or universally valid social structures).

Implications of Each Discourse for Practice and Agency

Each discourse focuses on different social practices and kinds of individual engagement. The juxtaposition of the three discourses entails a broad spectrum of action from the individual to the cultural to the institutional. At the conceptual level the three discourses together might be conceived as a complementary. However, the tensions among them must be acknowledged, and it is dubious whether they can coexist in actuality. Even though the transpersonal-collaborative discourse is the broadest, in the sense that it is the most inclusive of different disciplines, viewpoints, and domains such as biophysical, social, political, cultural, and personal, it is questionable whether it embraces or utterly includes and transcends the other discourses. Whatever the case, we believe that it is useful to differentiate among the strategies and actions resulting from each discourse [for place-based applications of this framework, see Manuel-Navarrete (2005) and Manuel-Navarrete et al. (2006)].

It is quite obvious that any change of human-nature relationships involves society and people. Therefore the assessment of such strategies must necessarily include the consideration of what kind of social organization we want at the global level and what our real needs and goals, as individuals, are.

The normative discourse is based on the belief that our current social organization "works" (or might work with some minor adjustments) to solve the global environmental crisis. All that is required from individuals is professionalism and an ethical commitment. Solutions to the crisis are derived from both scientific truths and morally correct analyses. It is argued that the only way that a real change can take place is through the effective implementation of policies to preserve nature. Therefore the main means for achieving EI is to influence decision makers as soon and effectively as possible. It is believed that decision makers require simple, user-friendly, cost-effective tools that permit the implementation of actions promoting EI. This belief of what is needed leads science to an emphasis on stability, unique equilibrium, normative states, and above all, the use of a set of simple measures each of which, independent of the others, indicate if the situation is better. This emphasis is consistent with structural definitions of biodiversity (Lister, chapter 6).

It is important to note that this reasoning provides a consistent, coherent, and, consequently, highly incomplete formalism. This incompleteness becomes quite evident upon asking some basic questions: What if scientific findings are controversial? What if political actions, once implemented, do not convey the results they were intended to? What if decision makers' agendas are not only guided by scientific findings? What if some people believe that integrity cannot be achieved without dealing with issues like, for instance, north-south inequalities or the improvement of the current mechanism of representative democracy? What if people do not feel comfortable with an exclusive professional kind of engagement because they are more concerned with, for instance, cultural evolution and spiritual awareness? These questions do not indicate that this discourse is wrong but simply that it is dramatically incomplete.

The normative discourse is built upon a formalism that embraces a small portion of reality, the perception of the biophysical world as only a set of objectively measurable phenomena. This simplification of reality[5] fits well with Western management and policy approaches. This discourse tends not to take into account issues related to different cultural views, aspirations, and ways of addressing human activity on the landscape. Consideration of social, political, institutional, or cultural changes tends not to be considered relevant. It assumes a relatively homogenous cultural landscape.

This relatively shallow approach gives rise to actions that seem universal in their applicability, as they ignore the complex dynamics of coupled social-ecological systems. They are presumed to transcend cultural boundaries and do not acknowledge the need to take into account different cultural settings or people's relationship with their environment (Manuel-Navarrete et al. 2006). By ignoring the complexity of the situation, these narrowly focused "scientific" actions give a false sense that they are a simple effective means, universal in application, for dealing with EI issues. In the long term this illusion works against the resolution of environmental issues. However their power is that they fit well with current Western societal myths and decision making and management approaches and therefore, in the short term, can be quite effective, particularly in dealing with immediate crisis situations.

The ecosystemic-pluralistic discourse is less consistent but more complete. In part, owing to the use of formalism and methods of implementation that try to account for uncertainty and complexity (e.g., complex systems thinking) and also in that they try to go beyond the biophysical to deal with some of the broader contextual and larger-scale societal issues raised above. Scientific uncertainties are put up front (thus allowing less consistency), whereas formal participatory processes and management widen the scope of decision-making roles and responsibilities to the whole society (i.e., more completeness).

The idea of incorporating a plurality of perspectives in decision making can be intuitively appealing. It requires, to some degree at least, a shift of power from technical and political elites to the "ordinary" people. However, control over decision making is seldom given up in real situations (Goodwin 1998). As a consequence, formal participation usually has complex dynamics derived from existing power structures. Eventually, its potential to challenge patterns of dominance may be transmuted into new avenues of power and influence that ultimately undermine collective decision making (White 1996; Barham 2001).

Collaboration requires a sense of community and a favorable political and institutional environment based on trust, goodwill, and power sharing (Saarikoski 2000). Serious challenges to the ecosystem-pluralistic discourse are presented by the mistrust, fragmentation, and perceived failure of institutions to meet social expectations that characterize some political arenas (Macnaghten and Jacobs 1997; Goodwin 1998). In addition, governmental agencies usually come to the process with an agenda that leaves little room for reflecting a real inclusion of people's needs. Participants' motivation is also critically important to fostering long-term collaborations. It is generally assumed that this motivation should arise from the expectation of gaining power. However, as discussed above, this seldom happens. Any voluntary sharing of power is particularly unlikely in institutional contexts characterized by impersonal and hierarchical structures.

More importantly, while they might be effective for fostering power sharing, little space is left for personal transformation and a sense of a larger community. "Empowered" individuals might be as unresponsive to EI as governmental agencies. In fact, the emphasis on instrumental forms of participation (i.e., when understood as a managerial tool to achieve a predetermined product) indirectly promotes the commoditization of the environment through strengthening private agendas rather than a relationship based on feelings of connection and respect.

The transpersonal-collaborative discourse is the most complete in the sense that it strives to take into account a broad spectrum of personal worldviews, cultural situations, and other considerations. This inclusivity makes this discourse the least consistent: different viewpoints will lead to internal inconsistencies and solutions that are restricted to very local applications. What works in one place (i.e., biophysical and cultural situation) probably will not in another. So a consistent universally applicable solution to EI cannot be supported by this discourse.

By emphasizing the changes occurring at the level of the noosphere (i.e., the sphere of consciousness), this discourse allows for a greater depth of consideration but at the cost of narrowing its breath of applicability. Consequently, its application can only be local, although it may allow for longer-term (i.e., more fundamental) changes. This discourse limits centralized institutions and does not include top-down policies or management mechanisms. However, this does not mean that there is no room for institutions or other organizational mechanisms.

In this last discourse, logic, deduction, and prediction are complemented with a narrative knowledge that prioritizes human intentionality, intuitive validation, meaning making, and finding truth in introspection rather than solely in what is externally (i.e., empirically) verifiable. Individuals assign values via inner mechanisms and judge the "worth" of information based not only on reason but also on their emotional (and spiritual) response to that information (Thompson 2001). Information without an effective response is simply information; it does not mean anything relevant to individuals until it has passed through interpretative, affective filters. Learning about other's interpretative filters may help individuals engage in practices that promote dialogue among cultures and ensure EI by taking into account the implications of their actions for all beings including their inner dimension. In short, consciousness evolution is depicted as the only "way out." According to this discourse, we cannot continue to grow into span (unfolding) but into depth (building up). We desperately need to integrate the biosphere into our psyche in order to put an end to their current dissociation and thus strive toward higher levels of complexity.

Epilogue

Overall, the normative discourse is quite narrow in the issues it focuses on and the people it includes; it also takes orthodox modern institutions and culture as given. This simplification will invariably result in important issues being ignored. Consequently, any solutions being proposed will inherently involve contradictions, and this will lead to its own unraveling. On the contrary, the ecosystemic-pluralistic and transpersonal-collaborative discourses make an effort to deal with the local context and the complexity of the situation. They include a broader spectrum of issues at different scales, strive to be inclusive of all people and their perspectives, and assume cultural diversity. This has the virtue of trying to deal with people, their issues and contradictions up front. The hope is that this will lead to a more effective and longer-term outcome. However, this is initially very costly in time and energy and requires social and cultural change. Therein lies the reason why it is not appropriate to identify one of these discourses as better than another. The trade-offs, between up-front investment, inclusivity (which implies messiness), context specificity, and robustness versus simplicity, generality, tidiness, efficiency, immediacy, fragility and ineffectiveness, constitute a dilemma that is inescapable when deciding how to undertake the quest for EI. Life is a trade-off!

Notes

[1] This could be understood by some as "ultimate anthropocentrism" in that it projects on the biotic realm the same sense of dualism as the one experienced by humans, that is, animals and plants being apart from humans.

[2] We are assuming here the possibility of other forms of democracy than Western representative.

[3] As indicated by Zack (2002), the use of deontological principles for preserving EI requires conferring personhood on natural systems and beings. As a consequence, depriving an ecosystem of life would be, in principle, no different from crimes of murder committed against human beings

[4] Not just political power, but power in the sense of skills, knowledge, economic and social status, and resources in general, which are available to the individual as they participate in the process.

[5] This is not stated in a pejorative sense. All approaches to reality are a simplification in some regard. The important thing is to understand this and the consequential limitations of an approach and to not get caught up in "misplaced concreteness."

References

Allen, W., O. Bosch, M. Kilvington, J. Oliver and M. Gilbert. 2001. Benefits of collaborative learning for environmental management: Applying the integrated systems for knowledge management approach to support animal pest control. *Environmental Management* 27(2):215–223.

Barham, E. 2001. Ecological boundaries as community boundaries: The politics of watershed. *Society and Natural Resources* 14:181–191.

Bavington, D. 2002 Managerial ecology and its discontents: Exploring the complexities of control, careful use and coping in resource and environmental management. *Environments* 30(3):3–21.

Daniels, S. E. and G. B. Walker. 1996. Collaborative learning: Improving public deliberation in ecosystem-based management. *Environmental Impact Assessment Review* 16:71–102.

De Leo, G. A. and S. Levin. 1997. The multifaceted aspects of ecosystem integrity. *Conservation Ecology* [online]1(1):3. (Available at: http://www.consecol.org/vol1/iss1/art3/) Duane, T. P. 1997. Community participation in ecosystem management. *Ecology Law Quarterly* 24(4):771–797.

Erickson, F. 2002. Culture and human development. *Human Development* 45:299–306.

Goodwin, P. 1998. Hired hands or local voices: Understandings and experiences of local participation in conservation. *Transactions of the Institute of British Geographers* 23(4):481–499.

Grey, C. 1999. 'We are all managers now'; 'we always were': On the development and demise of management. *Journal of Management Studies* 36(5):561–585.

Hester, L., D. McPherson, A. Booth and J. Cheney. 2000. Indigenous worlds and Callicott´s land ethic. *Environmental Ethics* 22:273–290.

Holling, C. S. (ed.) 1978. *Adaptive Environmental Assessment and Management.* New York: Wiley.

Kay, J. J. and H. Regier. 2000. Uncertainty, complexity, and ecological integrity: Insights from and ecosystem approach. In *Implementing Ecological Integrity: Restoring Regional and Global Environmental and Human Health*, eds. P. Crabbé et al., 121–156. Dordrecht, Netherlands: Kluwer Academic.

Ludwig, D. 2001. The era of management is over. *Ecosystems* 4:758–764.

Macnaghten, P. and M. Jacobs. 1997. Public identification with sustainable development: Investigating cultural barriers to participation. *Global Environmental Change* 7(1):5–24.

Manuel-Navarrete, D. 2005. Challenges and opportunities of metadisciplinary place-based research: The case of the Maya forest, *Environments: A Journal of Interdisciplinary Studies* 33(1):81–96.

Manuel-Navarrete, D., J. Kay and D. Dolderman 2004. Ecological integrity: Linking biology with cultural transformation. *Human Ecology Review* 11(3):215–229.

Manuel-Navarrete, D., S. Slocombe and B. Mitchell. 2006. Science for place-based socio-ecological management: Lessons from the Maya forest (Chiapas and Petén), *Ecology and Society* [online]11(1): 8. [Available at: http://www.ecologyandsociety.org/vol11/iss1/art8/)

Manuel-Navarrete, D., J. J. Kay, and D. Dolderman. 2007. Evolution of the ecological integrity debate. In *Sustaining Life on Earth: Environmental and Human Health Through Global Governance*, eds. C. L. Soskolne et al., 127–138. Lanham, Maryland: Lexington Books.

Michael, D. N. 1995. Barriers and bridges to learning in a turbulent human ecology. In *Barriers and Bridges to the Renewal of Ecosystems and Institutions*, eds. L. H. Gunderson et al., 461–485. New York: Columbia University Press.

Michaels, S., R. J. Mason and W. D. Solecki. 2001. Participatory research on collaborative environmental management: Results form Adirondack Park. *Society and Natural Resources* 14:251–255.

Richards, R. 2001. A new aesthetic for environmental awareness: Chaos theory, the beauty of nature, and our broader humanistic identity. *Journal of Humanistic Psychology* 41(2):59–95.

Richardson, F. C., A. Rogers and J. McCarrol. 1998. Toward a dialogical self. *American Behavioral Scientist* 41(4):496–515.

Saarikoski, H. 2000. Environmental impact assessment (EIA) as collaborative learning process. *Environmental Impact Assessment Review* 20:681–700.

Thompson, E. 2001. Empathy and consciousness. *Journal of Consciousness Studies* 8(5-7):1–32.

Wenger, E. 1998. *Communities of Practice: Learning, Meaning, and Identity*. Cambridge, UK: Cambridge University Press.

Westra, L. 2001. From Aldo Leopold to the wildlands project: The ethics of integrity. *Environmental Ethics* 23:261–274.

White, S. C. 1996. Depoliticising development: The uses and abuses of participation. *Development in Practice* 6(1):6–15.

Zack, N. 2002. Human values as a source for sustaining the environment. In *Just Ecological Integrity*, eds. P. Miller and L. Westra, 69–73. Lanham, Maryland: Rowman & Littlefield.

20

The Water or the Wave?

Toward an Ecosystem Approach for Cross-Cultural Dialogue on the Whanganui River, New Zealand

Charlotte Helen Šunde

Introduction: Whanganui River Reflections

July 18, 2001, was a crisp winter's day in Aotearoa, New Zealand. I awoke at Papakai Marae, under the shadow of the snow-cloaked mountains. By midmorning a full bus, followed by an ensemble of several other vehicles, pulled off the state highway and onto a minor road marked by the small signpost: "Whanganui River Intake." The bus occupants assembled on the concrete platform that is the intake structure of the Whanganui River headwaters. The group comprised several generations of an indigenous people united by bloodline and by river, and me—the only one not of Whanganui River Māori relation. The atmosphere was thick with reverence, and the air heavy with grief. In the silence that enveloped us, I stood looking toward the mountains. The river wound into the distance across rolling plains of mountain tussock and native grasses. Its waters were clear and clean, its flow strong; the feeling was one of a river with visible vitality and, indeed, mana (spiritual authority).

Yet, beneath my feet, the waters were suddenly hurried, sucked over the cusp of a metal grate and into a large pipe. The grate—a fortified grid for filtering debris—seemed almost offensive, like the personified toothless grin of "development" itself. From there, the waters churned noisily as their natural flow was disrupted abruptly and detoured at a right angle from their original path to the sea. I turned to face the direction in which the Whanganui River headwaters would have once flowed, prior to the Tongariro Power Development scheme. The scene now depicted a completely different future. Aside from the ochre-colored pool of still mud (discolored by iron oxides leaching from the surrounding soils), the Whanganui River was reduced to a series of disconnected puddles with only a trickle linking them.

I now understood that the watercourse that flows past my parents' house in Wanganui[1] is not that of the Whanganui River *proper* but, rather, a mixture fed and now sustained only by its tributaries (originating from the Taranaki region and elsewhere). The only time that the headwaters are granted permission to enliven the flow of the Whanganui River is during heavy rain when the intake structure reaches its full capacity (or when Lake Taupo is bursting at its seams). Then, and only then, do its headwaters spill over the boundaries of development in anything like their original volume.

From within the group rose a slow and mournful wailing. It was joined by voices singing to their tupuna (ancestor)—their Awa Tupua. Tears merged with mountain waters. I wished for a better future than the one that had beheaded the past.[2]

On the Water Again: Whanganui River, 2003

The river called me back. In February 2003 I was again paddling on the Whanganui River, this time with a group of thirty American students on the International Honors Program. As part of their university degrees the Global Ecology students were traveling the world to see first hand the devastating impact of Western-driven development on local environments and tribal peoples' way of life. Their journey had commenced in Washington, D.C., at the headquarters of the World Bank. A few months later, they were hosted by villagers in India who, as the result of a hydroelectric development project funded by the World Bank, had been relocated from their tribal lands (hence from their identity, livelihood, and ancestral future). In Wanganui I joined the students and, over the few days we shared, became a focus for their keen questioning. Yet I found myself increasingly unable to provide immediate answers about the Whanganui River beyond reciting scientific facts and historical dates. Furthermore, I felt increasingly reticent about doing that, especially in response to deeper questions. Rather, my feelings confirmed to me that it was through reflecting on lived experiences and returning to the source that my answers flowed more naturally.

At the end of our second day on the Whanganui River, we pulled our canoes ashore at Tieke Marae. We were formally welcomed onto the marae (tribal community meeting place) by the local people of the land, the tangata whenua. They had returned to Tieke in 1993 to live in the place of their ancestors. When Māori first reoccupied the hut and surrounding site in what was reported by the mainstream media to be an act of civil disobedience, relationships between Whanganui National Park managers (Department of Conservation) and the local people (Tamahaki; a hapū of the Whanganui River) were strained. Ten years later, those relationships have improved, although disparities over conservation and other issues remain. That evening we all settled in the main whare (house) and, in turn, stood to introduce ourselves. The students had come from rural towns, cities, and megalopolises scattered throughout the states of America. Nearly every one commented how beautiful the river is. At last the introductions were completed and a local Māori woman came forward to speak.

She agreed: "The river *is* beautiful." However, she qualified, that beauty is threatened. The river is polluted. The bush is poisoned. The birds and fish are greatly reduced in number and variety. Introduced animals and plants have taken over native species. The river's headwaters are diverted for electricity generation. The customary fishing has been seriously undermined. Faced with these problems and more, the local people are rendered powerless. And yet, despite all this, Whanganui River Māori continue to come back to the source, to the healing waters, to recenter and revitalize. Yes, the river is beautiful, but it is the beauty that extends beyond that which the eye beholds that draws the people back. In what seemed, to me, to encapsulate the difference between the long-term river resident and the recreational visitor, her relative left us with this health warning: "Do not drink the water."

The Dialectical Framework

In Whanganui (as for all New Zealand), every development proposal with a potential impact on the river and surrounding lands is dealt with through a resource management framework that is dialectical rather than dialogical (Panikkar 1999).[3] For example, the planning process forces a division between those supporting and those opposing a development to argue their respective cases dialectically through legal adjudication techniques. Under the Resource Management Act 1991, developers are required to apply for resource consent(s), which may (or may not) be granted by the relevant local government authority, subject to certain conditions. The Genesis power company recently applied for fifty-seven resource consents for the ongoing operation of the Tongariro Power Development scheme for an extended tenure of thirty-five years. The initial decision by the planning committee was to grant those resource consents. Appeals against that decision were then lodged by a number of Māori Iwi (tribal) authorities. The appeal hearings took place in 2003, and the decision was released in June 2004 (Ngāti Rangi et al. 2004). The Environment Court took the view that there had been insufficient dialogue or "meeting of the minds" between the Māori appellants and the scientific experts from Genesis and directed that such meetings take place. As a result, the request by Genesis for resource consents for a thirty-five year time period was reduced to ten years in order to reflect an outcome that would balance Māori concerns with the electricity generation. Predictably, Genesis has lodged a further appeal on a point of law to a higher court.

At one level, on the basis of the English legal and planning systems, the dialectical framing described above is democratic and just. For example, councilors who determine resource management decisions are elected by the citizenry. Furthermore, in Western capitalist countries, every individual (and corporation) is granted the "right" to pursue development of personal financial gain, just as much as any individual or group has the "right" to voice their opposition to that development. A developer's rights are restricted by environmental protection legislation and rules in regional and district council plans. Citizens have the opportunity to contribute directly to the formulation of those regulatory mechanisms (i.e., plans) and also during the public hearing of a development proposal through statutory consultation and submission processes. The dialectical framework emphasizes efficiency, transparency, and accountability when making decisions. Ultimately, those decisions on development proposals must be legally defensible. Dissatisfaction with the outcome may then be pursued further through higher legal channels.

From a different perspective the planning hearings for the Tongariro Power Development scheme could be described as a highly ritualistic form of "game playing." In this "game," two or more opponents enter the legal arena. An implicit assumption is that all contestants are equal before the law, and therefore symmetry between them exists at the outset. The reality, however, is that Genesis Power Corporation draws on financial resources far in excess of those available to their opponents (e.g., Māori tribal organizations, local councils, recreation groups, community citizens). As with modern professionalized sport, the legal game becomes more and more serious and competitive as the economic stakes are raised higher. Therefore the possibility that Genesis might lose (i.e., headwaters currently diverted to the power scheme be returned in full to the natural watercourses) is the first priority that the corporation's lawyers must eliminate.

In order to assert their authority in this legal game, Genesis has used a number of tactics. Their aim is ultimately to win the "legal race." When others stand in the way of the developer, they are seen as obstacles to progress, and therefore they must either be outcompeted or coerced out of the dialectical arena altogether. A key strategy in this legalistic game is that each competitor accumulates large amounts of evidence to convince the adjudicators. Thus Genesis employed consultants to prepare numerous technical reports detailing the technological and economic merits of the Tongariro Power Development scheme. In addition, consultants with scientific expertise were contracted by Genesis and arrived at conclusions that countered the evidence presented by the Department of Conservation scientists and others. Those "contestants" then enter into dialectical argument whereby each "expert" attempts to undermine the legitimacy of all others, especially that of the "nonprofessional."

What results is a dead-end argumentation between experts (in science, law, etc.) in which no competing party truly listens to the claims of the other and the real intent of each contestant is masked by legal terminology and endless dialectical bantering. Because the technical reports were produced by people with professional qualifications (called experts), they are elevated to an almost exclusive "truth." The reductionist technical detail, masses of quantitative data, computer-generated graphs, and bureaucratic jargon that characterizes those reports renders them impenetrable to the layperson. People who have fished and swum in, boated on, and been ritually blessed by the waters of the Whanganui River are not equally valued in this dialectical arena despite their enormous expertise derived from direct and lived experiences, nor, for that matter, are technical experts, scientists, economists, and lawyers appropriately equipped to pass judgment on spiritual relationships of great importance to Māori and others.

If abstract postulates and jargon are the tools of the bureaucrat and corporate executive, then experience and ingenuousness are gifts of the "witness," as defined by Raimundo Panikkar (1979:238): "The witness knows, understands, recollects, is anxious for, concerned about; he thinks, considers, is preoccupied with what he will manifest to another in his testimony." It is precisely the presence and conviction of the witness that exposes the susceptibility to economic ideology of those who approach the legal game as a dialectical contest. In the case of planning hearings for the Tongariro Power Development scheme, submitters come forward as witnesses to present their testimonies before an "audience." In this case, that audience is a planning committee whose role it is to judge according to the rules of the legal system. The very logic of that system is legitimized within the dominant European worldview. In its modern expression, that worldview regards rationality as the only basis of reason and therefore as all that is real. The dialectical framework excludes any other rationale, and yet, Māori (and others) claim that rational reasoning is not the only way of approaching reality.

The dialectical framework is fundamentally flawed because it operates like a game: it aims to produce a "winner" and therefore, by implication, qualifies all others as "losers." The framework incites confrontation among people by imposing a dialectical situation where they are set apart into opposing positions (e.g., submissions must either support or oppose a developer's application for planning consent). Even the more open processes of consultation may be seen in this light, given situations where the dominant player has a monetary monopoly and uses that power to strike deals with opponents to gain their support or, at least, to silence opposition. Evidence is presented in such a way as to convince or coerce those who are empowered in the position to pass judgment. Hence the dialectical arena is reduced to a battlefield for lawyers. What the dialectical framework blocks is the possibility for dialogue; it blocks the possibility of overcoming conflict and misunderstanding through cross-cultural dialogue characterized by a genuine openness to questioning the other and, in turn, to being questioned or "called into question" (Dallmayr 2002).

The way in which dialectical argument exacerbates conflict is exemplified in claims that today "we" enjoy an infrastructure that supports a lifestyle of comfort and ease with technological gadgetry that eliminates many of the mundane tasks of a bygone era. Indigenous people living within a modern, secular society often enjoy those same benefits: the speed of jet boat travel, the communication advantages of cellular phones, and the instant gratification of electricity at the flick of a switch. Therefore some might argue, how is it that Māori can justify a stance that jeopardizes the continued supply of electricity? The question, however, is specious because it assumes that an either/or situation exists: it assumes that respecting their Tupuna Awa (Ancestral River) is incompatible with enjoying the benefits of a modern technological society. Even from a technical point of view, the question is specious because there are other viable ways of supplying electricity that do not interfere with (i.e., dam and divert) the headwaters of the Whanganui River.

Māori and others (e.g., environmentalists) then become locked into a dialectical situation whereby opposing the diversion of the Whanganui River headwaters for hydroelectricity generation is viewed by some as antidevelopment, romantic, or even hypocritical. The reality, however, is that others have not only benefited from using the Whanganui River but some (such as Genesis) have also gained substantial profits through taking water from the river in a way that denies others their relationship with the river. The question (above) could be redirected: How can Genesis justify their stance when it continues to jeopardize others' spiritual, cultural, ecological, and recreational values of the Whanganui River? Of course, this question responds in dialectical fashion and therefore remains ultimately unsatisfactory.

Monism, Dualism, and the Reductionist Response

The factors that have contributed to environmental degradation and disempowerment of indigenous peoples and their traditional ways of life are not unique to the case of the Whanganui River or even to New Zealand. In fact, there is similarity among events throughout the world wherever a colonizing culture has imposed its worldview and enforced its authority over an indigenous landscape and autochthonous peoples (see, e.g., Berkes and Davidson-Hunt, chapter 7). A pattern has emerged that suggests that the situation characteristic of Whanganui today is not the result of random cross-cultural interactions. Rather, it can be attributed to a systematically executed process of British colonization based on socially enforced beliefs in European cultural supremacy. This chapter, however, does not attempt a detailed historical examination of the events and effects of colonization per se on the Whanganui River (see Waitangi Tribunal 1999). Instead, the intent is to expose the monistic attitude and related scientific paradigm of reductionism that underpins not only the historical relations among different cultures as a result of colonization practices, but also remains a powerful cultural force behind the modernist drive for social and economic development.

Monism is displayed in the attitude that there exists one authority, a common law, and a way of living and knowing that is upheld as the best and most truthful claim on reality. The belief in the supremacy of that one way is considered absolute and universal. That is not to deny the existence of a plurality who objects to the imposition of a monopolistic authority. However, as long as those who dissent from the majority do not become too great a threat, the monistic One will tolerate them until they can either be assimilated or extinguished. In New Zealand, a monistic attitude is evident in the politicized assertion that "we are all New Zealanders." The assumption is that there is general agreement on what constitutes the "common good"; dissent is the mark of the radical. Laws are drafted by the state and enforced through its agents (police and judiciary) to protect the rights of a collective majority (and the privileges of an elite minority). It is little wonder that Māori-led initiatives to determine their own ways of governing and educating their people are often met with suspicion, if not open hostility, by those who hold to monistic beliefs.

The Tongariro Power Development scheme is justified in monistic terms as development that benefits the national economy and therefore the so-called national interest. The reduction of the Whanganui River to economic rationale is evident in claims that, prior to the abstraction of the river's headwaters for power development, the river was an "unproductive" resource. According to that logic, the Whanganui River is now a "valuable" economic resource that contributes to the Western diversion of the power scheme with an estimated profit of fourteen and a half million dollars per year (Waitangi Tribunal 1999). Therefore the Whanganui River forms part of the development infrastructure that fuels the national economy. Economists argue that it is through such efficient engineering initiatives that "natural capital" is put to good use to enable the New Zealand economy to grow and prosper as a competitive export trading nation in the global economic market. As the political rhetoric goes, what is good for the economy is good for us all.

The Tongariro Power Development scheme is a product of the dualistic mind. That way of perceiving reality divorces the "subjective mind" from the so-called "objective world" and, in doing so, reduces all "others" to the status of objects. The Whanganui River is treated as an object when "it" is viewed simplistically as a body of water (potential kinetic energy) to be tapped for hydroelectricity generation and so transformed to produce economic goods and services. Furthermore, not only is the river perceived as simply a body of water but it is also dissected bit by bit into mechanistic units intended primarily for utilitarian purposes. This approach is classic reductionism: breaking a whole into its component parts, assuming that by doing so those parts will be more easily understandable (through analytical studies), predictable and therefore also manipulable. This reductionist approach underpins the entire Western legal framework, property regime, and prevailing scientific management rationale that act as external controlling constraints upon the river and Māori communities. It can also be recognized in the individualism of modern technocratic society. As some have argued, reductionism is a significant challenge, if not a fundamental barrier, to the ecosystem approach explored in this book.

Yet, there is something essential that the reductionist approach misses entirely. The gulf in understanding between a "reductionist economist" and a "holistic" Māori leader is relayed in the following personal account of an exchange that took place between them. The economist, who was working on property rights issues for the Treaty of Waitangi claim on the Whanganui River, sought from the Māori leader a precise definition of the physical extent and content of the Whanganui River. From a legal perspective, the focus of the case (the river) must first be isolated and described by an unambiguous set of necessary and sufficient conditions so that the case for (ownership of) the river may then be rigorously defended in a court of law. The economist recalled their conversation as follows:

"During this work, at an early point, the Chairman [of the Whanganui River Maori Trust Board] . . ., said to me something along the lines of 'we just want to own and manage the river as we did before the Pakeha came along.' I responded 'OK …, but what do you mean by "the river"?.' I went on to ask:

- Do you mean the bed of the river or something more?
- Do you mean the banks of the river?
- If so how far from the water line does the bank stretch?

- Do you mean the water flowing in the bed?
- Do you mean the flora and fauna living in and around the river?
- Do you mean the air space above the river and how far above?

I could have added more. In typically helpful fashion [the Chairman] replied "Yeah, all of that."—and he could have added 'together' and 'at the same time.'" (Wheeler 1999:2)

The economist's commentary continues with the conclusion that: ". . . western jurisdictions and management regimes. . . are arguably more simple in what they seek than those of Maori. . . ." However, from another perspective the case appears quite the contrary. In attempting to simplify the complex and interconnected whole, the reductionist approach unnecessarily complicates the explanation. What results is an endless list of questions (such as those above) that, in fact, avoid forming any relationship and therefore any real understanding of the river beyond superficial definitions of what appears to be only a physical entity. The very meaning of the whole is lost in such pedantic legal enquiry. Not only is the river rendered a mere object, but the questioner himself is diminished in the process: his search for objective definitions actually reduces him to an object (because he avoids relating to the river as would a subject).

A somewhat different illustration might better press the point home. Let the questioning be reversed so that the economist is now addressed and asked to define what he means by "his mother." Is she simply the female who gave birth to him? Is she a structure of bones and collection of organs linked in an assembly of arteries and veins circulating blood within a bounded layer of skin? Is she that individual member of the human species whose physical identity records her as age seventy with curly grey hair and green eyes? Is she his mother only until he turned five years old and began school? Or is she the woman who has provided him with healthy meals, a safe environment, a clean house, and nurtured him when he was ill and bereaved? Is she a person who is not only a mother but also a wife, daughter, sister, cousin, grandmother, neighbor, friend, employee, and citizen? Does the economist include in his definition of her the air she breathes or the food she eats and, if so, how much? What about the songs she sings, the cakes she bakes, the laughter, tears, hugs, the ways of disciplining and of loving? To these questions, many of which appear absurdly self-evident to the extent that the economist may consider them personally offensive, he could justifiably reply: "Yeah, all of those"; "together"; and "at the same time." Perhaps, most importantly, he might also add that she is: "so much more than all of those things (parts) described."

This reverse questioning is not intended as a personal attack on that economist per se nor are the questions raised in an attempt to imply that Whanganui River Māori relate to their Tupuna Awa solely as a respectful child does to his or her mother. However, the analogy suggests that there is some similarity in emotional depth, physical relationship, and spiritual connection to that which exists between a person and his or her maternal parent and the many who still relate to and care deeply for Papa-tū-ā-nuku (Mother Earth). Most importantly, it highlights the futility of such questions—for how can you attempt to disclose in a legally defined or physically tangible sense the fullness of the relationship to your parent? For Whanganui River Māori the inextricable tie between the people and river transcends words. The Tongariro Power Development might be reconsidered in a very different light if one were to borrow an economist's reasoning that your mother is "worthless" as she is and is only "valuable" when employed full time in a factory [although not all economists agree with this narrow neoclassical economics reasoning; see, for example, Waring (1988)]. For most people, such a thought is outrageously insulting, and yet modern civilization has progressed under just such a system of exploitation: not the very least in the transformation of nature into natural resources and natural capital (see Daly and Cobb 1994).

Resource management and utilitarian views of water in New Zealand (and other countries) have been dominated by the reductionist approach. However, from an ecological perspective, reductionist explanations are inadequate because the Whanganui River is not simply a collection of parts (a bed, a body, surface, banks, fauna, and flora) but is a complex system and, indeed, a living whole. The reductionist perspective is particularly unsatisfactory in that it fails to acknowledge the interdependency of humans with nature. In fact, that perspective actually encourages humans to view the river as an object to be exploited or a resource to be managed to meet human demands and expectations for economic growth. Given this anthropocentric failing, it is little wonder that Genesis and Whanganui Iwi find themselves in opposition, in a dialectical struggle stemming from essentially incompatible worldviews: one primarily reductionist and the other holistic.

An Ecosystem Approach to the Whanganui River

Conversely, the emerging "complex systems thinking" in science (as discussed in the first section of this book) makes explicit the interdependent relationships of humans with nature. It represents a more holistic conception of ecological interconnections between parts in a dynamic whole that is the ecosystem, and in that respect it is much more closely in tune with Māori understanding. From a complex systems perspective, the Whanganui River is a self-organizing ecosystem sustained by a throughput flow of high-quality energy and matter (e.g., as part of the hydrological cycle). The river ecosystem may be further understood as an open and dynamic, or SOHO system, subject to unpredictable behavior and responding adaptively. However, when the water flow entering this system was drastically altered by the diversion of the headwaters for hydroelectricity generation, the ability of the Whanganui River to maintain its ecological integrity became seriously threatened.

The resilience of the Whanganui River had already been undermined by the accumulated stress of incremental developments throughout the river catchment. The further interference with one of the key/critical "parts" of the river (i.e., its headwaters) has resulted in serious ecological (and other) ramifications throughout the entire ecosystem, many of which are still not understood, nor, for that matter, recognized. Although technical experts continue to persuade decision makers that ecological impacts may be mitigated, their self-assuredness is false given that uncertainty and at least limited predictability are inherent to complex systems. The imposition of a managed water flow regime may be placing the Whanganui River at even greater risk, given that the natural freshets critical to generating biological heterogeneity are now subject to a managed regulation and monitoring procedure (see Gunderson et al. 1995). Complex systems thinking, as discussed throughout this book presents significant challenges for resource management, not least in that it questions the very assumption that humans can "manage" and attempt to predict ecosystem behavior.

The emphasis on relationships (e.g., energy and material flow, networks, etc.) in ecosystem studies, as opposed to the reductionist focus on isolated objects, is similarly critical to indigenous people's epistemologies. While some scientists and indigenous peoples share a similar attitude toward nature as "holistic," indigenous ecological knowledge is neither identical with nor reducible to complex systems thinking. In Western philosophy, holism is understood as the process of whole-making; the emphasis is on the dynamics of the whole and the interconnections between things as constitutive participants of a whole that is "more than the sum of its parts" (Smuts 1926). An enlarged ecological perspective, some would argue, is necessitated by the ecosystem approach; this includes human beings as one species (albeit with dominant predatory habits) among many in the holistic web of life. However, with exceptions, systems thinking is still mainly restricted to intellectual arenas, and ecological awareness struggles to penetrate mainstream consciousness. There is a vast lag between the insights of holistic, systems-based theories and their expression in the fuller sense of community.

For indigenous peoples, the importance of relationships is not limited to ways of knowing but is essential to ways of living—culturally, socially, and spiritually. Māori regard whakapapa as the inextricable interrelatedness of all life-forms: the bloodlines uniting a person to their family, tribal groups, ancestors, mountain, river, lakes, and land. Those relationships are not only physical but also psychical and personal. Therefore "holism" refers not only to the interconnections between humans and nature but also with tūpuna (ancestors) and atua (deities). These three "cosmotheandric" strands (human-nature-divine) are inseparable in Māori cosmology, in which the spiritual world is interwoven throughout the physical world (Panikkar 1993). Therefore an awareness of and respect for the fullness of those tripartite relationships is critical to human use of natural phenomena. Humans must be cognizant of these interconnections and negotiate a harmonious balance between the sacred and the profane.

Nondual Awareness of the Awa

There are other ways of knowing the Whanganui River that do not take the dualistic categories of subject/object to be discrete or fixed entities. A nondual approach relates to the river as a living being—indeed, as life.[4] This nondual intuition, where the seer and the seen merge and unite, may only be approached through nonrational channels of perception. The rational intellect may guide in an understanding of the river, but a higher faculty than conscious, directed thought is needed for insight into the inseparability of the river and "me." While direct awareness of this relationship may be attained through intuition, it cannot be obtained through the logos, for thoughts and words are always secondary to experience: "the awareness *of* an experience is not the experience" (Panikkar 1979:292). In contrast to the normal scientific experiment, where methodological procedures are set out in advance, an experience is truly unique and cannot be proven through analytical reasoning. Only afterward may a spiritual experience be recalled to others as the awareness of a feeling or intuition of deep connection: "Ko au te Awa, Ko te Awa ko au—I am the river, and the river is me."

Seen from this nondual perspective, pollution is a manifestation of deep disharmony between humans and their spiritual relationship with nature. From a dualistic rationality, pollution of waterways is tolerated (within reason) as long as it does not adversely affect opportunities for human use and enjoyment or affect the physical health of riverine organisms and species. However, it must be understood as more than that, for ultimately environmental pollution is an abuse of the self. When we desecrate sacred places, like the Whanganui River, we damage the essence of our own spiritual being. For Māori, pollution is a cultural affront because it severely affects the mental, physical, and spiritual health of the many hapū and iwi (tribal groupings) on the Whanganui River. However, those connections between humans, the Earth, and the spiritual world are seldom regarded seriously by those whose understanding of the river is restricted to dualistic rationality. A Māori Member of Parliament, Tariana Turia (2001:8) points out: "Too often planners cannot see the connection between water and the life, soul and spirituality of communities. . . " The river life, which is placed in jeopardy, is not listened to by those who wield the power to destroy that life force. Instead, the river is talked about as if it is inanimate, as if it has nothing to say and humans have nothing to learn. Yet, no one who has been truly immersed in those waters could ever consider that the river is expressionless or without voice. To talk with the river is to enter a very different relationship in which the language is one of profound silence.

In secular society, spiritual experiences and mystical revelations are rejected outright as superstition or relegated to children's books to be retold as mythical monster stories. For more than a century, Whanganui River Māori have faced an onslaught of ridicule and humiliation by those whose rationality denies an opening to the sacred. Despite this, they have not forced their relationship with their Tupuna Awa onto others but have consistently requested that others respect their cultural traditions and practices by not exploiting the river. However, developments throughout the Whanganui catchment do threaten the spiritual (and physical) welfare of the river and its communities. In attempting to defend the river against these threats, Māori have been forced to pursue legal claims to ownership and management—concepts that are alien to their ancestral relationship with the river and that emanate from the Western, dualistic worldview. Until ways of understanding other than those limited to this dualistic perspective are accepted not merely in secular law but also in spirit, the prospects for resolution of disputes concerning the Whanganui River are bleak. With most, perhaps everything, to lose, the challenge for Māori is to find ways to open the "other" to these ways of understanding—ways that must be nondual.

The Dialogical Opening

Whenever we hold to fixed views, positions, and notions of truth as this or as thus, we do so because we stand on one riverbank and stare across an abyss to the other side. From my side, your riverbank appears blurry, incoherent, and lacking stability. Conversely, my riverbank makes sense to me. I can convince you of the superiority of life on this side; for here we have transformed a muddy backwater into a thriving river city that is no longer directly dependent on (and therefore vulnerable to) the river for provisions. I am moved to generosity such that I build you a raft so that you might sail across to my shore. Even if you refuse my gift, I am convinced that it is in your best interest to reside on this side of the river; therefore I throw you a lifeline to haul you across the waters. If you sink on the way, it is not the fault of the noose that I tied, but it is because you did not try hard enough to swim and thus submitted to the undertow. Once you are safely on my side, rebuilding your life with the materials that I have provided you with, you may also begin to see that your riverbank is but an erosion zone with little hope for concrete survival.

When we stand on one side of the river, the temptation is to take a fixed stance; the risk is getting stuck in the mud. Often, we are so busy making sure that our dwellings are free of dirt that we neglect the signs that warn of a flood until it is too late. Our actions then can only be reactions. The lesson taken is usually shallow: to build ever higher stopbanks (levees) to block out the possibility of another flood. However, in doing just that, we also block the view across to the other riverbank. We narrow our vistas and limit the possibilities for seeing new horizons. In building physical barriers to ward off the river in flood, we also project to those on the other side of the river that we do not trust nor listen to their cries when they try to alert us to impending danger. Thus we deny opportunities to assist each other in overcoming mutually dangerous situations. Similarly, it becomes more difficult to receive invitations to share in celebratory ritual feasts of fish. Only, perhaps, through individual acts of courage or by accidentally falling overboard can the other save us from drowning and bring us safely across to their side of the river. Is it only then that we, blinking, may open our eyes to the possibility of viewing our own side from the perspective of the other?

This scenario might now be grounded to illustrate the conflicting positions taken by Genesis and Whanganui Iwi. From Genesis' perspective, their existence as a financially viable corporation depends on the continued abstraction and diversion of water to the Tongariro Power Development scheme. However, for Whanganui River Māori, their cultural identity and survival as a people is entirely dependent on the ecological and spiritual health of the Whanganui River. The position that Whanganui River Māori assert—the full return of all headwaters to the Whanganui River—is intolerable to Genesis. Conversely, the continued diversion of the spiritual sustenance of the Whanganui River is totally unacceptable to Māori. It would appear that these two positions are incompatible. Thus a stalemate exists and will remain under the dialectical framework until one or the other convinces a third party (the planning committee or the judiciary) of the absolute validity of their own position or of the absolute invalidity of the other's claims.

There are other ways of outlining the situation of the riverbank dwellers. Rather than emphasizing their separation on either side of the river, the banks might also be seen as connected by the water that flows between them. While the river may meander to and fro and shift shingle from side to side, it owes its form and character (as "river") to both banks that constrain its otherwise entropic dispersal. However, this is not to say that the river owes its existence to its banks; that is another kettle of fish altogether. Because the river is and because water laps at both shores, there is the potential for people who live on different sides of the river to erect a bridge. Of course, there is also the potential that one side might design a war vessel to launch an attack and conquer the other side. However, this is not a study of war or peace. Rather, it is a call to recognize that the dual tensions that appear to divide may also be seen as creative tensions with potential to unite.

The "dialogical opening" is more than a conversational exchange of information, data, ideas, or opinions on a given topic of common interest. There must be an acknowledgement at the outset that the "partners" entering dialogue are from different cultural worldviews. In the case of the Whanganui River, the dialogue that is urgently needed is that which opens up a bridge of understanding between Māori and European worldviews. This extends beyond the dialectics of mediation among a plurality of groups with differing perspectives and calls instead for an awareness and positive acceptance of "radical pluralism" (Krieger 1991; Panikkar 1999).[4] When planning processes treat Māori testimonies as another set of interests to be considered alongside issues raised by interest groups, they reduce the Māori worldview to one group among a plurality within the Western rational framework. What the Treaty of Waitangi 1840 envisaged, however, was the basis for pluralism—an interrelationship between two peoples—a taonga (treasure) that binds people together.

Dialogue requires openness in that it is a "talking through" or "thinking through," as the etymology of *dia-logos* suggests (Barnhart 1988). Openness in dialogue is essential because it encourages listening with an ear to understanding and learning with the potential for growth.

In authentic dialogue, there must be a willingness to bear witness to one's own testimony with a sincerity that flows naturally when one speaks (or writes) with the "optimism of the heart" (Panikkar 1979:243). Thus dialogue does not stem from the rationalized order of the mind, indeed: "If you can prove with reason or furnish evidence, you are not, strictly speaking, testifying, you are not witnessing but demonstrating" (Panikkar 1979:240). In witnessing, the kaumātua (Māori elder) speaks with the wisdom and clarity of the heart of his or her concern for the

Whanganui River. In contrast, the expert employed for Genesis does not testify as such but uses the force of dialectics: he or she delivers information, points out "facts," and "proves" through critical argument. The former bears witness to an intrinsic relationship with the river; the latter sets out to convince the audience of the corporation's right to take from the river.

Through dialogue, people listen to and learn from each other. What may emerge from such a process is a broader dimensional understanding of the Whanganui River from a variety of perspectives and experiences that each speaker communicates. Through listening with an openness to receive, each partner in dialogue might also come to better know the other and their connection with the river. Yet cross-cultural dialogue has an even more essential role—it offers a way of "knowing myself." As Panikkar explains: "Dialogue is, fundamentally, opening myself to another so that he might speak and reveal my myth that I cannot know by myself because it is transparent to me, self-evident" (Panikkar 1979:242).[5] In communicating his or her experiences, the other (perhaps quite unwittingly) raises questions that challenge the very presuppositions that I take for granted. Ultimately, the challenge is one of rejecting claims to universalism—universalism being, inevitably, self-destructive. This remains a key challenge of the ecosystem approach as it has been set out in this book.

While dialogue with the other draws me out of myself, the river brings us back together. The other gives us the ability to see ourselves as fully human. The river makes us whole. Yet, the present condition of the Whanganui River is not "whole," not healthy.[6] Both Western-trained scientists and Whanganui River Māori testify to this. From a Western rationale, science and technology may offer practical solutions to restore the river's physical (environmental) health. At one level, ecological remedy is urgently needed and is a goal currently pursued by resource management agencies, environmental groups, and Whanganui River kaitiaki (guardians). However, ecosystem restoration is only one part of healing the whole. A nondual approach to the Whanganui River alerts us to our most vital calling—that of healing ecological and spiritual connections—and, in so doing, making ourselves whole.

Riding the Wave: Whanganui River, 2003

Following the formal Māori farewell (poroporoaki) from Tieke Marae, we (the group of Global Ecology students, teachers, and myself) returned to our canoes to resume paddling what was the final leg of our journey on the Whanganui River to destination Pipiriki. My new canoe partner elected me to the position at the back of our boat; in charge of navigation. Initially, my unfamiliarity with this role sent us snaking first this way and then that way, making me frustrated with my own inefficient paddling stroke. Even a slight distraction off center could result in a turn in the opposite direction: alas, enantiadromia! Misreading the flow of water could spin us into a back eddy. Yet, with helpful advice from others and through practice, I relaxed and began to keep a more even keel. Then I could look up again and take in the beauty of the bush-clad embankments and the reflections in the water all around me. The large stretches of slower moving water between rapids invited such moments for quiet reflection of a more personal and profound nature.

Our friends from Tieke passed us in a jet boat that left our canoes rocking unsteadily in their wake. They waited for us in the slipstream of the Ngaporo rapid for the entertainment that we would provide. The turbulent whitewater was to test each one of us. Some made it through via the safe side current and thus avoided the rapid's rough ride. Others only just hung on throughout the roller-coaster ride; not by any special talent but by throwing caution to the wind (or the wave). Others, it seemed, were fated for a swim. Only an elect few really rode the wave. They were rewarded with a standing ovation from the jet boat onlookers whose exhilaration had become contagious. As the boat then roared into motion and began to pull away from us, the passengers waved in a reciprocated departing gesture. Over the noise of the engine, almost drowning it, a woman's voice called out strongly in a karanga; a cry that penetrated beyond the water to acknowledge the spirit of the tupuna whose wave had carried us.

When I reflect back to the health warning we received at Tieke, "Do not drink the water," I see in those words a message that contains within it the seeds for its own healing. After a couple of days paddling along with the river's rhythm, the students' almost unanimous expression that the river is beautiful must be appreciated for more than mere frustration with the inadequacy of words (and the logos). The river mirrored back to each of us the physical beauty of a landscape in which water is the essence. Yet the beauty of the river experience cannot be captured in any physical description or rational analysis. This touches on the ultimate struggle of the poet: how to give expression to that which is truly inexpressible. The beauty is not only in the water, but in the wave—the realm of mystery, of Beauty, that washes through every being. If I can no longer drink the water, should I quench my thirst by drinking the wave?

> "'Utram bibis? Aquam an undam? 'Which are you drinking? The water or the wave?'"[7] (Fowles 1977:188)

Notes

[1] Confusion over the spelling and pronunciation of "Whanganui" and "Wanganui" derive from European misunderstanding of the local Māori dialect. In 1991, the Minister of Lands announced that the official spelling of the Whanganui River should revert to its correct original. The city, however, remains misspelled as Wanganui. According to Whanganui River Māori, the absence of the letter "h" takes away the true meaning of the name. The perpetuation of this misunderstanding remains a point of contention that sometimes manifests in race relations disputes.

[2] The phrase "beheaded the past" may seem dramatic/controversial. For Whanganui River Māori, the headwaters signify the "head" of their ancestor, so the diversion of the headwaters is, for them, literally a beheading of their ances-

tor (hence the spiritual basis of their people). The impact of this action is so much greater in Māori and Polynesian societies because a person's head is revered as tapu (sacred). This (poetic) phrase also has a temporal aspect: to say "beheaded the past" is to imply that the diversion of the headwaters has destroyed not only the future of the river (and its people), but also present and future generations connection to their past.

³In this chapter, "dialectical" is contrasted with "dialogical" (see Panikkar 1999). For example, Whanganui River Māori and Genesis Power Limited are currently locked into a dialectical framework: the legal arena forces them into opposing positions and restricts their interaction to that of a battle (debate: *de-battle*) of the intellect/logos; the critical realm of reasoned argumentation. A dialogical opening allows for talking through (*dia-logos*) and piercing the logos to attain a truth that transcends it. In the present situation, a dialogical opening would allow for the possibility of mutual fecundation in understanding across cultural worldviews.

⁴The spelling "nondual" is preferred over "non-dual," where the latter might be misunderstood as "*not* dual" (i.e., the opposite or negation of dualism) and, paradoxically, categorized dualistically. Nondual is closer to the advaitic meaning "not divided."

⁵"Plurality" is used here as the numerical multiplicity, whereas "pluralism" goes beyond the acknowledgement of plurality (and postcolonial tolerance of cultural difference) to the acceptance that no single group embraces the totality of human experience (i.e., beyond universalism). Krieger's (1991:15–16) term "radical pluralism" is of interest: "a *radical* pluralism in the sense that it is a matter of different worldviews, that is, different ways of thinking and forms of life which are constituted by their own criteria of *meaning, truth* and *reality,* and which, therefore, claim absolute validity. Radical pluralism refers to the unique historical situation of the late 20th century wherein humanity finds itself on the threshold of a global culture."

⁶Panikkar continues: "Dialogue is a way of knowing myself and of disentangling my own point of view from other viewpoints and from me, because it is grounded so deeply in my own roots as to be utterly hidden from me. It is the other who through our encounter awakens this human depth latent in me in an endeavor that surpasses both of us. In authentic dialogue this process is reciprocal." (Panikkar 1979:242–243)

⁷"Health" is formed from the root *hāl*, such that "to heal" means "to make whole." Of particular interest, the etymological connection may be extended to the word "holy," suggestive that: "The primary meaning of the word [holy] may have been 'that must be preserved whole or intact, that cannot be transgressed or violated,' which would support its relationship to Old English *hāl* whole." (Barnhart 1988:487)

⁸In *The Magus,* a novel by John Fowles, the question is posed by Maurice Conchis as a challenge to Nicholas Urfe, the young man whose quest is a search within for the path between both extremes. Conchis adds: "We all drink both. But he

meant the question should always be asked. It is not a precept. But a mirror."
(Fowles 1977:188)

References

Barnhart, R. K. (ed.). 1988. *Chambers Dictionary of Etymology*. New York: Chambers.

Daly, H. E. and J. B. Cobb Jr. 1994. *For the Common Good: Redirecting the Economy toward Community, the Environment, and a Sustainable Future* (2nd ed.). Boston, Mass.: Beacon Press.

Dallmayr, F. R. 2002. *Dialogue Among Civilizations: Some Exemplary Voices*. New York: Palgrave Macmillan.

Fowles, J. 1977. *The Magus: A Revised Version*. London, UK: Jonathan Cape.

Gunderson, L. H., C. S. Holling and S. S. Light (eds.). 1995. *Barriers and Bridges to the Renewal of Ecosystems and Institutions*. New York: Columbia University Press.

Krieger, D. J. 1991. *The New Universalism: Foundations for a Global Theology*. New York: Orbis Books.

Ngāti Rangi Trust, Tamahaki Inc. Society, Whanganui River Māori Trust Board et al. v Genesis Power Limited (2004) Decision A 067/2004.

Panikkar, R. 1979. *Myth, Faith and Hermeneutics. Cross-Cultural Studies*. New York: Paulist Press.

Panikkar, R. 1993. *The Cosmotheandric Experience: Emerging Religious Consciousness*, ed. S. T. Eastham. Maryknoll, New York: Orbis Books.

Panikkar, R. 1999. *The Intra-Religious Dialogue* (rev. ed.). New York: Paulist Press.

Smuts, J. 1926. *Holism and Evolution*. London, UK: Macmillan.

Turia, T. 2001. Water is sacred—Turia. *Massey News* 12:8.

Waitangi Tribunal. 1999. The Whanganui River Report, Wai 167. Wellington, New Zealand: GP Publications.

Waring, M. 1988. *Counting for Nothing: What Men Value and What Women Are Worth*. Wellington, New Zealand: Allen & Unwin.

Wheeler, B. 1999. Water—Use and abuse. WATER: Values, Uses, Rights, Laws. Conference. Wellington, New Zealand: Foundation for Indigenous Research in Society and Technology (FIRST).

A Tribute to James J. Kay

As presented by David Waltner-Toews on July 11, 2004, at the Welcome Sessions of the International Society for Ecological Economics, Montreal, Canada.

James Kay died, at 11:00 p.m. on May 30 of this year, 2004, just a few weeks short of his fiftieth birthday. He died with his eyes open, both literally and figuratively.

For many, James was an exquisite physicist, theoretician on complexity and thermodynamics. In his early work with Eric Schneider, he reinterpreted the second law of thermodynamics, applying it to understand how exergy gradients induce self-organizing structures and how living systems organize so as to destroy exergy gradients at the fastest rates possible. His work was featured as a cover story in the *New Scientist*. His 1994 paper with Schneider was recently identified as one of the twelve most important papers in ecology in the 1990s and is included in the Oxford University Press *Readings in Ecology*. More recently, he extended and expanded on this work with Royden Fraser. Some have argued that in the unpublished papers on thermodynamics he left behind, there is more than enough material for a Nobel Prize. I cannot say I understand that part of his work well enough to make such a judgment. I do know that he never stopped at exquisite theory.

I first knew James as a teacher, when, as a veterinarian and public health epidemiologist, I was making my first tentative forays into the debates on sustainable development. I discovered what a wonderful teacher he was, translating the complex theories of thermodynamics into plain English and simple object lessons and diagrams. The irreducible uncertainty of complex life was to him not just a source of frustration, but of humility, and great wonder and amazement and good graphics. It was certainly at the core of how he approached the immense problems of science in the public domain, science for the public good.

He could move from thermodynamics to science fiction to municipal politics with ease, although some might argue that the latter two are not so different after all. He was the founding Chair of University of Waterloo's Greening the

Campus Committee and a founding member of the City of Kitchener's Environment Committee, which developed a Strategic Plan for the Environment and an ecosystem-based master plan for the Huron Natural Area. He sat on the committee that developed the award winning (Canadian Institute of Planners) bicycle master plan for Kitchener, was on the City's committee for the transition to a hydrogen economy. He advised the Ontario Ministry of the Environment on how to develop indicators of ecological integrity and delivered special guest lectures to the National Ministry of the Environment. He served on the Long Term Ecosystem Research and Monitoring Panel of the Royal Society of Canada and was a member of the Royal Swedish Academy of Sciences, Beijer Institute, Working Group on Complex Ecological Economic Systems Modeling. He was a very active member of the U.S. National Science Foundation Advisory Committee on Environmental Research and Education and had a profound influence on the shape of its ten-year outlook, *Complex Environmental Systems: Synthesis for Earth, Life and Society in the 21st Century*, published in 2003.

As academic friends will, we would sometimes commiserate about the immense inertia of formal institutional structures such as governments, granting councils, and universities, but he wouldn't let us stop at just complaining. When we founded the Network for Ecosystem Sustainability and Health, James saw it as a way to reinvent the university, to create a forum for intellectual debate, teaching and practice, a critique of business as usual, a way to use new Web-based technologies to create that extended peer group demanded by post-normal science.

In the last ten years, many of us had the privilege of working with him on a kind of globalization of this parallel, publicly engaged university. Since the early 1990s, he had been part of an ad hoc international group of scholars studying uncertainty, complexity, and managing for sustainability. An earlier name for the group, the Cali Cartel, named after the working place for one of our members, seemed to create some traveling problems. Then, we had a memorable meeting at La Faloria Convent in Cortina, in the Italian Alps, in which we flew people across the Atlantic with no clear agenda and no committed funds, our only sponsor an Italian winemaker who agreed to provide wine. It was one of the greatest scientific experiences of our lives. Like the fictional detective Dirk Gently, created by Douglas Adams, that hitchhiker of the galaxy, we sought to solve the whole crime, to find the whole person, to find the whole solution to our global problems. Served pasta by the nuns and drinking heavenly Italian wine, we read poetry and argued complex systems theories and explored the reasons for epidemics and the nature of agricultural development. When I suggested we might call ourselves the Dirk Gently Group, James, with typical impishness, insisted on referring to us as the Dirk Gently Gang.

Later, at the Ecological Economics meetings in Boston, and subsequent panels and special sessions in Cali, Colombia, Toronto, Guelph, James was always a central figure. Even this last November in Milan, and in March in Alexandria, when he was not there physically, we could feel his presence. Then again, there were the sessions he organized dedicated to theoretical ecology in the Biennial Workshop in Advances in Energy Studies of Porto Venere, Northern Italy, in which James always played the crucial role of the "skipper fighting in the typhoon." More than once, he told me, after some memorable engagement at a workshop on exergy or energy or integrity or health, there had been blood on the floor. More than once, he confided in me that he was distressed at the way some scholars seemed to believe that one side had to win or lose these debates. Did we not believe in the reality of complexity? In the necessity for well-argued, well-articulated multiple perspectives? In the reality of trade-offs?

In the last year, he was annoyed by the fact that there is a lot of work still to be done, and his health problems were preventing him from doing all he would have liked to do. At the beginning of May, I visited him in the hospice shortly after he had moved in. It was an unsettled spring day, with temperatures around 20°C, bits of rain, bits of sunshine. The hospice is in a small gulley, with a wetland nearby, and James had some bird feeders outside his floor-to-ceiling windowed doors. He immediately started talking about energetics and feed availability and species competition. We watched the young red-winged blackbirds, chipmunks, a flock of goldfinches. James was taking pictures with a new digital camera to document the ecology outside his window. The world never ceased to intrigue and amaze him. Until a couple of weeks before he died, he was still debating with Martin Bunch and I on the fine details of diagrams of self-organizing systems to be included in a book we had been working on, it seems to me, almost forever, entitled, *The Ecosystem Approach: Complexity, Uncertainty, and Managing for Sustainability*. For some of us, like me, who are at home with broad visions of health and sustainability, he could be an annoying bear of a man to work with, precisely because he was as demanding of his colleagues as he was of himself. He complained at how sloppily supposedly good scientists used words like attractors and phase changes, did not tolerate the kind of loose thinking or sloppy logic which some of us use to comfort ourselves with our cleverness. He wouldn't let us off the hook just because we were friends. And I think it is exactly because of this that I remember him as a friend, and not just as a colleague.

Looking over the edge of eternity, he was worried that his work might not amount to what it should, that it would fall short of its potential. On May 19, just a couple of weeks before he died, I dropped by his hospice on my way home from work and talked with him over supper, and then we wheeled out to the little wetland and looked at water and listened to the red-winged blackbirds. It's all going so fast, he said. I told him again about how he had connected so many diverse people from so many parts of the world, the marvelous influence he had. Only half joking, I think, he told me to keep saying that, as it made him feel better. He was pleased that Jona and Lise, his children, now both had their drivers' licenses; they were growing up and out into that world he so loved. I went home and worked in the back yard until dark and I had a sore back. I guess I just wanted to feel my body, knowing how easily it can slip away, how easily, if we let it, this world can slip away from us.

In spite of his regret at not having finished his work here, he left a huge legacy to us, in terms of books, papers, videos, tapes, intuitions, enthusiasm, contacts, friendships, shared experiences, memories, ongoing projects, and personal example. Visit his web site (accessible through www.nesh.ca; he can be found in the pioneers section); it is a treasure. James would want us to share it, to use it, to let it multiply in the minds of students and scholars everywhere. It is time for us, now, to use our intellectual drivers' licenses, and get out there, not to let it all slip away, this world full of misery and wonder, immense problems and immense resilience. With James' passing, the world lost a champion of good science, good ecological economics, and good citizenship. But most of all, for many of us on this strange and wonderful journey of life, we lost a traveling companion and a good friend.

David Waltner-Toews for Joan Martinez Alier, Tim Allen, Bruna de Marchi, Silvio Funtowicz, Gilberto Gallopin, Mario Giampietro, Nina-Marie Lister, Giuseppe Munda, Jerry Ravetz, Henry Regier, Joe Tainter, and Bob Ulanowicz.

Appendix
Hierarchy and Holonocracy
Henry Regier

The term *hierarchy* has long brought discomfort with it when used in discussions among and between natural and social scientists within an ecosystem or other systems-based approach. Etymologically in its purest sense, a *hierarch* is a *chief priest with the power to rule*, implicitly from the top in a rectilineal sense. A secular *monarch*, who has been invested in an office through the services of a priest, may feel sanctified or sacralized and thus also entitled to be accepted as a *hierarch*.

Biological ecologists with a reductionistic bent may use *hierarchy* as a code word for *systemically autogenous organizational nesting* with the exercise of power welling upward from the lowest level and not cascading downward from the highest level of organization. Etymologically, it seems silly to use the term *hierarchy* is such a context.

Antihierarchic revolutionaries may be termed *anarchists,* but few of them may commit to a completely chaotic *anti-archic* or *laissez faire* exercise of power. Some writers of the extreme Right in recent politics may have supported ethically unconstrained *anarchic* capitalism but may stealthily foster *autarchy* or *plutarchy* by libertarian capitalists, perhaps domiciled in a tax haven.

At an Alpbach Symposium in Austria in 1968, Arthur Koestler tried to transcend such confusion about *hierarchy* (Koestler and Smythies 1969; Koestler 1979). First, he used the term *holon* as a dialectical meld between *hol* for *whole* and *on* for *part*. Thus *holon* implies transcendence of the holistic versus reductionistic contrast. Then he inferred that evolutionary phenomena emerged autogenously as *Self-Organizing, Holonically nested, and Open (SOHO)* entities or beings, but Koestler may have created new confusion by using the term *holarchic* for a *holonically nested power relationship*. Social ecologists might take *holarchic* to mean top-down rulership of the whole nested phenomenon, as with strongly centralized state socialism.

In an attempt to free ourselves of code words that I find to have been used ambiguously, I suggest the term *holonocracy* for the kind of nested self-organization and power sharing that we perceive in ecosystems that exhibit *integrity*. I understand that the Greek root of the suffix *cracy* (as in *democracy*) may imply softer action than that of *archy* (as in *panarchy*).

References

Koestler, A. 1979. *Janus—A Summing Up*. New York: Vintage Books.
Koestler, A. and J. R. Smythies (eds.). 1969. *The Alpbach Symposium 1968: Beyond Reductionism—New Perspectives in the Life Sciences*. London: Hutchinson of London.

Contributors

Timothy F. H. Allen
University of Wisconsin, Botany
    Department
430 Lincoln Drive, Madison,
    WI 53706-1381, USA
Tel 608-262-2692 Fax 608-262-7509
tfallen@wisc.edu

Fikret Berkes
Natural Resources Institute, University
    of Manitoba,
Winnipeg, Manitoba, Canada R3T 2N2
Tel 204-474-6731 Fax 204-261-0038
berkes@cc.umanitoba.ca

Stephen Bocking
Environmental and Resource Studies
    Program
Trent University
Peterborough, Ontario, Canada K9J 7B8
Tel 705-748-1011 ext. 7883
    Fax 705-748-1569
sbocking@trentu.ca

Michelle Boyle
University of British Columbia
Institute for Resources, Environment
    and Sustainability
2202 Main Mall, 4th Floor
Vancouver, BC, Canada V6T 1Z4
mboyle@ires.ubc.ca

Martin J. Bunch
Faculty of Environmental Studies
York University, 4700 Keele Street
Toronto, Ontario, Canada M3J 1P3
Tel 416-736-2100 ext. 22630
    Fax 416-736-5679
bunchmj@yorku.ca
http://www.yorku.ca/bunchmj

Dominique Charron
Program Leader, Ecosystem Approaches
    to Human Health
International Development Research
    Centre
150 Kent St. PO Box 8500
Ottawa Ontario, Canada K1G 3H9
Tel 613-236-6163 ext 2079
    Fax 613-236-7293
dcharron@idrc.ca

Iain J. Davidson-Hunt
Natural Resources Institute, University
    of Manitoba,
Winnipeg, Manitoba, Canada R3T 2N2
Tel 204-474-8680; Fax 204-261-0038
davidso4@cc.umanitoba.ca

Silvio Funtowicz
Institute for the Protection and
  Security of the Citizen (IPSC)
European Commission, Joint Research
  Centre
Via E. Fermi, 2749
I-21027 Ispra (VA), Italy
Tel +39 0332 785934
silvio.funtowicz@jrc.it

Mario Giampietro
ICREA Research Professor
Universitat Autònoma de Barcelona (UAB)
Institute of Environmental Science and
  Technology (ICTA)
Edifici Q –(ETSE) Escola Tècnica
  Superior d'Enginyeria—ICTA
Campus de Bellaterra
08193 Cerdanyola del Vallès,
  Barcelona, Spain
mario.giampietro@uab.cat

Thomas Gitau (1967-2005)
Department of Public Health
Pharmacology and Toxicology
University of Nairobi, Kabete Campus
Box 29053, Nairobi, Kenya

James J. Kay (1954-2004),
Environment and Resource Studies,
University of Waterloo
Waterloo, Ontario, Canada N2L 3G1
www.jameskay.ca

Clive Lightfoot
Lightfoot Consulting
3 Court Close
Bray, Berkshire, SL6 2DL UK
clive.lightfoot@linkinglearners.net

Nina-Marie E. Lister
Associate Professor
School of Urban and Regional Planning
Ryerson University
350 Victoria Street
Toronto, Ontario, Canada M5B 2K3
Tel 416-979-5000 ext. 6769 Fax 416-979-5357
nm.lister@ryerson.ca
http://www.ryerson.ca/surp/

David Manuel-Navarrete
Department of Geography
King's College London
Strand, London, United Kingdom,
  WC2R 2LS
david.manuel_navarrete@kcl.ac.uk

Joan Martinez-Alier
Department of Economics and
  Economic History
Universitat Autonoma de Barcelona,
  Spain

Dan McCarthy
Department of Environment and
  Resources Studies
University of Waterloo
200 University Avenue West
Waterloo, Ontario, Canada N2L 3G1
Tel 519-888-4567 ext. 33065
dmccarth@fes.uwaterloo.ca
http://www3.sympatico.ca/
  dkmccarthy/

John McDermott
Deputy Director General—Research
International Livestock Research
  Institute (ILRI)
P.O. Box 30709
00100 Nairobi, Kenya
Tel 254 20 4223207 Fax: 254 20 4223001
j.mcdermott@cgiar.org

Tamsyn Murray
42 Turners Track
Kerrie, Victoria, Australia 3434
Tel 03 5427 0995
tamsyn@softtissuecentre.com.au

Cynthia Neudoerffer
School of Environmental Design and
  Rural Development
University of Guelph
Guelph, Ontario, Canada N1G 2W1
rneudeor@uoguelph.ca

Reg Noble
Academic Coordinator for
  Postgraduate Program in Food
  Security
Ryerson University, Centre for Studies
  in Food Security
350 Victoria Street
Toronto, Ontario, Canada M5B 2K3
rnoble@ryerson.ca

Ernesto F. Ráez-Luna
Centro para la Sostenibilidad
  Ambiental/Environmental
  Sustainability Centre
Universidad Peruana Cayetano
  Heredia/Cayetano Heredia
  University
Armendáriz 445, Miraflores, Lima 18,
  Peru
Tel 51-1-4470317
eraez@csa-upch.org

Ricardo Ramirez
Communication Consulting
(Adjunct Professor, School of
  Environmental Design & Rural
  Development, University of Guelph)
44 Caledonia St, Guelph, ON, N1G 2C9
  Canada
Tel +1 519 824-5519
rramirez@uoguelph.ca

Jerome Ravetz
111 Victoria Road
Oxford, United Kingdom OX2 7QG
Tel +44 1865 512247
jerome-ravetz@tiscali.co.uk.

Henry A. Regier
Professor Emeritus, Zoology
University of Toronto
10 Ernst St.
Elmira, Ontario, Canada N3B 1K5

Tel 519-669-5552
hregier@rogers.com
José G. Sánchez Choy
Consultant Agronomist
Professor, University Private of
  Pucallpa
Tarapaca 1019 Pucallpa, Perú
Tel 51-61-572847
  Mobile 51-61-961843519
jose.sanchezchoy@yahoo.com
www.fortunecity.es/losqueamamos/
  gordo/64/

Felix Sánchez Zavala
Consultant Nutricionist
Los Jardines Mz B L 20
Yarinacocha, Perú
Mobile 51-61-9617372
fsanchez07@yahoo.es

Charlotte Helen Šunde
Resource and Environmental Planning
  Programme
Massey University
Private Bag 11-222
Palmerston North
Aotearoa New Zealand
chsunde@gmail.com

Julian van Mossel-Forrester
Graphic Designer
Van Mossel Forrester Art & Design
Kitchener, Ontario, Canada
julian@vanmf.com

David Waltner-Toews
Department of Population Medicine
University of Guelph
Guelph, Ontario, Canada N1G 2W1
Tel 519-824-4120 ext 54745
  Fax 519-763-3117
dwaltner@uoguelph.ca

Index

Boxes are indicated by b, figures by f, and tables by t following the page number.

approach to sustainability and, 11–12; systems thinking and, 3–4, 8;
see also emergent complex systems
complication: complexity vs., 79, 310; definition, 40
conflict resolution, 240–241
Conklin, H. C., 115
consciousness transformation, 337
conservation: adaptive management of, 100–102, 102f; of biodiversity, 83–103; complexity and, 97–99; ecological contexts for, 88–89; of energy, 15–16; management choices, 98; as participatory and cooperative endeavor, 98–99; sociocultural values, 84, 99
Conservation Authorities (Ontario), 187, 188
Consultative Group on International Agricultural Research, Eco-Regional Program, 214
context: in adaptive ecosystems approaches, 242–243, 292f, 305; ecological, 88–89; for open systems, 52; social, 150, 151; in sustainability, 241; in systems description, 24–25
contradiction, 313–315, 319–320
Convention on Biological Diversity (U.N. Environment Program), 85
Co-operative Inquiry, 132
Cooum River (India) case study, 157–172; alternative domains of system organization, 169–170; CATWOE analysis, 161, 163; core system structure, 163, 164f; decision support system, 159; interventions, 170–171; problem identification exercise, 159, 160b, 161; research program overview, 158–159; rich picture for, 159, 161, 162f; self-organization around attractor states, 170–171; socio-ecological problem description, 159–163, 170; soft systems methodology, 158, 168–171, 169b; subsystems, 163, 166–167, 166b; urban character of, 161, 163; visioning of desirable future states, 165–168, 165b
corn production, 180, 184, 185–186
Cotton, C. M., 116
Cox, P. A., 116
Cox, Robert, 326, 332nn4–5
creative destruction, 310, 314
creative tension, 314
critical link species, 93
cross-cultural dialogue, 345–360
cross-scale problems, 28
cultural transformation, ecological integrity and, 338
cybernetics, 7–8
cyberspace, 311

Daniels, D. E., 143
Darwinism (natural selection), 95
data collection and storage, 303–304

Davidson-Hunt, Iain J., 240
DDWSC (Department of Drinking Water Supply Corporation), Kathmandu, 263, 266
DeAngelis, D. L., 4
decision making: adaptive ecosystem monitoring and, 290; environmental, post-normal science and, 10; evolution in complexity and uncertainty, 99–100, 100f; multiple perspectives in, 340; observer decisions and system type, 42–44; open approach for conservation, 102; participatory, complexity and, 80
decision support system (DSS), 159
Declaration of Belem, 116
deforestation, 213, 214–215
deontological principles, 336, 342n3
Department of Drinking Water Supply Corporation (Kathmandu), 263, 266
design process, 314–315
destructive conflict, 314
development: environment impacts, 346; monism and, 350–351; sustainable, 319, 323
dialectical framework, 347–350, 360n3
diamond diagram, 243–245, 244b
Dirk Gently Gang, 239, 246
dissipative structures, 57, 60–62, 61f, 95
diversity: organization and, 317
Dixon, J., 153
Dolderman, Dan, 99
domination, in emergent complex systems, 309
dreamtime, 114, 114t
dualism: biocentric, 335–336, 342n1; in Whanganui River development, 351
dynamical quality of complexity, 40–41, 47–48, 310

Earth Summit (Brazil, 1992), 90
echinococcosis, 258
ecological economics, 72–74
ecological integrity, 52, 335–342; deontological principles for preserving, 336, 342n3; discourse implications for practice and agency, 339–341; ecosystemic-pluralistic discourse, 336–337, 340–341, 342; human-ecosystem relationship, 335; negotiation and, 336–337; normative discourse, 335–336, 339–340, 342, 342nn2–3; social practices and, 338; transpersonal-collaborative discourse, 337–338, 341, 342
ecological systems: definition, 64b; self-organizing nature of, 11–12
ecology: biodiversity and, 90–97; conceptual shifts in, 109; environmentalism and, 90; paradigms in, 89–90; sacred concept of, 112; schism in, 90
economics, ecological, 72–74

CPSIA information can be obtained at www.ICGtesting.com
Printed in the USA
LVOW03s1051010814

397081LV00010B/55/P